Acclaim for Peter Steinhart's

THE COMPANY OF WOLVES

"Steinhart is . . . capable of evoking the mythic qualities that we associate with the wolf, the sheer poetry of its form and function, along with the scientific details." —*Los Angeles Times*

"*The Company of Wolves* is a must read. What big teeth it has." —*Outside* magazine

"Steinhart is able to assess and explain the emotional as well as the scientific debates over a complex piece of wilderness engineering. It is a rare talent." —*The Economist*

"In *The Company of Wolves* Peter Steinhart provides an insightful look at the duality of our view of wolves as well as a comprehensive review of their natural history." —*Cleveland Plain Dealer*

"Our response to this wildest of creatures provides a measure of our own humanity. . . . *The Company of Wolves* is worthy of our serious attention." —*The New York Times Book Review*

"A far-ranging, well-balanced portrait." —*Seattle Times*

"Peter Steinhart's *The Company of Wolves* is remarkable for its detail and comprehensiveness." —*The New York Review*

Peter Steinhart

THE COMPANY OF WOLVES

Peter Steinhart was a columnist for *Audubon* and is the author of *Tracks in the Sky*, *Two Eagles/Dos Aguilas* (both with photographer Tupper Ansel Blake), and *California's Wild Heritage*. He lives in Palo Alto, California.

Books by Peter Steinhart

Tracks in the Sky:
Wildlife and Wetlands of the Pacific Flyway
(with Tupper Ansel Blake)

California's Wild Heritage:
Threatened and Endangered Animals in the Golden State

Two Eagles/Dos Aguilas:
The Natural World of the United States–Mexico Borderlands
(with Tupper Ansel Blake)

The Company of Wolves

THE COMPANY OF WOLVES

THE COMPANY OF WOLVES

Peter Steinhart

VINTAGE BOOKS
A Division of Random House, Inc.
New York

FIRST VINTAGE BOOKS EDITION, JULY 1996

 Published in the United States by Vintage Books, a division of Random House, Inc., New York, and simultaneously in Canada by Random House of Canada Limited, Toronto. Originally published in hardcover by Alfred A. Knopf, Inc., New York, in 1995.

The Library of Congress has cataloged
the Knopf edition as follows:
Steinhart, Peter.
The company of wolves / by Peter Steinhart.
p. cm.
Includes bibliographical references.
ISBN 0-679-41881-4
1. Wolves. I. Title.
QL737.C22S74. 1995
599.74'442—DC20 94-26913
CIP
Vintage ISBN: 0-679-74387-1

Random House Web address: http://www.randomhouse.com/

Printed in the United States of America
10 9 8 7 6 5 4 3 2 1

For Judy

CONTENTS

ACKNOWLEDGMENTS

A number of people gave me invaluable assistance in this journey, and I would like to thank them for having made this book possible. Thanks for opening up your stores of knowledge, your hospitality, and your patience to: L. David Mech, Diane Boyd, Rolf and Candy Peterson, John and Mary Theberge, Ed Bangs, Lloyd Antoine, Durward Allen, Charles Jonkel, Paul Joslin, Peggy Graham, Robert Stephenson, Anne Ruggles, Robert Wayne, Ronald Nowak, Norman Bishop, Renée Askins, Terry Johnson, David Parsons, Harry Frank, Victor Van Ballenberghe, Robert Ream, Mike Jimenez, Pat Tucker, and Bruce Weide. The errors that no doubt will creep into this book are not in any way theirs. But whatever good sense may creep in owes especially to their experience and generosity. Thanks to the Peninsula Conservation Center Foundation for its willingness to sponsor parts of this project, and to Dieter Walz for translating some difficult German. A special thanks to Neil Soderstrom, whose faith and persistence made this book possible.

INTRODUCTION

No two species have a more tangled, more intimate, and more shadowy set of relationships than *Homo sapiens* and *Canis lupus*. In ancient times, we seem to have admired and perhaps even worshipped them. For centuries after that, we persecuted them, put out poisons for them, shot them on sight, told stories about them that frightened children and grown-ups alike, and all but exterminated them in the lower forty-eight states. Today, we argue over wolves. While we argue, wolves are returning on their own to Washington, Idaho, Montana, Wyoming, North Dakota, Minnesota, Wisconsin, and Michigan. There are controversial plans to reintroduce them in several other states. For more than a decade, people have been surreptitiously trying to reintroduce wolves into the wilds of California, Oregon, Utah, Montana, and possibly other states. There are bitter disputes over government programs to shoot wolves in order to increase the number of moose and caribou for the sake of sport hunters. Some individuals ill-advisedly keep wolves or wolf-dog hybrids, thinking they are helping to conserve the wolf. Dozens of books have been written about them, and dozens of television programs filmed about them. Even so, they still dwell in the realm of myth, for rarely does any human catch more than a furtive glimpse of a wolf in the wild—even in Canada or Alaska, where wolves are relatively abundant. In the absence of real encounter, there is invention. And we bring those inventions to our rare encounters—with real consequences for wolves and humans alike.

Consider the following:

Lash Callison is a seventy-eight-year-old retired hunting guide who lives in Fort St. John, British Columbia, rugged, mountainous country deep in the Canadian Rockies. Callison had a hunting territory on the Little Toad River in the Muskwa Range, where he took dudes out to hunt sheep, grizzly, mountain goat, caribou, and moose. He doesn't like wolves. "As far as big-game hunting, they'll run the game right down," says Callison. "If you ever was a big-game hunter or a trapper and had to make a living and you went in there after the wolves, you'd know what I'm talking about. You don't see wolves very often. But if I saw them around here right now, I'd shoot 'em."

Callison felt that his clients weren't getting the big trophies they had been getting when he started guiding. He would tell local game officials that the government ought to get rid of the predators. "But the more you'd complain," he recalls, "the closer they'd watch you. I said to my brother, The only thing for us to do is don't complain, don't say anything, and just take care of these wolves ourselves. That's what we did. We got poison."

They injected tallow or pieces of moose liver with strychnine and put it on the ice of a winter lake or along a mountain pass where wolf tracks had been seen. In summers, they buried bait where a wolf would smell it and dig it up—if they just left it on the ground, crows or jays would take it. In the snow, they would put baits under strips of moose hide and throw a little water over everything. The water would freeze, and the ice would keep crows and jays and smaller mammals off the baits, but wolves could pry them up. Callison carried out his own predator controls from the 1940s through the 1960s.

"Sure it was illegal," says Callison. But wolves seemed so evil to him that legality hardly entered into it.

Today, Callison no longer hunts, but he still has the same sentiments about wolves. He sees the wolf as a competitor, and thinks it would be foolish of him to let the wolves take away the source of his livelihood. "Kill all these wolf off in the ranching country and the hunting country," he says. "There's plenty of other places for that wolf to go. Let 'em live out in the muskegs."

Hugh Walker grew up in Shageluk, on the Innoko River in Alaska. A native American, he is working on a degree in social work at the University of Alaska. He tells a story his father told him. In the 1960s, his father and his uncle were shooting wolves from airplanes, hoping to collect bounties and sell the furs. One winter day, he said, after following wolf tracks, his father and his uncle herded a pack of wolves out of the forest and into the open on a valley floor. Says Walker, "The wolves were heading for the timber and they got almost halfway over and they realized the plane was going to catch them. They turned around and faced the airplane and were jumping off the ground at it. Over the sound of the engine, my father and my uncle could hear the wolves barking and snapping. They were barking and screaming because they knew it was over. My uncle did shoot them. But from that day, they could not do that any more. They didn't have the stomach for it any more after that."

Max Lipinski lives on the farm where he was born in 1917, just outside of Bonnechere, Ontario. His parents were Polish immigrants who brought their own traditions of wolves with them. He is an austere man, a man without fancy, a man who never married and lived alone through long, silent winters. His voice is like wind on rock, almost a whisper, almost a groan. "I know wolves," he says. One February morning, he looked up from his porch and saw a wolf in his field. "It wasn't bothering anything," he admits. He long ago stopped raising livestock or poultry. "But I want to get even. I know what they can do. They took a lot of sheep. They took calves. And they didn't do it because they were hungry. There were many deer, beaver, moose—lots to eat." So he got his gun and shot the wolf.

"I'd shoot them every time," he says. "They got no business being here."

Jason Badridze is a small, balding man with dark, heavy-lidded eyes and an aquiline nose. Trained as a neurophysiologist, he is director of the Vertebrate Behavior and Ecology Section at the Institute of Zoology in Tbilisi, Republic of Georgia. As a boy, he hiked in the nearby Caucasus Mountains, and, perhaps once a year, he would hear wolves howling. His father had taught him to respect animals. He saw parallels between the communist occupation of Georgia and government persecution of wolves. In 1974, he began to buy wolf pups from local hunters, paying them the bounty they would have gotten from the government. He studied them in his lab, and concluded that wolves have their own languages, and that they can count up to seven. But when the captive wolves got older and Badridze realized he could not keep them in the lab, he had to choose between euthanizing them and releasing them into the mountains. He chose the latter. Using homemade electric-shock collars, he conditioned them to avoid livestock and humans other than himself, then released the wolves in the mountains and began studying them in the wild. He could get them to come to him by howling, and he continued to feed them even after they proved capable of catching their own prey.

By 1992, he had released fourteen wolves in the mountains. Because they were hard to observe in the forest, he would take some to a desert valley near the Trialete Mountains, release them, observe them, and bring them back to his laboratory. In order to follow them, he constructed his own radio collars and his own radio receiver. One day, he was sitting between two overhanging rocks to stay out of the hot sun, fiddling with the controls on his radio receiver, when a vehicle full of soldiers rumbled up to him. The soldiers had chanced upon mysterious radio transmissions out in the desert, and had followed them to this curious anchorite, turning a radio antenna suspiciously between two rocks in the middle of nowhere. They thought him a spy, for only a spy could, in their imaginations, possess such a radio, but they could not begin to guess what sort of spying he might be doing out there. When he told them he was studying wolves and had equipped them with radio collars, they didn't believe him. He said he would call the wolves so they could see the radio collars, but his sympathy with wolves seemed so preposterous that they took him away. They detained him for three days, repeatedly asking to whom he was giving information. At last, unable to make sense of the incident, they relented. He took the soldiers to where they had found him and howled; the wolves came in, he returned them to his laboratory, and the army took the collars away. But, he says, "The government doesn't know about the wolf releases."

. . .

In none of these stories are the actors able to explain adequately what wolves mean to them. But their feelings are intense, and the stories argue that wolves exist as much inside our minds as outside.

It is not news that we use animals as symbols—as mascots for athletic teams or national emblems. But sometimes we use animal symbols in very profound ways. We don't only think *about* animals: we think *through* them. They become mental forms around which we wrap ideas, hopes, fears, and longings.

For hundreds of thousands—perhaps millions—of years, we lived as hunters and gatherers. Our survival and the survival of our offspring depended upon seeing deeply into nature and into the behavior, not just of prey and competing predators, but of all the smaller creatures that animated the world around us. For millennia, human awareness of the world was attuned to the cadence of bird song and insect buzz, the movement of shadows through the trees, the flutter of leaf in the wind, and our minds must have been selected on the basis of how acutely we observed other animals.

Just as our bones and teeth are shaped by what we eat, our minds are shaped by what we think about. Our thought processes arose in part through consideration of the immense variety of form in other living creatures. Animals have come to embody concepts for us. They give life to abstract qualities, such as quickness, strength, cunning, timidity, and aggressiveness. We think about human character by thinking about animals, because it allows us to define sharply personality traits that may be confusing in the people we know. Humans are complex and paradoxical. A man may be a fierce warrior and a warm and loving father. He may kill without remorse one day and cry at a wedding the next. The concept of fierceness is clearer when it is cloaked in animal skins, so we describe such a man as a wolf or a bear.

We almost daily use animals as metaphor or simile to communicate human qualities to one another. We are lion-hearted or eagle-eyed, proud as peacocks, sly as foxes, big as whales, timid as mice, cross as bears, stubborn as mules. We'll worm our way into a conversation, hawk our goods, or weasel out of a deal. We still use animals in advertisements to try to convey ideas of integrity, freedom, power, and simplicity, but there is more going on here than mere metaphor. We keep in our minds a menagerie of types which we have drawn from animals, and we use them to formulate thoughts about human behavior. Since the 1930s, scientists have sought more formal insights into human nature by studying the behavior of elephants, lions, monkeys, gorillas, and chimpanzees. They have brought new rules to the custom. But the habit of understanding and explaining ourselves by thinking about other creatures is older than science.

We are increasingly estranged from nature, in cities where walls and pavement banish most other forms of life. Our mental menagerie grows smaller and smaller, because we see fewer and fewer wild creatures. Even those of us who are birdwatchers or hunters or hikers spend far less time out of doors than our ancestors did, and if we step out at all, it is into an intensely altered and managed landscape. What we see out there are glimpses, slivers of life, mere hints that there is a wider, deeper world.

We still have need for animals to think upon. Despite what we have learned from the biological and social sciences, at no time in human history have we been less certain of who we are as people and what we are as a species. Many of us are drawn to animals, we say, because of their beauty or their liveliness or their faithfulness to ancient form. But I suspect we are really drawn, not by their shapes and textures, but by what we think through them. And what we think through wolves is never simple.

We are drawn to wolves because no other animal is so like us. Of all the rest of creation, wolves reflect our own images back to us most dramatically, most realistically, and most intensely. We recognize chimpanzees and gorillas, which are more like us in body structure and which show capacities for language and tool-making, as our closest evolutionary relatives. But long ago we diverged from chimpanzees and gorillas, and we have been shaped by different habits. As a result, though we are in many ways like chimpanzees, we are in some ways more like wolves. Like wolves, we evolved as hunters; we have long legs and considerable powers of endurance, adaptations to the chase rather than to hiding; we have minds that are capable of fine calculation, not just of spatial relationships, but of strategy and coordination. Like wolves, we band together to kill larger prey, and that has given us a different social system and a different personality from the chimpanzee; we have long childhoods, strong social bonds, complex social roles, and status differences; we tend to claim and defend territories; we have complex forms of communication; we are individuals; we have strong emotions. Humans and wolves are so much alike that they take an unusual interest in one another. Wild wolves have often followed humans with what the humans felt was friendly curiosity, perhaps even a desire for company. And humans have, for thousands of years, adopted wolves and felt with them a mutual sense of companionship.

Wolves do little to resist our mythologies. The real facts of wolf life remain hidden, because wolves in the wild have evolved a shyness of humans. When humans appear, wolves vanish. Our glimpses of them, with rare exceptions, have been fleeting and incomplete. In thirty-five years of studying wolves, except in the Arctic, biologist David Mech saw wolves only fifteen times without the assistance of radio-telemetry devices or airplanes, and most of those sightings were of wolves that dashed across a road in front of his car. In 1986, he began observing wolves on Ellesmere Island, in the Canadian Arctic, where there are no trees and where he could follow the wolves all day. Such day-to-day familiarity between human and wolf is something almost wholly new, something denied to humans other than the Eskimos and Indians who ventured into the far north, where trees do not obscure the view. For the most part, we have seen only a fraction of the animal's nature, and we have let our imaginations fill in the gaps.

What do we really know about wolves? What do we merely imagine? In a world molded so much by human activity, the future of wolves depends on our understanding of how fact shapes symbol and symbol shapes fact. It is a problem that all wildlife conservation faces: before we can save biological habitat for birds and butterflies, we must take steps to manage the habitat of the human heart. In the following chapters, I hope to explore the tangled and shadowy relationships between fact and feeling, biology and mythology, wolf and human.

THE COMPANY OF WOLVES

1

THE COMPANY OF WOLVES

Diane Boyd came to Montana for the wolves. In the 1970s, as a wildlife-management student at the University of Minnesota, she had worked for L. David Mech (pronounced "Meech"), whose studies on Isle Royale and in northern Minnesota had made him the dean of wolf researchers. It was a period of remarkable new interest in wolves. Mech had published *The Wolves of Isle Royale* in 1966, and then his detailed account of wolf biology, *The Wolf,* in 1970. These and the writings of a handful of other researchers had already turned aside centuries of folklore and apprehension of wolves. And while Boyd was at the University of Minnesota, Barry Lopez's *Of Wolves and Men* had edged American and European views of wolves into spiritual and ethical realms.

It was a time of change for wolves, too. Since 1973, Dr. Robert Ream, a biologist at the University of Montana, had been collecting reports of wolf sightings in Montana and Idaho. Wolves had been declared an endangered species in all the United States outside of

Alaska. They had not reproduced successfully in Glacier National Park for fifty years, and in that time there had been only sporadic sightings of wolves in Montana. The closest known breeding population was 150 miles away, in Canada's Banff National Park. Ream believed wolves might return to Montana. In April 1979, Joe Smith, working with Ream and with Canadian bear-researchers, had trapped a female wolf in British Columbia, not far from Glacier. Smith put a radio collar around the wolf's neck and released her. The presence of a radio-collared wolf was a fresh research opportunity for Ream, and he wanted the animal watched. When Diane Boyd enrolled as a graduate student at Montana, Ream hired her to work on the Wolf Ecology Project.

Tall and blonde, with the high cheekbones and pale complexion of her Nordic ancestors, Boyd would become an anomaly in the world of wolf research, a woman in a field dominated by men, and a person of searching curiosity in a science often dominated by reductionist skepticism. The job might have gone to a male researcher had others than Ream believed that this might be the beginning of a return of wolves to Glacier National Park. Other wolf scientists pointed out that the one radio-collared female did not constitute a breeding population. They saw it as an outlier, an oddity that would probably disappear.

Boyd proposed to study the relationship between wolves and coyotes and to write a master's thesis on the results. She intended to track radio-collared coyotes, the radio-collared wolf, and any other wolves she could find. There were wolves just north of the international border, which hunters and timber cruisers saw now and then. She would collect and analyze scats and search for carcasses of wolf-killed elk, deer, and moose, find out what coyotes and wolves were eating, and judge whether they competed for food.

Boyd caught dozens of coyotes, but she didn't see signs of another wolf. The lone radio-collared female she sought to study was a traveler, a crosser of ridges and rivers, a consumer of distances. Wolves don't stay in any one place; motion is their characteristic state. They must move around as packs in search of prey, or move as individuals to new locations in search of unoccupied ground or a pack hospitable enough to welcome them. The long-legged, loose-jointed trot of a moving wolf is as much a defining quality as the creature's teeth. A wolf traveling across the landscape bounces slightly. Its big, splaying paws glance off the earth and curl as they rise, and they seem to whip the body forward. The lithe backbone coils and releases like a spring, but the bend is almost unnoticeable, so fluid is the motion. The hind feet tread in line behind the forefeet, unlike dogs, whose hind feet slap down *beside* the forefeet. When a wolf is walking, its hind feet step right into the impressions left by the forefeet; the movement is spare and economical. A wolf effortlessly travels thirty miles in a day. David Mech followed a pack on Isle Royale that covered 277 miles in nine days.

The gray female was a disperser, a wolf that had left her natal pack somewhere far to the north and gone wandering. In this mountainous environment, wolves may disperse as far as five hundred miles. It is still not well understood why they sometimes leave their packs. Perhaps this wolf was driven south by some inherited longing to breed and have her own pack, and couldn't wait for the older females in her pack to die off. Perhaps she was chased out of her pack by more aggressive siblings. Perhaps she was simply not social enough to stick with the pack and one day drifted into the pines and never looked back. No one can say what may have led her south. Wolves tend to go where wolves have gone before. There are den sites and trails in the Arctic that have apparently been used continuously for centuries. Her journey may have followed paths taken by generations of wolves, or she may have followed the borders of existing wolf territories until she stopped where there was no fresh scent of other wolves. Perhaps she was led by an ingrained ability to recognize good country, by some inherited mental pattern of the relationship between slope and tree cover and meadow and lake surface that over the millennia programmed wolves to settle where deer or elk or caribou or bison would also find sustenance. For whatever reason, she came into the drainage of the North Fork of the Flathead River, on the northern edge of Montana, and there she stayed.

Why a lone wolf should have arrived in 1979, and not ten or twenty years before, may be easier to guess. Wolves had been persecuted in southern British Columbia and Alberta until the 1970s. They were poisoned until the 1960s, and no game laws restricted the hunting of wolves until the 1970s. In Alberta alone, more than fifty-four hundred wolves were destroyed in a rabies-control operation between 1952 and 1956. Wolves had been exterminated even from Banff National Park. But the poisoning stopped, and by the 1970s wolves were beginning to reappear there. It is likely that the gray female came from at least as far away as Banff. Habitat change might have helped draw her south. When logging companies clear-cut large swaths of forest in southern British Columbia and Alberta in the 1970s, they opened up new habitat to white-tailed deer. The deer proliferated, offering the wolves a food source they may have found newly sustaining. And in northern Montana, a series of mild winters gave an additional boost to the white-tailed deer population.

For fifteen months after the wolf was trapped, the radio collar beeped out the solitary wolf's locations in Glacier National Park and north of the border, in Canada. Skiing in winter or flying in small airplanes in summer, Boyd would locate the radio signal and plot the wolf's movements on maps. Ream named the wolf Kishinena, after a creek in the northeastern corner of the park. From the airplane, Boyd occasionally saw the wolf, a gray shape loafing in the snow or heading for the cover of trees to hide from the airplane. "She was extremely elusive," Boyd recalls. "Nobody ever saw her from the ground. *Nobody.* In winter, a lot of wolves, when they hit your ski tracks, will walk for a while in them, because it's easier travel. She would never do that. When she hit your tracks, she would go back the way she came and go around them. Or she would jump over them."

Still, Boyd was getting to know the wolf, and getting a feeling for her engagement with the wider world. One winter day, Boyd happened onto her tracks in the snow. Usually she backtracked such trails, lest she scare the wolf or accustom it to human presence. That day, she followed the tracks and came upon the carcass of a moose calf minutes after Kishinena had killed it. She didn't see the wolf, but the carcass was still steaming. The wolf had gnawed a small hole between the ribs of the moose and was starting to pull the entrails out when it heard the skidding sound of Boyd's skis and darted into the pines.

Realizing she had frightened the wolf away, Boyd took a quick look at the kill and left immediately. She waited four days before going back to the scene. When she got there, it was clear that Kishinena had not returned. Smaller tracks in the snow indicated that coyotes had eaten the moose calf.

She found that coyotes often trailed the wolf, and even urinated on the same rocks and tufts of grass the wolf marked as she traveled around her territory. They didn't accompany the wolf, but regularly, if the wolf left a kill to go off and sleep in the woods, the coyotes moved in and finished it. "They kept her hungry," says Boyd.

In July 1980, Kishinena's radio collar stopped transmitting, probably because the batteries died. Unable to get the wolf to step into a trap so she could replace the batteries, Boyd concentrated her study on the radio-collared coyotes. Every once in a while, she would come upon the tracks of the wolf, and backtrack them. She found very few of Kishinena's kills, and saw no sign of another wolf. There were reported sightings of a black wolf in the fall of 1981, but no one managed to snare or trap it, and its presence was so far simply a rumor. With no additional wolves, interest in Ream's Wolf Ecology Project flagged, and by 1982 Ream's funding had dried up. Boyd took a job as a fire lookout and stayed on the North Fork as an unpaid volunteer.

Then things changed suddenly. One overcast, ten-degree February morning in 1982, rangers Jerry Desanto and Steve Frye skied out from a patrol cabin in the northwest corner of Glacier National Park and crossed the fresh tracks of two wolves. They skied the four miles to Boyd's cabin on the North Fork, near the Canadian border, to tell her about it. "I just dropped what I was doing," says Boyd. She strapped on her skis and went out with Desanto and Frye to see the tracks.

They picked up Rosalind Yanishevsky, a volunteer bear researcher who was a winter neighbor, skied east along the international boundary to Sage Creek, and followed it south into the park. Along the creek, they struck the tracks of the two wolves. Desanto guessed that one set of tracks belonged to Kishinena, the only known wolf in the neighborhood, and the larger, accompanying tracks appeared to belong to a male. The male's prints were distinctive: the outer toe on his left front foot was missing.

The wolves were clearly traveling together. It was midwinter, the time when courtship and mating occurs. Desanto told Boyd that he had seen places where the two wolves urinated in the same spot. That excited Boyd even more. She had seen such double markings in Minnesota and knew that courting males and females urinate over each other's marks in the snow. Not far down the track, they all saw double marking, the sign of two wolves bonding. This was a mating couple.

For two years, Boyd had been tracking a creature she seldom saw, and which, because it was alone, was biologically incomplete. "It seemed all along like such a natural place for wolves," says Boyd. "But with only one, it seemed so bleak. These marks foretold a future for the wolves. I felt great excitement."

There was something more. The trail of the wolves led them south on Sage Creek to its confluence with the North Fork. With Yanishevsky, Boyd followed the tracks north along the North Fork. They found that the wolves had passed just across the river from Boyd's cabin and stopped for a long while, looking at the cabin as if considering at length something that was going on there.

After her two years of rare and sporadic sightings of the solitary wolf, Boyd was elated by the prospect that there might soon be a family. She was not to be disappointed. In June, Bruce McLelland, a Canadian biologist, drove up to a recently logged clearing in a remote part of the forest a few miles north of the international border and saw seven wolf pups. Boyd and Desanto hiked to the spot and found the litter. One gray pup poked along through the weeds and wildflowers, and a second pup darted into the woods so quickly that she could not decide what color it was. The other wolves had already fled. She did not see any adults. She could not be certain that this litter was Kishinena's, but she knew of no other wolf in the area. Both Boyd and Desanto were wildly excited.

With pups, the pair of wolves had become a pack, a complete and functioning unit of wolf society. When the pups were two months old, the three-toed male was accidentally killed in a bear snare. But the female successfully reared all seven pups without the male, and by winter the pack was traveling south across the international boundary. For the first time in a half-century, a family of wolves hunted, howled, and tested the winds of Glacier National Park.

The event pumped new interest and new funding into Ream's Wolf Ecology Project, and Diane Boyd was again employed. She would hear of hunter sightings of the pack, and now and then she would see their tracks. She never identified Kishinena or her tracks again, but eventually she would trap and radio-collar many of her presumed descendants. In 1984, a lone gray male Boyd believed was a member of this original litter was trapped and radio-collared. Boyd tracked him for two years as he wandered northeast to Waterton Park and south to Lake McDonald, and then circled back and found a mate in the vicinity of the Wigwam River, just northwest of Glacier National Park. Boyd guessed the female had dispersed from somewhere to the north. In 1985, McLelland caught a lactating female, probably one of the 1982 litter, put a radio collar on her, and released her. Boyd followed her back to the den, a few miles north of the Canadian boundary, and found a pack of thirteen wolves. That pack wandered back and forth across the international boundary. Boyd, Ream, and Mike Fairchild, who joined the project in 1984, named the group the Magic Pack, because it appeared and disappeared as if by sorcery. By the end of 1985, they estimated that there were fifteen to twenty wolves in Glacier National Park.

In 1986, the Magic Pack left Canada and denned in the park. It was the first known reproduction in Montana in over fifty years. That same year, the pair of wolves on the Wigwam River produced a litter a few miles north of the border. There were two packs to follow, Magic and Wigwam, and more wolves to trap. In summers, Boyd would set steel-jawed traps along unused park and national forest roads. The jaws of the traps were offset to minimize injury to the wolves, and Boyd checked the traps twice daily. The pups were not shy of traps, and Boyd was able to catch several and put radio collars on them. Though she found it relatively easy to trap adult females, the big feet of the large males would straddle the jaws of the traps, which would throw the feet out as they snapped shut. She collared only a few big males, but she trapped enough wolves to begin to know the social structure of the packs.

For fourteen years, Boyd kept the company of wolves. In summers, she would fly to locate radio-collared wolves and trap and radio-collar new ones. In winters, she would ski to tracks she had found by radiotracking from the ground and backtrack. She generally would not follow the wolves on the ground, because she felt it important not to accustom them to humans. Wolves are legally hunted just across the border, in Canada, where they are not considered an endangered species. In the United States, there are stiff penalties for those who harass or kill them, but they are nevertheless shot clandestinely by ranchers and outfitters who consider them dangerous to livestock or competition for hunters. Boyd didn't want to let the wolves get used to her presence only to be shot by someone else. Besides, backtracking would reveal what the wolves had done. Boyd would find where they rested, where they played, and where they hunted. She would see their scent marks in the snow in winter, and scats and scrape marks, where they had scratched the dirt, by the sides of roads in summer.

Over the years, she followed hundreds of miles of wolf tracks in the snow. She seldom saw the wolves. What she saw regularly were impressions of wolf life stamped upon the ground in a kind of code, the individual unit of which was the footprint. A wolf's paw is much bigger than a dog's, and Boyd could not completely cover a large one with her whole hand. A wolf print shows four big toe pads, each larger than a human thumbprint, and a big triangular heel pad behind them. There may be claw marks, but wolves tend to wear down their claws traveling, and prominent claw marks are more characteristic of dogs. The toes of dogs tend to splay outward, whereas wolf toes and claw marks tend to point ahead, as if the wolf were more focused on its goals. The footprints in the snow or the soft mud of a little-used dirt road telegraphed meanings the way letters cluster into words and sentences on a page of print. Here, where a wolf ran, prints were far apart, and the toes dug in. There, where a wolf walked, they settled down close and left even impressions all around the edges. Big wolves left deep impressions, younger animals daintier prints. A shallow depression showed where a wolf slept, or sat and scratched. Here they passed in single file through the deep snow of a winter meadow. There a playful wolf jumped out of ambush on an unsuspecting companion. Here a pack moved along a seldom-used dirt track at the edge of the park, focused and businesslike as it left scent marks and scats to notify other wolves that this territory was taken. Boyd could read the passage of wolves as if their images still hovered over the ground.

She often thus reconstructed hunts. One day, she and Ream were skiing on the Magic Pack's trail in British Columbia. The wolves had been trotting along in single file when they suddenly veered off the path they had taken. From the way the tracks turned, Boyd guessed they had scented a small herd of elk some distance away, in a clearing below a forested ridge. The wolves moved quickly through thick timber. Before they reached the end of the timber, the wolves began to lope. The elk, less than a hundred yards away, had probably not yet detected their approach. The wolves were running when they burst through the timber and into the clearing. The elk fled toward the opposite side of the clearing, but the snow was deep, and the wolves made directly for a cow elk that was having difficulty breaking a trail through it. The chase was short; they brought her down within a few yards. There was little sign of struggle. By the time Boyd and Ream arrived, all that was left was the skeleton, part of the face, and tatters of meat on the ribs. Ravens had already begun to pick the carcass clean.

Most of Boyd's views of wolves were from airplanes. She would see their long purple shadows strung in a line on a frozen lake in the low sun of winter. Sometimes, the shadows would be eight times taller than the standing wolf and would loom up like billboards as she flew toward a snow-covered meadow or lake. A couple of times each winter, she would ski to a location determined a few days earlier by flyover, and happen to reach the spot just as the wolves were passing back through it. "If they detected me, they would disappear," she says. Or she might see them far ahead, loafing in a meadow or sleeping around a kill, and would stop and watch for twenty minutes or an hour, until she became too chilled to stand still.

Boyd was witnessing a relatively unstudied aspect of wolf nature: the developing edge of a wolf population, the pioneering explorers and settlers of untenanted ground. Ream's collection of reports had suggested a trickle of wolves coming down into the Glacier area from Canada. Now the wolves were claiming the territory, settling down, having young and forming packs. Boyd tracked and watched. Nowhere else had human observers recognized the pioneering movement and studied it from the first wolf. And as she watched, a distinctive picture of wolf life emerged.

Glacier wolves are big wolves, comparable in size and weight to Alaskan wolves. An adult female weighs 80 to 100 pounds, an adult male as much as 125 pounds. Wolves of desert regions, such as *Canis lupus arabs,* the wolf of the Arabian desert, weigh 50 to 60 pounds and are small enough to be confused with the jackals or coyotes that share their range. The largest wolf taken in Alaska weighed 175 pounds. Hunters glimpsing such wolves in the woods have sometimes mistaken them for moose. Size is an advantage in bringing down larger prey, defending a territory against rival packs, or contending for standing within a pack, and, to some extent, in retaining body heat in a cold climate; but it is a disadvantage in that it requires more food to meet individual energy needs. In desert regions, prey is not large enough or densely populated enough to sustain large wolves.

Wolves vary in color from pure white to soot black. There are silky gray wolves the color of wood ash, and creamy tan wolves with gray brindled backs. Arctic wolves are mostly whites. The wolves of Glacier are mostly grays and blacks.

Wolves are social creatures, no less convivial and attached to each other than humans. The life of the individual is inextricably woven into the life of the pack. Some wolves travel and hunt alone, but their journeys are probably temporary solitudes that will eventually bring them back to pack life. Or, like Boyd's original subject, they are dispersers, gamblers on the future of the species, like humans who go off to cross unknown seas or explore remote frontiers; they are, in an evolutionary sense, probes sent off to try to establish wolf genes on new ground. Nineteenth- and early-twentieth-century literary naturalists focused at length on lone wolves. Perhaps the idea derived from our parting view of wolves as we shot and poisoned the last of them into oblivion. Stanley Young's *Last of the Loners,* a collection of accounts of legendary stock-killing wolves and the trappers who ultimately exterminated them, gave them names and characterized them as individuals alone and at odds with the world. A popular print by Alfred von Kowalski Wierold Zowalski, entitled *Lone Wolf,* shows a wolf standing in the snow on a hill at night, overlooking a small, low-roofed winter cabin with smoke rising from its chimney and a pale light blinking from its window. The picture is allegorical, not of the ravenous wolf on the prowl, but of the outsider looking in on the snug and private comforts of other beings. It hung in ranch houses and back-country cabins all over North America. The pictures and the literary wolves suggest that the lone wolf has become a human idea, a myth that says something about will or fortitude or

one's own tragically civilized loneliness. Th
metaphor for the alienation of modern lif
the nineteenth-century frontier, but it
wolves, tucked as they are into the c
do not much experience.

Most wolves live in packs of two to t
ing largely on the size of available prey. In M
the chief prey, packs number five to ten individua
which feed on smaller, more widely dispersed drylan
not to occur in packs larger than a pair plus their young o
In Alaska, where wolves prey on moose, packs may number te
twenty. Packs of more than forty wolves have been reported in bison range in northern Alberta, but whether these aggregations are long-lasting is not known.

Though wolf packs are cooperative, they are not democratic: they have rank orders among each sex. Typically, the dominant, or alpha, male lords it over all the other males, and a dominant, or alpha, female asserts her will over the other females. There may be a beta and a gamma, second- and third-ranked wolves of each sex, each imposing some will on the lesser-ranking wolves down through the Greek alphabet to omega. Boyd could read the dominance rankings even from the air. When wolves were spread out in single file in the snow, the dominant males and females carried their tails high, while the subordinates carried their tails low; the lower the tail, the lower the rank of the individual. There were other signs, too. Subordinates would approach higher-ranking wolves and lick at their faces, their heads low, their ears flattened, their tails pressed to their bellies. Or they would roll over in the snow and present their bellies and groin areas to be sniffed. Dominant males and females standing over subordinates in such encounters had stiff, unyielding postures.

The dominance rank order of a pack changes as wolves age or get injured and younger wolves mature and gain strength, confidence, and understanding. Typically, dominance orders are most unsettled early in winter, when there seems to be more fighting. Old dominants may be chased out of the pack or beaten and brought down to submissive roles. Younger challengers may arise, assert themselves, be beaten, and sink back into submission.

Hierarchies are a part of the dynamics of pack splitting and re-

In January 1987, Boyd found that Phyllis, the alpha fe-
e Magic Pack and the wolf collared by McLelland in 1985,
vn off by herself, either to the edge of the pack or wander-
tirely apart. Phyllis had evidently been displaced as alpha fe-
. Boyd had radio collars on a half-dozen wolves in the pack, and
was able to follow Phyllis as she left the pack and went north into
anada, found another mate, and denned. One by one, former packmates peeled off and joined her. Eventually, about half the founder wolves left the Magic Pack to join Phyllis. With the Magic Pack thus split into halves, Boyd and Ream retired the Magic name and designated the two new packs Camas and Sage Creek. In time, she would see at least ten packs form as wolves dispersed from existing packs.

The life of the pack keeps an annual rhythm. Females normally mature at twenty-two months. They come into estrus, or breeding condition, only once a year, in winter. The onset of estrus is gradual, and early in the season, before females are fully fertile, subordinate males may mount even high-ranking females, but they do not actually breed, because the females have not yet ovulated. As the dominant female becomes fertile, both she and the alpha male become increasingly aggressive toward subordinates. Something—perhaps the stress from this aggression—keeps subordinate females from becoming fertile. As she becomes ready to breed, the alpha female secretes a pheromone that causes the alpha male to become possessive and to guard her from subordinate males. Around the end of February, the alpha male will mount her, and during intercourse the bulbous end of his penis swells so that they are locked together or "tied" for up to thirty minutes. After they have tied thus once or twice, a firm bond seems to form. Once that has happened, the pack begins to calm down. There is less competition for status, and roles are generally settled for the time being. One way of viewing the returning peace is to say that the fighting is not so much over rank as over the right to reproduce. Subordinate males may mount and copulate with the alpha female after that, but the likelihood of conception so late in estrus is small. Usually, a pack produces only one litter each year, and that litter comes of the mating between the alpha male and the alpha female, but there are exceptions: in captive packs, subordinate females have borne litters alongside of or in place of the litters of alpha females. The bonded alpha pair may remain together for years, until one of them dies or is displaced from the top ranking by another wolf.

Gestation takes sixty-two to sixty-three days. In April or May, the dominant pair may draw off and den together, perhaps using a den that has been used many times before. The Magic Pack's 1986 den was in a thick stand of willows at the edge of a marsh, and the entire pack remained with the alpha pair. There the wolves dug tunnels and nesting chambers twelve feet into the hillside. The opening was so narrow that Boyd could only push her head and shoulders in and measure by pushing a flashlight and steel tape measure ahead of her. She thought it remarkably clean and odorless, with no scats or bones, no sign of wolves other than a few hairs rubbed onto the tree roots on the ceiling. The Glacier Park litters were born in April and May and numbered five to seven black, gray, or multi-colored pups. Wolves in the Arctic give birth as late as June, and those living in places other than Glacier have had litters with as many as thirteen pups. While the young are unable to hunt, one parent may stay with them as the other hunts (either parent may hunt), or the pups may be left alone, even in bad weather. The mother, however, must return to nurse pups in the den for thirty-four to fifty-one days. The successful hunter brings food back to the den in its stomach and regurgitates it to the waiting parent. Occasionally, a subordinate adult may stay with the young while both of the parents hunt, a habit some wolf biologists refer to as "aunting." Boyd found that widowed males or females like Kishinena successfully reared the young when a mate was killed, even without the help of aunts and uncles.

A pack is likely to consist of the alpha pair, its young of the year, and other adults, presumably the offspring of previous years. While the mated pair den, the other adults may disperse and hunt in smaller groups or as solitary individuals. Spring and early summer are times of plenty for wolves, since deer, elk, moose, and caribou are calving and the young herbivores are easy prey. David Mech found that in summer wolves on Ellesmere Island hunted as individuals as much as 80 percent of the time.

At about two months of age, the pups emerge from the den, and the adults move them to a meadow or an open area near dense forest or other cover, which biologists call a "rendezvous site." With moose, caribou, or deer calves now large enough to be more challenging prey, the pack may now reassemble around the rendezvous site, and leave the young there, sometimes with a subordinate adult, while adults go off together to hunt. And when parents or other adults return from a successful hunt, the young mob them, whining and licking their faces, thus stimulating them to regurgitate partially digested meat for them. A pack may change rendezvous sites several times during the summer as they follow prey or are disturbed by humans. By the end of summer, the young are able to go out hunting with the adults, and the pack is on the move as a group again.

Packs possess and defend territories from other wolves. We think wolves keep territories either to protect den sites or to conserve hunting opportunities, and that the abundance of prey generally defines the size of a territory. On Arctic islands where musk oxen and snowshoe hares are the prey, a pack may cover several thousand square miles; one Ellesmere Island pack covered 5,000 square miles in a six-week period. In northern Minnesota, where deer are more densely populated, a territory may be 50 to 100 square miles. In Alaska, where the prey is moose, a territory may be 800 square miles. In northern Montana, where prey is elk and deer, Boyd found territories are typically 300 to 400 square miles. In winter, the wolves may concentrate in small parts of those territories, because the wintering deer and elk concentrate in the river bottoms and adjacent hillsides. Territories change size and shape from year to year. Resident wolves scent-mark heavily at the boundaries of their territories, and trespassing wolves mark assiduously when they stray inside another pack's domain. It is as if a wolf thus leaves a string of boasts and threats on the ground to taunt and intimidate other packs and keep them out of its larder. Wolves in Alaska and Minnesota have been known to kill neighboring wolves that strayed into their territories. Boyd has, in fourteen years, found only one case of wolves killing wolves in the Glacier area. "I think that's because there is a low density of wolves, a high density of prey, and a lot of unoccupied range," she explains.

In the wild, wolves probably do not often live more than ten or twelve years, although in captivity wolves have lived as long as sixteen years. In Alaska, the average age of an adult wild wolf is three to four years, and only a few live to be ten. Normal birth rates can double the number of wolves each spring, but by winter natural mortality has eliminated 50 percent—chiefly young wolves and old ones. Some die of starvation when they are so badly injured by moose or caribou that they can't hunt. Wolf skulls have been found with holes kicked in them by deer. Some starve when hard winters reduce prey populations to levels that won't sustain the pack. Some kill each other in territorial and status battles. Bears have been known to dig into wolf dens and kill their pups. Wolves have also died in avalanches or falls through ice. Then there are human-caused fatalities. Wolves die of diseases such as rabies, distemper, and canine parvovirus, brought to their world by domestic dogs. Humans shoot, poison, snare, and trap them. In 1989, all of the members of the Wigwam Pack suddenly vanished, presumably poisoned. There were no operating radio collars left in the pack, and repeated searches for tracks and howls failed to turn up any trace of wolves. In 1991, the Headwaters Pack, which had formed in 1987 in Canada, disappeared: Boyd retrieved five of their bodies and concluded that they, too, had been poisoned. About 25 percent of the wolves born into the Glacier-area population die of trapping, shooting, or poisoning.

The picture emerging from Boyd's experience with wolves was different from the one people had held before the middle of the twentieth century. Before 1950, wolves were generally viewed in narrowly restricted terms. The standard text about wolf life at that time was Stanley Young and Edward Goldman's 1944 *The Wolves of North America,* a compendium of tales from the diaries and journals of explorers, trappers, ranchers, and historians of the West, combined with Young's own lifetime of experience with federal bounty hunters. Goldman, a taxonomist, contributed to the work chiefly by identifying the characteristics and ranges of various subspecies. Young wrote the narrative and descriptive parts of the work. He had himself made a living trapping wolves, and had worked since the 1920s for government predator-control agencies. Young drew very little from scientists, because few had ever studied wolves in the wild. In *The Wolves of North America,* he described a creature "symbolizing power, ferocity, sagacity, courage, fighting ability and ruthlessness." It was "a menace to human life," and "everywhere so destructive to domestic stock that constant warfare had to be waged against it." Because the wolf was being exterminated in the United States, Young professed to see it as a tragic creature, akin to the outlaw gunmen of the old Wild West. That merely reinforced the old myth of the wolf as a malevolent will with large teeth and glowing yellow eyes. At midcentury, one could hardly think of a wolf without thinking of the slaughter of innocent sheep or attacks on sleigh riders in winter snows.

The actuality never really fit Young's picture of murder and mayhem. In North America, accounts of wolf attacks on humans seem generally to be tall tales or accounts of rabid wolves. Randolph Peterson of the Royal Ontario Museum reported an attack in 1942 by a wolf on a Canadian railway section foreman, who was knocked, along with his handcar, off the railroad tracks. For half an hour, he fought the wolf with an ax. The wolf growled and gnashed its teeth and refused to flee even after the railroad man had hit it with the ax. Finally, a freight train came along and stopped. The engineer, fireman, and brakeman all came to the foreman's aid and killed the wolf. Though the animal was never tested for rabies, its behavior strongly suggests that it was rabid. Naturalist and editor George Bird Grinnell said he had looked for years for an authentic case of a healthy wolf attacking a human and found only the story of an eighteen-year-old Colorado girl who met a young wolf while herding milk cows at dusk. She called out and threw a stone at it, and the animal took her by the shoulder, knocked her down, and bit her on the legs and arms until her brother came to her rescue and killed the wolf. That wolf was not tested for rabies, either. J. W. Curran, editor of the Sault Ste. Marie *Daily Star,* for years offered a $100 reward to anyone who could prove he had been attacked by a wolf in the Algoma district of Ontario, but never had to part with his money.

The fact that wolves kill has always colored humankind's view of them. We have a hard time separating killing as necessary predation from killing as moral outrage. Real reappraisal of the myths of wolf ferocity did not come until scientists began to think of them in ecological and evolutionary terms, and thus changed the moral basis of the question of killing.

An ecological view states that, if nothing died, all the earth's available materials would be locked up in a kind of carbon freeze. If nothing ate plants, they would simply become huge masses of wood and coal sitting glumly over the millennia. At some point, all the available carbon would be tied up in living plants, and nothing more would be born. Death is nature's way of making things continually interesting. Death is the possibility of change. Every individual gets its allotted lifespan, its opportunity to introduce change through mutation or culture, its chance to try something new on the world. But time is called, and the molecules which make up leaf and limb, heart and eye are disassembled and redistributed to other tenants.

A considerable portion of the creation is devoted to the disassembly and redistribution of organic materials. Soil-dwelling fungi take apart the wood of trees, and bacteria consume what the beetles and fungi don't get. Many creatures don't wait for others to depart voluntarily: they kill and eat living organisms. Predation is a fact older than mammals, older than reptiles, older than the tooth or the claw or blood itself. Single-celled organisms in primeval pools hunted and gobbled each other with abandon. Predation is today the rule, rather than the exception, among vertebrates. Few fish, amphibians, reptiles, or birds live exclusively on plants. Among North American mammals, the only vegetarians are beaver, porcupine, hares and rabbits, pikas, manatees, pocket gophers, elk, deer, antelope, moose, bison, bighorn sheep, and mountain goats. Even creatures we expect to be meek have predatory moments: chipmunks have been known to kill mice, and one observer has seen deer consume fish.

An evolutionary view sees the wolf as the product of a long line of evolutionary choices.

Two hundred million years ago, reptilian ancestors began to connect the lower jaw to the skull by means of a cheekbone hinge, enabling an individual to chew food into small bits, exposing more surface area of food to digestive enzymes, and thus allowing quick digestion. That development enabled creatures to heat their own bodies efficiently and gave rise to warm-bloodedness. That ushered in the age of mammals.

The earliest mammals were small nocturnal creatures, living in a world of fierce, quick, predatory dinosaurs. They probably relied more on scent than sight, and were furtive, secretive, and probably drab in color. They were more frequently prey than predator.

But, as often happens, the drab get even. About sixty-five million years ago, the large dinosaurs disappeared. Many scientists believe that an asteroid hit the earth, throwing up huge dust clouds that blotted out the sun and reduced photosynthesis to the point where food resources for the larger vegetarian reptiles crashed. Creatures that lived in earth burrows and manufactured their own heat when abroad were more likely to survive. The age of reptiles came to an end, and the age of mammals began.

Even without an asteroid collision, the days of the cold-blooded dinosaurs were probably numbered. Sixty million years ago, the continents were drifting apart from their earlier union in a single supercontinent. The continental masses moved away from the earth's equator, and as they drifted, they rose and folded. Mountains uplifted and ocean currents changed. Cooler local climates came with the uplands and the sea currents. In place of the tropical forests, new kinds of plants were evolving. Given the cooler climate and the new plants, dinosaurs would have been hard-pressed.

In this changing world, grasses evolved. Savanna and prairie spread over large sections of the continents. And with this new habitat came a burst of evolutionary activity among mammals. Both hunters and hunted evolved dramatically. Earlier plant-eaters were slow and clumsy creatures with broad heads, big bones, and elephantine feet. To take advantage of the new grasses and shrubs, some plant-eaters developed the ruminant digestive tract, which allowed cattle, deer, and antelopes to eat quickly, then go off to some safer place, regurgitate the unchewed food, and quietly chew and swallow it. That put a greater premium on the ability to escape, and deer and antelope developed longer legs, smaller feet, and relatively large bodies, to contain larger digestive systems and larger lungs. They developed a wrist bone that hinged top and bottom and so gave them greater flexibility and extension. The long bones of third and fourth digits fused to form the cannon bone. Gradually, the radius and the ulna fused into a single bone. Plant-eaters were increasingly designed to eat and run. Escape and wariness became the great advantages, and deer, camels, llamas, pigs, sheep, goats, bison, and cattle became the common herbivores.

As the herbivores grew faster, the slow, broad-headed, short-legged creodonts and mianids that had preyed on their ancestors died out. The surviving predators developed longer limbs and bigger brains. Out of these trends emerged the modern carnivores. Carnivores are defined by the presence of long sharp canine teeth and steeply ridged molars and premolars known as "carnassials," which are used for shearing meat. Carnivores include bears, cats, hyenas, weasels, skunks, raccoons, and dogs. And the dogs, or canids, include foxes, coyotes, jackals, and wolves.

The first wolflike canids emerged about three million years ago. In the Eastern Hemisphere, they gave rise to the various species of jackal; in the Western Hemisphere, to coyotes. The first wolves probably split from the coyote line about one million years ago in the New World and then migrated to the Old World. Perhaps seven hundred thousand years ago—perhaps more recently—the gray wolf emerged from the evolutionary mists in the Eastern Hemisphere. There is much debate about the exact line of descent. Ronald Nowak of the U.S. Fish and Wildlife Service believes the ancestors of today's red wolf were in North America a million years ago, and that one descendant migrated to South America, where it evolved into the dire wolf, and another went north to Alaska and the Old World, where it evolved into the gray wolf. The gray wolf recrossed the land bridge to America sometime between three and six hundred thousand years ago. By the time of the Roman Empire, the gray wolf ranged over most of Eurasia and North America and boasted the largest range of any land mammal.

Cats split off the line that would lead to dogs about forty million years ago. The differences between dogs and cats tell much about the kinds of choices that were being made. Dogs are relatively long-limbed and slender-bodied. The lower part of the leg is generally longer than the upper part, for greater leverage. Dogs evolved in open country, where hunting requires pursuit. They are built for speed and endurance. Cats, on the other hand, are designed largely to lie in wait and take prey by stealth. Because they are built for speed, dogs walk on the very tips of their toes. Cats are slightly more flat-footed, a trait that allows them to maintain sharp claws, whereas dogs wear theirs down. Cats have wrists that turn, and with their extensible claws they can grab a prey and hold on to it while biting.

Long legs alone don't make a runner: runners need long noses. Long-distance running produces a lot of body heat, and, to vent that heat before it forces the animal to stop or damages the animal's organs, dogs have long muzzles. Carnivores, like most of the mammals, don't sweat. They cool down by panting. A large portion of a dog's blood is pumped through the blood vessels of the nose. The *rete mirabile,* a network of blood vessels at the base of the brain, carries this cooled blood from the nasal passages across the arteries conducting blood from the heart to cool the brain. Thus, a dog's long face is there not just to smell things but to chase things.

Dog skulls tell yet more about evolution's choices. Cats are shorter-muzzled and have, for the size of their jaws, a more powerful bite than dogs. A lion, for instance, can bite through the neck of a buffalo. Cats tend to hunt alone, coming together solely to breed. Only lions and cheetahs hunt in groups. The solitary habit probably explains, too, why the largest cats are larger than the largest dogs. A single lion or tiger or leopard can bring down a large prey that even a pack of dogs might be unable to tackle.

Since dogs can't deliver the quick fatal bite, they tend to hunt in packs. Their jaws are designed to nip and tear. When wolves attack a moose, one wolf may grab the moose's nose and hang on while the much larger moose shakes him like a rag doll, trying to dislodge him. Meanwhile, other wolves are diving in from the rear to slash at the moose's flanks. The operation may take hours. The wolves will withdraw and wait while the wounded moose stiffens or weakens from loss of blood. Death can be slow for the prey, and because this seems cruel and inefficient, it has earned wolves the scorn of humans.

Much is made of a wolf's fangs, but it is worth noting that a wolf's canine teeth are not really all that distinctive. Big canines are relatively common among mammals. Several species of deer have diabolically overdeveloped canines that protrude over their lips. Pigs elaborate their canines into huge tusks that curl back toward the eye. Hippopotami have extravagantly developed canines curving into sharp tusks. It's not the canines that make wolves special—it's the teeth behind them, the carnassials.

The teeth of dogs may also tell something about their social order. Dogs have four premolars, cats only two. Once a cat has stabbed or suffocated its prey to death, it can drag the prey to a tree or some safe place and eat without hurrying. But because they hunt in packs, dogs cannot. In a pack, there are fights over meat, and a slow eater among fast eaters may be doomed. Premolars are useful for holding food, and for shearing it off quickly so that it can be bolted down. The dog's extra premolars might have evolved as a way of coping with the darker side of sociability—the greed, jealousy, and envy that are the shadows of cooperation, love, and care.

Coordinated pursuit requires greater intelligence. The wolf's ancestors learned to track, to communicate with one another, and to read strengths or weaknesses in prospective prey. Wolves learn what to hunt and how to hunt it. There are wolves that live today among livestock but never seem to think of cows as food. Near Churchill Bay, Ontario, a wolf pack learned to distract mother polar bears long enough to kill and eat their cubs. In this century, Alaska's Brooks Range wolves were almost eradicated by poisons, traps, and disease, and in the absence of predator pressure, moose moved into the region. After wolf controls ended, wolves returned to the area, but fed on caribou. Twenty-five years passed before they started preying on moose again.

As intelligence becomes more important, youth becomes more protracted. Nature decrees dependency and helplessness as a way of keeping an individual out of trouble until it has sufficient knowledge and experience. To equip an infant with adult bulk and power is to make a destructive monster, a creature that poses danger to its kind and lacks the experience needed to live long enough to pass on its genes. Large carnivores require long schooling. Bear cubs may stay with their mothers two years. Wolves may stay with their parents all their lives.

That long childhood leads to strong social bonds, affection, loyalty, self-sacrifice, and the love of touch. Twenty-nine of the thirty-five canid species have been studied well enough for us to understand how they rear offspring, and in all twenty-nine the male helps the female rear the young. In nine of them, other individuals are also involved in the rearing. Red foxes may have an individual other than the mother or father feeding the young. Unmated jackals of either sex act as nannies while the parents go off to hunt. Both male and female domestic dogs have a propensity to adopt orphaned pups.

All that tightly knit sociability also leads to more complex forms of communication. Canids develop languages of gesture, posture, voice, and scent, and these languages are most complex among the most social carnivores. Wolves are extremely expressive. Their faces are designed to convey quick and subtle changes of mood. The color and shape of their bodies allow them to transmit messages with changes in posture. They have supracaudal glands which secrete substances that impart information about an individual's readiness to breed, and anal glands which impart odors to feces that indicate the individual's identity and probably a great deal more. They constantly urinate and defecate around their territories, and other wolves are presumably able to smell in these scent markings indications of the depositor's sex, age, physical condition, and emotional state. So they are able thereby to share with one another and intruding canids essential information about who they are, what they are eating, how healthy they are, perhaps even how they feel about strangers walking into their territories. They have a variety of vocalizations—barks, howls, growls, whines, and whimpers—each of which may have subtle and complex meanings.

Predatory habits led also to strong emotions. Predators must be a balance of calculation and action. The action must be intense, moved at times by desperate need, and emotion is what triggers such action. In the wolf, emotion seems to run strong. Some individuals get carried away with rage or aggression, some can be paralyzed with fear or resignation. To see wolves at play, leaping over one another, backs arching sinuously, bodies almost hovering in air, with necks craned and eyes sparkling, is to see clear expressions of joy. Several observers have concluded that wolves howl in a particularly mournful way when a beloved companion has died.

Social predation also led to strong tendencies toward individual variety in temperament and character. When wolves hunt, they make use of disparate special abilities. Some wolves are good at finding prey, some at reading the strengths and weaknesses of prey, some at chasing prey, some at killing prey. A wolf pack is a collection of individuals. At the Folsom City Zoo in California, curator Terry Jenkins built a life-sized caribou out of cardboard and crepe paper for a wolf celebration at the zoo. She put the creature into the wolf enclosure. Each of the zoo's four wolves treated the caribou differently. Sage, the big white dominant male, focused entirely on the caribou's antlers, and seemed oblivious to the rest of it. Onyx, the black dominant female, sneaked up behind the caribou and bit mouthfuls of crepe-paper hair out of its flanks. Terra, a subordinate female, came over to Jenkins and jabbed her with her nose to ask for help. And Lupine, the lowest-ranking female, lunged at the caribou and knocked it down. Jenkins has since put out two more cardboard caribous for two more wolf celebrations, and the roles remained the same. Lupine made the knock-down and Terra asked for help.

Those who study wolves today see them as a collection of different personalities. There are gentle wolves and rough ones, shy ones and outgoing ones. There are clowns and scholars, bullies and pals. A pack is well served by having such variety. Whereas one wolf's aggressiveness may be valuable in bringing down large prey, another's timidity may serve the pack if the moose and caribou vanish and the hares and birds it has sharpened its skills on must feed the pack. There is probably call, too, for different styles within the pack—for nurturing wolves to bring pups along, for aggressive wolves to keep neighboring packs off the hunting territory, for studious wolves to read the prey better, for friendly wolves to stimulate the greater solidarity of the pack. "Wolves are individuals," says Boyd. "There are playful wolves. There are smart wolves. Some are born to be leaders. Some are not. It's just the same as people."

The wolf Diane Boyd observed was, at first glance, very much in the mode of this new ecological view, and a creature of more complex and intriguing character than the beast of Young and Goldman's views. But Boyd's experiences with wolves pushed her even further, into realms the more cautious scientists regard at best as softheaded and at worst as perilous.

Boyd talks about experiences she can't explain. Sometimes they are merely ironies. For example, she says, she may spend a long day crawling through tangled thickets of lodgepole pine without seeing a wolf or locating a radio signal. "Then you come home and find wolf scat outside your cabin door." At times she suspects wolves are laughing at her.

Sometimes, however, she has the impression the wolves are talking to her. Time and again, wolves have come to her cabin on the

North Fork and sat there, watching. She recalls a wolf that came down to the river near her cabin and sat there in plain sight howling, while she approached and took photographs. Whether it meant anything, and if so what, she cannot say.

Boyd tells a story she would not share at scientific meetings. For two of the early summers while she was trapping wolves, she would occasionally dream at night of trapping one. "When I woke up in the morning, I knew I had caught a wolf. I knew the color of the animal I'd caught, because I'd seen it in the dream." She would tell Mike Fairchild that she had had the dream, and they would go out to the trap line and find that she had indeed caught a wolf. She believes that 90 to 95 percent of the time she caught wolves those two summers, she had had such dreams. But it happened only during those two summers, and then stopped. "It sounds so wacky that I don't tell people about it," says Boyd.

To some extent, the ground on which such experiences fall was prepared by Barry Lopez. In *Of Wolves and Men,* Lopez dashed the myths of wolf rapaciousness and savagery, and showed that most of what people said and wrote about wolves consisted merely of deflected views of humans. The wolf "takes your stare and turns it back on you," he wrote. "People suddenly want to explain the feelings that come over them when confronted with that stare—their fear, their hatred, their respect, their curiosity." Lopez dealt especially with mythic and spiritual—sometimes even mystical—connections with wolves. He drew much from Eskimo and Indian views, in which wolves were spiritual, even supernatural beings. He focused a great deal on the fear and hatred in those who had defined the wolf before, and he showed that the real wolf bore little resemblance to the wolf of their views.

Lopez based his work on good science and close personal observation of wolves. He pointed out that many of the things wolf defenders said weren't true, either. He maintained that wolves did kill more than the old and the weak, that they sometimes killed more than they would eat, and that they had indeed been known to make unprovoked attacks on humans in North America. Readers sympathetic to wolves often ignored such statements, because, when it comes to wolves, we see what we want to see.

Lopez offered a wider context, in which views other than science might obtain a hearing. He wrote, for example, that there was more to hunting than killing, and that, in the context of hunting for food, dying might be as sacred as living. Because science could not respond to such an idea, he asked us to view the creature with our hearts as well as our minds. Many scientists, old trappers, and ranchers, people who based their view of wolves on the old understanding, did not like the book. Environmentalists and city dwellers who felt something was very wrong with the way humans approached nature embraced it. *Of Wolves and Men* threw the subject of wolves open to new approaches. And into the opening rushed a new generation of controversy.

Scientists would still claim the high ground in debates about wolves, but there would be other voices in the fray, voices informed by other kinds of perception. Says Boyd, "In the early years, I was careful to have no feelings about wolves on the outside, and to keep them as data points. Now I don't have to worry about that stigma any more. Now I know it with my heart and my brain."

Boyd is no less a scientist for recognizing that there are different ways of seeing wolves. She is careful to limit her scientific papers to observations and conclusions that other scientists can reproduce, and she is at least acquainted with her biases, whereas many more "objective" scientists have no idea how much their eyes are trained by culture. Moreover, she believes that keeping her mind open keeps her eyes open to unexpected possibilities in the lives of wolves.

In 1989, the Camas Pack denned in a hollow log in a stand of lodgepole and ponderosa pines, but the pups all died. Boyd guessed from the fact that the adults spent less time at the den that some disaster had befallen the litter, but she didn't want to endanger any surviving pups, so she waited a month before going in to inspect the den. Not far from the den, she found a scrape where an animal had dug something out of the soft duff of rotted wood and soil. The excavation was five inches deep, with the excavated earth piled to the side. In the excavated dirt she found some skin and part of the jawbone of a wolf pup. "It was apparent that it had been buried," says Boyd. "And it was apparent that something had come along and dug it up." Boyd returned the next day with her dog Max and followed Max's nose around the site; it led her to five similar scrapes, one of which held the remains of a pup. Max responded to all the scrapes with the same keen interest. Blood samples taken from trapped wolves that year showed that the wolf population had been infected with canine parvovirus. Boyd concluded that the pups had probably died of parvovirus and that some solicitous wolf had buried them. No other evidence of wolf burials has ever been recorded. Humans are the only species supposed to bury their dead. If a wolf inters her pups, what does that say about wolves? What does it say about us? Boyd reflects, "You can take that and run with it forever."

Boyd has become one of the rare researchers who are capable of encompassing the wide range of views about wolves. She hunts deer and elk for food, because she feels hunting allows a deeper relationship with the landscape than one may find shopping at Safeway. "Why would I want to eat a cow any more?" she says. "They don't have any spirit. A deer is a gift from the land." And the land holds secrets to our own nature. Watching wolves has made her more aware. "When I'm out there, I can track a wolf and see where they paused and looked out over a valley at something, and I'll do the same thing." She has learned to listen more carefully to what people say, and to be more tolerant of opposing views.

At the same time, she cares deeply about the wolves. In December 1992, Phyllis, who had been the alpha female of the Magic and Sage Creek packs, was shot to death by a hunter who had often seen her before and who knew her status in the pack. Phyllis had been an alpha female for seven years. She was an old wolf, wise in the ways of the wild and matriarch of a large and dispersing clan that might one day repopulate Montana's wilderness with wolves. "Of all the wolves," says Boyd, "I felt strongest about her." When she heard that Phyllis had been killed, Boyd grieved. She was angry. "I felt this no-good jerk went and plugged her."

But a year later, on her Christmas vacation, Boyd went into Canada and visited the site of the shooting. She spent half a day talking with the hunter who had shot Phyllis, and she came away feeling better. "The hunter is neither a bad person nor a wolf hater," she says. "I wish he hadn't shot her. I'm not saying what he did was right or wrong. But after talking to him and thinking about the ideas he grew up with, I understand why."

The new science struggles to be wholly empirical, objective, based entirely on the eye. But, however its practitioners may trim the heart out of their writings, they are still moved. It's a shadowy world out there, and unseen things murmur and scuttle under the leaves of fact. Those who understand this are better able to understand their own vision, and understanding our vision is part of understanding wolves.

2

LAST OF THE BOUNTY HUNTERS

It is hard to talk about wolves in North America without talking about the campaign of eradication we waged against them. But other views of wolves preceded the era of organized slaughter, and the mania for wolf control, laid against the broader backdrop of human experience, appears to be an aberration, a temporary sickness that afflicted only some of our species, and which even some of the most avid wolf hunters came to regret.

Earlier cultures looked more favorably upon wolves. Among North American Indians, clans were identified with particular animals, and a member of the bear or badger clan might look to that animal for guidance or inspiration. A clan member might be prohibited from killing his or her totem animal, lest the animal spirit take offense and abandon the mortal. Wolves were often the totems. The Moquis of the American Southwest, for example, divided into wolf, bear, eagle, and deer clans and believed that at death the spirit of the departed returned to the body of a living bear or wolf or eagle or deer. The Niska of British Columbia divided into wolf, bear, eagle, and raven clans, and all had specific prohibitions connected with the totem animals.

The sanctions of a totem animal could be forceful. Frank Glaser, who trapped and poisoned wolves in Alaska early in this century, told a story in *Outdoor Life* magazine about a rabid wolf that attacked an Eskimo named Punyuk. In the middle of January, Punyuk was camped in a stove-heated tent between the Kobuk and Selawik rivers in Alaska. It was late in the day; the sun was down and it was dark. His dogs, tethered to willow clumps outside, began to growl and bark. Punyuk went out to see what was going on. In the darkness he saw what he took to be one of his dogs running loose. He threw a chunk of ice at the animal and ordered it to come. It leapt on him, knocked him down, and bit him about the head, tearing open his scalp. He managed to rise, open a pocket knife, slash the animal, and then choke it into unconsciousness. But he could not kill it, because he was a member of the wolf clan, and the spirit within the wolf might belong to an ancestor. He called it Grandmother and told it to go and leave him in peace. But the wolf revived, renewed the attack, and bit Punyuk on the thigh, laying bare the bone. It knocked him down again and bit into his shoulder. Punyuk passed out. When he awoke, the wolf was gone.

Punyuk managed to harness his dogs and sled his way to medical care. His wounds, though serious, did not seem to be mortal. Meanwhile, the wolf entered a village and killed some dogs, and was shot. Glaser was able to get the wolf's head, and sent it to a laboratory to be tested for rabies. The test came back positive, but it took the laboratory a month to report its findings. In the meantime, Punyuk, apparently healed from his wounds, went back out to camp, where he collapsed and died of rabies.

Wolves were sometimes invested with special powers by whole societies. A Kwakiutl creation myth tells how the antediluvian ancestors of the people took off their wolf masks and became humans. The Mongols viewed themselves as "sons of the blue wolf," descended through Genghis Khan from a mythical wolf that came down from heaven. Men dressed in wolf skins and ran through the streets beating people with leather thongs to purify Sabine cities. Roman soldiers wore wolf helmets to honor a wolf-god.

Even among people who respected and admired wolves, wolf tales tend to reflect and focus on the conflicts between our humane and destructive impulses. A Sioux woman named Brings the Buffalo Girl told the story of The Woman who Lived with Wolves to Royal B. Hassrick, who recounted it in *The Sioux: Life and Customs of a Warrior Society.* A young woman fought with her husband and ran away into the winter plains. She walked for days without eating, determined that her people not find her and return her to her husband. She climbed a hill and found a cave, crawled into its darkness, and went to sleep. When she awoke, in the dim light of morning, she could see she was in a den full of wolves. They spoke to her in human voices and told her not to be afraid. They brought her fresh deer meat and she ate. She stayed with them. The wolves hunted for her, and she cooked and dried the meat and made pemmican with meat and berries. She tanned hides and made dresses. But after two years, the wolves told her she must return to her people. The great wolf told her to walk to a herd of wild horses near the cave, and they would lead her back. He warned her that the stallions would try to force her to stay with them, however, and that she must not let that happen, for her people would get her back anyway. The woman left the wolves and found the horses. The stallions courted her, and she succumbed to them and refused to go back to her people. She ran with the herd. Her clothes turned to tatters, and she was covered with dirt and unrecognizable as a human. One day, hunters from her people came upon the herd and captured horses. They found her among them, roped her, tied her up, and dragged her back to camp. When they cleaned her up, they recognized her, but though they combed her hair and dressed her in clean clothes, she would never tame down. She lived with her people, yet apart, as a creature half wild.

The story trained the listener's attention on the line between sociability and individuality—between the need to cooperate and the urge to vent one's passions—that divides our lives. Many cultures made similar uses of wolves. Among the Nootka, boys were initiated into adulthood by being ceremonially killed and carried off by men dressed in wolf skins. The ritual represented, said a nineteenth-century observer, "the killing of the lad in order that he might be born anew as a wolf." It reenacted a tribal myth in which wolves taught a chief's son their rites and carried him back to the village, to teach humans how to live. The ceremony implied that to be an adult was to have lethal powers that one must learn to live with.

Romulus and Remus, the founders of Rome, were the issue of an illicit union between Rhea, the daughter of the deposed king Numitor, and Mars, the god of war. The twins were condemned to be put into the Tiber by Rhea's uncle, who had deposed Numitor. A wolf found them and suckled them. The implication in the story is that being suckled by a wolf gave the twins a wolfish nature. Raised by a shepherd, the twins returned to take revenge on their uncle, and eventually to rule the kingdom. They founded the city of Rome on the spot where the wolf had fed them. As in American Indian tales, the Roman wolf seems to be a lens through which we view the dual and conflicting nature of humans, for Romulus was fierce and suspicious, Remus deriding and jealous. Romulus forbade his brother to go past the frontiers, Remus disobeyed the order, and Romulus killed him.

When humans begin to fear their own predatory nature, wolves come in for very much darker imagery. At the festival of Lupercalia in Sabine cities, men dressed in wolf skins, slaughtered goats, and ran through the town beating whomever they met. Among the Greeks, there was a cult that imitated wolves and practiced ritual cannibalism. In Navaho society, witches were identified with wolves. Navahos believed in "skinwalkers," men or women who dressed in wolf skins, climbed on top of a neighbor's hogan at night, and dropped pollen or the ground-up bones of children down the smoke hole. It was said that they gathered in caves and sang songs backward to create chaos out of order, that they ate the flesh of the dead and had intercourse with corpses. They broke the boundaries between life and death, good and evil, community and selfishness.

Wolves seem to grow more fearsome as human conduct becomes more fearsome, and that may explain why Western culture takes such a dim view of wolves. European history over the past two millennia has been a progression of ever-widening wars. As the scale of battle grew, the number of dead left on the field also grew. And wolves, being scavengers, fed on the corpses. The Hundred Years' War between the Armagnacs and Burgundians left thousands of dead on the fields. At night, wolves came to feed on them, and, having acquired a taste for human flesh, the wolves came into towns and attacked people. In 1423 and again in 1438, wolves came into Paris, seizing dogs and children. On a December day in 1439, wolves ate four Parisian women. On the following Friday, they attacked sixteen more, eleven of whom died. Wolves would continually appear in Paris until the seventeenth century; the Louvre Museum is so named because it sits on a spot once frequented by them.

This kind of scavenging did not occur in North America, where Indian battles left few corpses behind. Plains Indians, for example, carried off their battlefield dead and put the bodies on platforms to keep the wolves from getting them. Only occasionally in North America did wolves feast on human flesh. They were said to have attacked and eaten Indians of Delaware Bay who were dying of smallpox in 1781. And when cholera struck the emigrant trains heading west along the Platte River in 1849–51, hundreds of trailside graves were dug into by wolves. But in general, wolves in the Western Hemisphere did not scavenge on human corpses, and that may account for the rarity of wolf attacks on humans in North America.

In Europe, those who saw wolves scavenging on battlefield dead associated the gruesome sight with the underworld and Satan. By A.D. 500, the Germanic word *wargus* was used to refer both to the wolf and someone who desecrated the dead. In time, it would also mean "outlaw," "evil one," a human possessed. Throughout Europe, wolves became associates of war gods. Artemis in her capacity as destroyer of life was accompanied by wolves. The wolf Fenrir accompanied the Teutonic god of war, Odin, and according to myth it was the breaking of the chain that restrained Fenrir that set in motion the end of the world. That myth may well have seeded the twentieth century's view that it would be war, rather than disease, overpopulation, famine, or pollution, that ultimately destroyed humankind.

By the fifteenth century, Europeans widely believed that there were wolves that prowled about human habitations and could not be killed. They were, according to the writings of religious scholars, emissaries of the Devil. Between 1598 and 1600, a French judge sentenced six hundred people to death, believing the Devil had rubbed their bodies with a satanic unguent, turned them into werewolves, and sent them to torment the countryside.

In spasms of civic spirit, European communities launched organized wolf hunts. The French army designated *louvetiers,* officers charged with organizing ordinary citizens to hunt wolves, and the office persisted into the twentieth century. The British felt so beset by wolves that in 1652 Oliver Cromwell forbade the export of Irish wolfhounds, lest Ireland have an inadequate supply of the dogs.

The war against wolves came to the New World as a virus in the mind of the first Europeans to settle North America. Wherever it found farmers keeping livestock, it grew deadly. The first domestic livestock arrived in the New World in 1512. In 1609, cattle, pigs, and horses arrived at Jamestown. Early settlements carved pastures out of the almost continuous forest of the Eastern Seaboard. That concentrated cattle and sheep onto open ground, where it was relatively easy for wolves to attack. At the same time, the settlers so reduced the deer and other native prey of wolves, that wolves were *compelled* to feed on livestock. The result was that, within twenty years, the colonies were establishing bounties on wolves. Massachusetts installed a bounty in 1630, Jamestown less than two years later.

We can get some idea what indiscriminate trapping, poisoning, and shooting might have done to wolf populations by looking at what happens to coyote and mountain-lion populations today in the western states. Not all coyotes attack sheep. Resident breeding pairs of coyotes tend to hold a territory and to drive out young dispersers, who have left their birthplaces and gone looking for unoccupied space to claim as their own. If the resident coyotes do not eat sheep, there is little or no predation on the neighboring ranchers' flocks. But if the resident pair is eliminated, the territory becomes a sink into which young wolves disperse from neighboring areas. Moreover, constant shooting lowers the age at which females bear young and raises the number of young in an average litter. So, where coyotes are shot, population density is apt to be high, competition for food intense, and the loss of sheep more likely.

A similar process may have occurred among mountain lions in California. In a study of mountain-lion predation on deer fawns in the Sierra National Forest, researchers found that mountain lions took a very high toll on the young deer. But they had trouble keeping the radio-collared lions on the airwaves, because poachers were shooting them. One of the collars ended up broadcasting from the bottom of a lake. With the poaching, there may have been no resident lions available to keep dispersers moving on. The researcher, perhaps not coincidentally, reported the highest density of mountain lions known to science.

Wolves, like coyotes, reproduce at earlier ages and have larger litters when their population is dramatically reduced. To control wolf numbers, the population must be reduced at least 70 percent each year. Early on, bounty hunting didn't destroy that many wolves, and probably did little more than give ranchers the impression that they had to keep up the pressure to prevent wolves from simply overwhelming them. In 1925, Henry Boice of the Chiricahua Cattle Company in Arizona observed that, although bounty hunters were taking 15 to 125 wolves a year from ranch properties, "the number of wolves running on our range remained about the same."

In the East, predator controls combined with alteration of habitat and elimination of deer to cause the extinction of the wolf by the early part of the century. In the West, the land wasn't cleared for farms, and extinction took longer.

Before the coming of the railroads, western ranching coexisted with wolves. The West was arid land and could not support cattle in concentrated pastures typical of the eastern states and Europe. A rancher in the West turned large numbers of cows loose on the open range, and left them unattended until roundup. Mexican ranchers turned wild longhorn cattle onto the range, and they probably held their own against wolves, because they were aggressive and stood their ground against wolf attacks.

But with the coming of the railroads in the 1860s and 1870s, that changed. A new kind of livestock industry appeared in the West. With access to eastern markets, investors poured large numbers of cattle onto the western range. Instead of the wild Mexican steers, they substituted more docile Herefords, Durhams, and Anguses, breeds that clustered around water sources and ran from wolves. Drought recurrently decimated these herds, and the investment ranchers looked to trim what adversities they could. There wasn't a thing they could do about the weather. But they could do something about predators.

In the last quarter of the nineteenth century, western ranchers set up local bounty systems, and many of the larger ranches hired their own "wolfers." They bought wagonloads of strychnine. Anywhere a cowhand found a dead cow or deer or dog, he would get down from his horse and lace the meat with poison. Stanley Young wrote in *Last of the Loners,* "There was a sort of unwritten law of the range that no cowman would knowingly pass by a carcass of any kind without inserting in it a goodly dose of strychnine, in the hope of killing one more wolf."

The poisoned wolves could then be submitted for bounties. Montana initiated a bounty in 1884, Arizona and New Mexico in 1893. In thirty-five years, more than eighty thousand wolves were submitted for bounty payments in Montana. Between 1895 and 1917, bounties were paid on more than thirty-six thousand wolves in Wyoming. The programs invited behavior we can today only consider insane. From 1905 to 1916, a Montana law required the state veterinarian to inoculate captured wolves with sarcoptic mange and turn them loose. Cowboys roped wolves, strung them between horses, and spurred the horses until the wolves were torn apart. They doused wolves in gasoline and set them afire.

The madness required justification. There were in these years fantastic estimates of the carnage supposed to be caused by wolves. Ben Corbin's *The Wolf Hunter's Guide* in 1901 calculated that there were more than a million wolves in North Dakota, that each consumed two pounds of beef a day, and that feeding them cost North Dakota over $44 million a year, and that it would take more than twelve thousand men killing a hundred wolves apiece to get rid of the problem. Arizona stock-growers estimated in 1917 that predators cost them $2.7 million a year. Curiously, seventy years later, when Dan Gish, a former wolf trapper himself, collected the annual reports of the Arizona federal wolf hunters from 1915 on, he found that the pages reporting stomach contents of the wolves taken had all been removed. In the 1918 report from New Mexico, of 189 wolves taken by the federal trappers, only 13 had beef in their stomachs.

The extravagant claims pushed counties and legislatures to keep the bounties coming. Ranchers pressed the economic argument, claiming that, since they leased grazing rights on public lands, those lands ought to come free of the costs of predators. They persuaded the United States Congress in 1915 to give the U.S. Department of Agriculture's Bureau of the Biological Survey funds with which to eradicate wolves on the public lands. The law established the Predatory Animal and Rodent Control Service, which employed hunters to trap and poison wolves, mountain lions, coyotes, and bears.

By the time PARC was established, poison had already eliminated all but the wariest of wolves, the wolves that were unlikely to take poison or step into a trap or the vee of a gunsight. Frustrated bounty hunters attributed to these remaining wolves extraordinary qualities. A wolf in the Chiricahua Mountains was said to have taken a cow every four days for at least four years. Old Aguila, a wolf that eluded Arizona trappers for eight years, was said to have cost $25,000 in stock. Individual wolves grew legendary, and before the war years were over, wolfers were selling tales of mythical wolves to popular magazines. Stanley Young, for example, in *Last of the Loners,* described Three Toes of the Apishapa, a wolf that was noble by day and diabolical by night: "Happy, carefree, led by the spirits of the wild," Three Toes and his mate "played in open sunny places, and then on dark nights trotted through the dusk, slaughtering and gorging in a bacchanalian orgy of blood feast." When Three Toes' mate was poisoned, the wolf howled in anguish, and then "there was new hate, new viciousness, new striking back at man." As the trapper approached, Three Toes waited: "Keen wolf teeth, bloodthirsty, ready to snap the veins in the throat of man, were at the end of his quest, and he could not come within their reach until they were made harmless." The trapper, the model of hard work and intelligent progress, always won.

In practice, the world of wolf control was less heroic and less dutiful. Bounty hunters sometimes left the dens untouched in order to ensure continued production of wolves on which to collect paychecks and bounties. They sometimes submitted the ears or noses of dogs or badgers or coyotes as those of wolves. And they sometimes turned in parts of animals for bounties in one county, sneaked them out the back door, and sold them again in another. In 1909, Vernon Bailey, a biologist with the Bureau of the Biological Survey, wrote a *Key to the Animals on which Wolf and Coyote Bounties are Paid,* to help county and state agents determine whether they were getting badger or dog parts from a professed wolfer. To keep the federal PARC trappers from abusing the bounty systems, the government required them to turn in the pelts, which the service then sold.

Because they were pursuing the last and wariest of wolves, PARC hunters became more skillful, more patient, more persistent than the run-of-the-mill bounty hunters. They set their traps around wolf-killed livestock, hoping the wolves would return, or set them on wolf runways, where they had found wolf urine or scats and knew wolves would return to sniff and leave scents. They spent months in pursuit of a single wolf.

But the traps merely finished off the toughest cases; it was poison that did the wolf in. Cowboys and trappers alike sprinkled strychnine crystals or inserted capsules of it into incisions in a carcass. Strychnine had a bitter, quininelike taste that caused wolves to spit it out, so trappers molded the tablets into balls of fat from the back of a wild burro or horse and scattered the baits around a wolf runway or a dead deer or cow. PARC hunters treated thirteen million acres of Arizona with such baits in 1923. They killed ravens, coyotes, foxes, wolverines, weasels, eagles, dogs, and human children with their poisons, but the justice of their cause was deemed unassailable. Mark Musgrave closed his newsletter to hunters, "Remember our Slogan, 'Bring Them in Regardless of How.' "

And his hunters did. Within three years, PARC hunters eliminated the last fourteen hundred gray wolves from Texas. By 1925, it was concluded that no wolves were rearing young in New Mexico, and in 1926, Musgrave reported from Arizona, "There are no more wolves left inside the borders of our state."

When wolves grew rare, coyotes kept the federal control agents in the field. After 1940, when the Bureau of the Biological Survey was merged into the new Fish and Wildlife Service in the U.S. Department of the Interior, federal trappers put out traps and poison for predators. (In 1974, the program took the name of Animal Damage Control.) Federal agents developed new methods of coyote control. One was the "coyote getter," a .38-caliber shell fixed to a firing-pin mechanism and mounted on a stake in the ground. When a coyote or wolf took the exposed end in its teeth, the device fired a charge of sodium cyanide into the animal's mouth, bringing death within seconds. A later, spring-loaded version of the device was called the M-44, because it had a mechanical firing mechanism and an enlarged shell-bore of .44 caliber. Coyote getters were not always carefully used. In 1952, PARC hunters set some at a rest stop along a major highway in Arizona, and several dogs were killed.

The fate of the wolf in the West was sealed in 1948, when a new poison, sodium monofluoroacetate, came into use. The chemical had been developed by the Army during World War II to kill rats in Southeast Asia. Modified by U.S. Fish and Wildlife Service scientists to kill predators, the formulation they arrived at was the 1080th compound they tried, and it became known as Compound 1080. Agents injected the tasteless, odorless substance into a freshly killed horse or burro, where it diffused to the whole body of the corpse, and remained potent for months. It was said to be selective, to kill only canids when used in proper doses. But in the field it was impossible to use only proper doses, and it killed bears, bobcats, badgers, foxes, skunks, raccoons, and carrion-eating birds such as eagles, magpies, and condors. In 1950, Fish and Wildlife Service officials went to Mexico to show U.S. ranchers there how to use it. Before the end of the decade, U.S. officials claimed complete control of wolves in northern Chihuahua.

These chemical predacides allowed predator-control programs to cover vast areas with little manpower. And though they didn't eliminate coyotes, they eradicated wolves. The last wolf was taken from Utah in 1929, from Colorado in 1943. The last "documented" wolf in New Mexico was a dead wolf, probably the victim of poisoning, found in 1970. Two wolves were taken in southwestern Texas the same year. The last wolf trapped in Arizona was taken quietly for a bounty offered by ranchers around 1976. Nobody was willing to say when the wolf was taken, because by then the wolf was on the endangered-species list, and killing it would have been illegal.

There was something both colorful and tragic about the era of wolf control—colorful because the stories the bounty hunters told romanticized the wolf and lent to the country a deeper mystery and a moral import, but tragic because the policy eradicated the wolf and damaged whole ecosystems. It also damaged the West's view of itself as something new, innocent, and pure of heart. The trappers would lose more than anyone: with the loss of the wolf would come the loss of their own identities.

. . .

Dan Gish was one of the last of the wolfers of the Southwest. I visited him at his home, a low, earth-colored house on a dirt street just outside Mesa, Arizona. The house is shaded with palo verde and other desert trees, and three large saguaros grow in the yard. Outside, it is leafy and overgrown. Inside, the walls are paneled with composite. The name "Jesus" is spelled in large cutout letters on one wall, and old illustrations are framed on another. Gish lives on a small pension and the house makes little boast of the American dream.

Gish has diabetes. Confined most of the time to a wheelchair, he loses sensation in his legs and feet. He gets up from the table in the living room and has walked across the room before he stops to look down and see if he still has his slippers. One of them has fallen off. "I'm not a failure," he says in answer to a question he has asked himself. "I succeeded in everything I undertook," he says. But it's clear he's struggling.

Gish has small, tight blue eyes set in a broad, beefy face. His hair is still full, dark, and wavy, graying only at the sides. Gish can be offhand, but he is not a man who smiles easily. He is watchful, perhaps apprehensive; he asks if I mind if he tapes our interview. Though he is passionate about his ideas, his sentences tend to swirl off into eddies of anger and bitterness.

His life, as he recounts it, runs a trail like that of a wolf pursued through a hard and unforgiving world. As a young man, Gish had trapped foxes, skunks, and weasels in the Midwest. When he worked in a factory in Milwaukee, he saw in it only an affirmation that humankind is greedy and corrupt. He joined the service in World War II, and at the end of the war he went west and found a job as an information officer for the Arizona Department of Game and Fish. Within a year, he was putting out strychnine drop baits to kill coyotes, in an effort to restore the deer population in the White Mountains. And in the course of that project, he started working with wolfers.

In the 1940s, wolfers still bore a tinge of romance. They worked alone in remote areas. If they were skillful, ranchers valued them highly. Stanley Young would write of them in *Last of the Loners,* "Dogged in a hunt, untiring in upbuilding of communities, quiet, high-powered, these men have been the bulwark of western progress." Gish still wears across his ample midsection a silver belt buckle given to him by a Sonoran rancher, General Alfonso Morales, whom he helped to dispatch wolves.

He speaks of the other trappers he knew as a select group of men. "These old trappers were peculiar individuals," he says. "They were cocky about themselves. They wouldn't share information. Dore Green, the head of PARC in Chicago, would try to get these guys together for meetings to demonstrate methods that were successful, and they would not do it. They would be squatting around a campfire at night, and if anybody tried to get them to share their method, they would just disappear."

In 1945, Gish began to work with Bill Casto, who had been trapping wolves since 1909. "He was an incredible naturalist and a loner and the best man at the job. It was an unbelievable education. He dragged me down into Sonora and Chihuahua. I participated with him in the capture of twenty-six wolves, all of them Mexican."

Canis lupus baileyi, the Mexican wolf, is the subspecies of gray wolf that ranged from southern Texas, New Mexico, and Arizona south into central Mexico. "These Mexican wolves have longer ears than the northern wolves," says Gish. "A Mexican wolf has longer legs. They're rangy. They have a deep but narrow chest, and the shoulders are very close together. From what I have observed in the field, with Casto and on my own, I think the range of a Mexican wolf extends three or four times more than a wolf on Isle Royale or up in Alaska, where the vegetation is contiguous and prey factors are different.

"Most of the Mexican wolves we found in Arizona were young, and most of them were males. But almost no reproduction. No denning. We only found one den, and that was on the east side of the Huachucas." Wolves then were highly migratory. The resident wolves had been exterminated, but, says Gish, "There was quite a bit of migration up out of the northern third of Mexico into this state. The movement of these wolves was in the vacuum created by the lack of wolves here. And they were exploratory and circuitous. Those that survived the trappers and ranchers followed a path back into Mexico.

"Bill knew the routes of these wolves. When word came of a kill, Bill went to where they would travel next. Wolves on the move will travel the highest ridges, the kind we called 'military ridges.' In the area here, the mountains were islands, and the desert between them didn't provide much food. Wolves also have a tendency to want to look down on the area they are in. Bill would determine where a wolf was likely to trail and he would set traps."

Gish trapped with Casto and on his own. When he got a wolf in a trap, it would act with submission and resignation, "as if it was almost in a trance." He would walk up to the trapped wolf and club it to death.

There were no biological studies of wolves in the Southwest, and no resident populations for a biologist to examine. Gish was asked by Charles Voorhies of the Izaak Walton League to do a study of the characteristics of these transient Mexican wolves. He was to accompany trappers like Casto, for they were the reigning authorities on wolves. It was an era in which old PARC trappers were cashing in on their experiences by telling tales to magazines of great wolf hunts and supernatural wolves. Charlie Gillham, who had poisoned a wolf that inhabited the low desert along the Gila River, turned that wolf into a legend in magazine articles about the Gila Wolf. "Every wolf was about forty feet high and consumed tons and tons of cattle and sheep, chickens, snakes, and acorns," says Gish.

Gish prided himself on being able to distinguish between story and lore. "These Mexican wolves never would try to pull down a cow on a range unless they turned it around and stampeded it. They almost always attacked its flanks." He guesses he has seen "maybe half a hundred" wolf kills, largely in Mexico, and all were of livestock. Most of the reports wolfers got of wolves were complaints from livestock owners, but they also followed up on other reports. "I've got a report of an old woman sitting on a privy in an outhouse in the rain and a wolf came and bumped the door open." In another report, "a male and a female wolf came down to a ranch down in the Chiricahuas, a couple of years ago. Their dogs barked at the wolves. Then the dogs ran. One of them lost its tail to one of the wolves, and hid under the porch of the house, where it died." In another, "a guy walked up to Big Lake in the spring, when it was still muddy. He saw a wolf jumping up and down in those thick weeds that come with the rains, jumping up and down, catching mice.

"Even when you're hunting wolves, the admiration for the animal's activities is tremendous. They have something no other animal has. They actually challenge you. They can come out with things that will astonish you." A trapper might pursue one wolf for months and see nothing more substantial than its tracks and the bones of its kills. Wolves would dig up the traps and leave them sprung and empty, as if to spite the trappers. "Bill Casto gloried in it. And he got a big kick out of it if they got the best of him."

Wolves back then didn't exactly thrive: Gish found that, when wolves managed to den, they had an average of five pups, but only an average of two survived the first six months. The rest, he guessed, starved. "I got the sense not all attacks by wolves were successful, even on domestic cattle. I think that the wolf's fear of being put out of commission entered into a lot of decisions in preying situations. I saw very little evidence of ruthless attacks. They just didn't take many chances." He describes seeing tracks indicating that wolves came upon cattle at a watering place at night and "walked around and around just keeping the cows restless. You could see where one of 'em would tramp around in one spot and circle and circle and circle. Then you can see where the steer kicked out and ran and the wolf followed," and made the kill.

The more he talks, the more Gish's face softens, the more his mind takes his body back to the days when it did not betray him. He recalls a wolf that killed cows in the Canelo Hills, and as he recounts finding the wolf's track and following it into the hills, he seems to rediscover an old truth, and he breaks into the story to exclaim, "They so intrigued you that you couldn't let go. You realized you didn't know half what you thought you did, and you wanted to keep going on. You couldn't let go.

"The Mexicans themselves never bothered to fool with the wolves. It was just like the water and the sky." Only when modern ranchers moved in and a European tradition took hold did they start to trap. He recalls Seminole Indian refugees he met while trapping along the Bavispe River in Sonora. He says, "They were very sharp people. They had an awful lot of respect for these wolves. It was a part of the folklore that the wolf was ethereal almost. That they had properties and capabilities that were not controllable by people. It was something that existed up there in the sky.

"I share their appreciation for the qualities of the wolf as he really was. I've got photos of wolves that were killed, and it's sad to see such a beautiful thing dead."

Behind Gish's house, there are low fences built to contain Destiny, a hybrid wolf-dog with a very aggressive view of strangers. And there is an old sheet-metal-sided trailer, built in the early 1950s by Gish's father-in-law, where Gish keeps his files, clippings, typewriter, and fax machine. The sheet-rock ceiling crumbles under a long-unrepaired leak in the roof. He hobbles in and begins to pick up books and papers to show his visitor.

"I have dug stuff out of the history that you wouldn't believe," he says. He picks up a tattered paperback copy of Allen and Allen's *Pioneering in Texas,* and, perhaps quoting something from the book, announces that a "Colonel Dodge" claimed to have seen forty thousand wolves crossing Nebraska's Platte River at night. He hefts a thick typescript that appears to be a compendium of short summaries of the lives of every Indian-fighter in Texas. Pointing to a copy of Young and Goldman's *Wolves of North America* on his shelf, he praises it as the best book on wolves. "You can bank on it," he says of the book. "You can bank on it because it was our people and that kind of people working with their own hands out in the field that he based it on." He waves a copy of Barry Lopez's *Of Wolves and Men* and says, "This is all bunk. This guy wouldn't know a wolf from a Volkswagen."

He says there has been no study, "nothing ethical, since Young." He believes, "A major amount of misinformation is trickling down from David Mech." He talks of other books about wolves, and is sour on all of them. "I'm sick and tired of these new biologists making up a biology which doesn't exist."

Gish did two studies of the Mexican wolf in his lifetime. The first was the study Voorhies commissioned him to do in the 1940s. The second was an unpublished historical account of the eradication of the Mexican wolf, entitled "An Historical Look at the Gray Wolf in Early Arizona Territory," commissioned by the U.S. Fish and Wildlife Service in 1970. David E. Brown drew heavily upon it in *The Wolf in the Southwest,* and listed Gish as a "major contributing author," but Gish feels his own study should have been published and given credit.

He feels he ought to have a say in the future of the wolf, the way he had had a say in its past. "I felt at least I was entitled to be considered as knowledgeable in the field, and at least deserved to be considered." But he is not. Though there is an effort to restore Mexican wolves under the leadership of officials of the U.S. Fish and Wildlife Service and the Arizona Department of Game and Fish, they do not ask Gish for his views.

If he were asked, he would say the effort is a sham. He believes fervently there are still wolves in Arizona. He says a woman biologist working north of the Animas Mountains in New Mexico ran into a wolf that she saw at a range of fifty yards, "in clear sight." When she went back to the New Mexico Game and Fish Department and reported it, he says, "she was transferred to a desk job." He shows three pictures taken by Ross Kane in 1991 in the Canelo Hills, of a wolf crossing a road. They show the landscape of southeastern Arizona, dry grass, low oak or scrub on the hills in the background, a dirt road, and a wolf crossing it. The light is low—late day, perhaps just before sunset. The wolf is deep-chested, with a broad head and long legs. In one picture its head and tail are both held low as it steps into the road, looking nervously toward the photographer.

He thinks reports of wolves are silenced, dismissed, put down the way UFO reports are put down, by bureaucrats who have a vested interest in denying the existence of wild wolves. "They want you to believe there aren't wolves here. They have declared the Mexican wolf extinct in the wild wherever he existed, and they made that determination without any fieldwork in the wild." They want instead to reintroduce captive-bred wolves, which he doubts will survive in the wild.

His face reddens, and his tone grows angry. "That's the ecofascists taking over total control of the resources of the United States under government mandate. Bureaucrats already have hundreds and hundreds of officials, and the program isn't even approved. They want a bureaucracy. They don't want a wolf. It's a land grab." A cascade of symbols issues forth. "Riparian water rights. Spotted owls. Such-and-such a minnow. Whatever. It's all designed to create a bureaucracy and bring a range of programs with no accountability. They are creating meetings and organizations all over the U.S. It's as phony as a three-dollar bill!

"They have to get title for the land for them to be on, and they have to keep people out. So who's going to enjoy it? Only the bureaucrats. The helicopter fliers and the dart nuts."

If the authorities are really interested in wolves, he says, "all they have to do is allow the wolves that are still alive in Mexico and in southern Arizona to come back in, and set up an effective compensation system for stockmen who lose stock to them.

"I have a relationship and an understanding of the wolf that these people don't have," he says. "I cannot be so far off base after all these years. I have no problem with wolves. I know what they do.

"I would like to see wolves in a lot of places where I see people," he says. Instead, what he sees creeping across the desert and up the mountains of Arizona is "the encroachment of rules, regulations, license fees, and arbitrary decisions on the world. I doubt that, anywhere in these United States and maybe in Mexico, these wolves can very long survive. I recognize the overpowering dominance of the open free market. And I have seen thousands and thousands of acres become asphalt and cement." When he bought the lot he lives on, he bought three lots together, so he would have space on all sides, but houses have crept up to his fences. At the time he moved out here, there were irrigation canals and citrus groves. "Every one of those irrigation canals used to have gigantic cottonwood trees, all of which have been cut. The citrus groves are gone. The wolf is not built for that kind of world. God never made him that way."

He would probably agree that God didn't make him to work in factories or to smile as the suburbs sprawled out from Phoenix, either. Gish watched the Old West turn into the New, and, like the wolf, he has not been master of the transformation. "I never made much money in my life," he declares with an air of defiance. "We haven't accomplished much in the way of getting on top of the economy." Asked if he means this implied comparison between his history and the wolf's, he looks off, through the piles of papers and the sheet metal walls of the trailer, at something beyond, and he nods and says in a soft voice, "Our time is past."

It is both a lie and a cliché to say that killing is a kind of love for the wolf. Dan Gish was certainly not a wolf hater, though he worked for wolf haters, for men who were so sour on life that they wanted to reduce it to numbers, to profit and loss, to what they could suck out of the marrow before the bones bleached. It's an awful paradox:

he identifies with what he spent his life chasing. The wolf has meanings for him as powerful, deep, and true as those cherished by the most ardent of wolf defenders. And when the last wolf vanished, he lost something more precious than he knew. It cut his bond to the earth and severed connections between eye and heart. There is a lesson in this for us all.

3

KILLING

In February, a pilot flying over the snowbound stillness of northern Minnesota has spotted a wolf kill in the snow below and radioed its location to wolf biologists at the North Central Forest Experiment Station, thirteen miles from Ely. Hours later, a group of wildlife enthusiasts and schoolteachers participating in a Vermilion Community College weekend wolf seminar have skied a mile over the flat surface of a frozen lake toward the kill. The skies are so gray they make the spruce-and-balsam-fir-forested ridges seem almost black in color. There are four inches of fresh powder on the ground from the previous night's storm. It is quiet but for the clatter of ski poles and the zipping of skis over ice crystals. The kill is away from the lake, in a spruce bog, a low area with dwarfed and twisted trees. The croak of ravens off in the woods at the edge of the swamp announces our arrival.

There is a strong catbox odor of foxes. Fox footprints weave through the spruce, looping around in wide, gossipy, intersecting arcs in the snow. They stop to sniff at something, then hurry on. A wolf kill is an event in the woods. It is food for ravens, foxes, and eagles, and they attend it like the opening of opera season.

To us, a wolf kill is a window on an unseen world. Wolves are not watched; they are glimpsed. One may see a flash of yellow eye or a movement in a thicket of branches, and then it is gone. Much of our impression of the animal has come from reading its tracks in the snow. And much of what we imagine comes from having found its kills.

By the time we find this one, there is nothing left but a mat of dun-colored fur and a welt of frozen blood. It was once, judging by the fur, a deer. The night's snowfall has blotted out the record of the hunt, but the surface of the snow around the hair is tattooed by the feet of ravens and foxes. There are delicate featherings in the snow where ravens landed or darted out of the way. No bones are left at the kill, but fox trails lead to a wolf bed, a hollow in the snow under a cluster of sheltering spruces where the wolf has curled up to sleep after its feast. Wolves sometimes cache pieces of meat near or under these beds, to hide them from scavengers. Digging with a bowie knife, we unearth a few chips of bone and a frozen chunk of blood. The bed has already been excavated and the meat removed by the foxes. Ravens must have watched excitedly, judging from the fanlike tracks of wing feathers around the fox trails.

The wolf's trail leads into the densest part of the spruce bog, where we find another bed already dug up by foxes. Right next to that are the deer's hind legs, rank with the scent of fox urine. The tracks out of this thicket are fresh, possibly made this morning—possibly when we arrived. Ravens croak and caw from the trees, but not from the kill site. The wolf may be nearby, wondering what all this clatter and exclamation and heavy breathing are about.

From the length of the leg bone, the deer appears to have been a yearling. We break open the bone. The gelatinous marrow is dark pink, almost crimson, indicating that the deer has been drawing on its last fat reserves. If the deer had been healthy, the bone marrow would be firm and white. Several of the group are fascinated by the bones and blood and tracks in the snow. We dig around and puzzle over the footprints, putting the events of the day together. Did a single wolf ambush the deer? Or did a pack run it down? What did the wolf see in the deer that urged it to attack? Did the deer limp or look skinny or hang its head in exhaustion? Was it obvious to the wolf that this was a deer failing its first test of winter and likely therefore to fail the test of wolf? After the wolf ate, did it cache food here and there while ravens and foxes darted in to gnaw and peck at the corpse behind its back? Did the wolf protest their lootings or resignedly concede them a share? Did the wolf stay for days by the kill, alternately eating its fill and sleeping, until the clatter of skis and the rasp of tree branches against L. L. Bean parkas and Jansport packs spooked it out of a full-bellied slumber and sent it skulking off into some darker recess of the forest?

What have we seen here? Tracks in the snow. Hieroglyphs of struggle. Tufts of hair. Crystals of blood. What is the nature of the beast that left them? What is the nature of the world inhabited by such a beast? What's going on here?

Predation has always fascinated humans. Perhaps we are interested because we are the world's preeminent predator, killing more kinds of creatures in more kinds of ways and for more kinds of reasons than any other species. Perhaps we have an ingrown interest in killing that we sublimate for the sake of society, but indulge with abandon when we look at other species. Perhaps we have a long history of perfecting our arts by borrowing from other species. Perhaps we're just irrepressibly curious about how others live and die.

Sometimes, the fascination discomforts us. A few members of the group hang back at the edges of the circle—looking for chickadees in the branches, following the passage of a gyrfalcon as it flies just over the treetops. "I'm not excited about seeing the kill," confesses a woman from Indiana. "I mean, it's the natural process and all . . ." Her voice trails off and the sentence remains unfinished. Her expression, however, says, "Sure, the wolf kills to eat, but it's something we ought to regret."

We have regretted the killing for centuries. Early in American history, when settlers cleared pastures in the forest and concentrated livestock in a few small areas, where wolves easily attacked them, they quickly developed an antipathy for wolves. William Bradford, who arrived in Massachusetts on the *Mayflower,* declared in 1624, "The country is annoyed with foxes and wolves." Roger Williams, founder of Rhode Island, wrote, "The Wolfe is an Embleme of a fierce blood-sucking persecutor." Mark Catesby, the early naturalist, held, "The wolves in Carolina are very numerous, and more destructive than another animal. They go in droves by night, and hunt deer like hounds, with dismal yelling cries." More than a century later, in 1904, William T. Hornaday, longtime director of the New York Zoological Park and a man who ought to have seen beyond the stereotypes, wrote, "Of all the wild creatures of North America, none are more despicable than wolves. There is no depth of meanness, treachery or cruelty to which they do not cheerfully descend."

We have made much of the cruelty. Roy McBride, an experienced wolf trapper, examined many cows killed by wolves in Mexico and the American Southwest. He says, "In many cases, I don't think the cow was dead when they ate them. I think these wolves just start eating on them. I find some of them alive, laying on the ground, with their back end eaten."

The heart of our quarrel with wolves is that we humans have developed a different view of death. We think of it as an event staged by and in the service of the supernatural. We die to receive judgment—to ascend to heaven or descend to hell. It is God's will, what we deserve, what we are entitled to by the sum of our moral qualities. We are called away to reflect upon virtue. Even the deaths of other creatures are computed with this kind of accounting. Native Americans believe that an animal presents itself to a hunter because that hunter has been caring, hardworking, and respectful, and the animal gives its life as a reward.

Predation is a challenge to this view. It is death unwarranted by divine intention; it is morally unscheduled suffering. Our view of death as moral summation is of little help in understanding nature's workings, and biologists have had to search for evolutionary and ecological meanings in death.

Real study of predation is a recent thing. Even fifty years ago, studies were being made in the half-light between science and woodlore. On a winter night in the late 1930s, Sigurd Olson, the great wilderness writer, was snowshoeing over the ice of the frozen Kawishiwi River near Ely, Minnesota. The moon cast the shadows of twisted branches over the silvery snow. It was silent, except for the swish and creak of his snowshoes, the hiss of his breath, and the crackle of branches in the brush beyond the river's edge.

He could hear the wolves on both sides of him, following him. "I knew I was being watched, a lone dark spot moving slowly along the frozen river," he later wrote. He recalled the long war with wolves, the poisoning and shooting, and tales of wolves stalking humans hovered in his mind. Olson had once seen a wolf kill a doe near Basswood Lake. The wolf had loped easily behind the doe, then leapt on it and grabbed it by the nose. The doe had somersaulted, and the fall had broken her back. Olson thought of this as he approached a narrow point in the river, and he became afraid that he might be attacked. He knew of no authenticated instance of an unprovoked attack by wolves on a human being, but, he wrote, "In spite of reason and my knowledge of the predators, ancient reactions were coming to the fore, intuitive warnings out of the past."

The river narrowed between dark, forested hills, and there two shadows left the brush and came rapidly toward him. He stopped and removed his pack. The wolves stopped fifty feet away, and looked him over. "In the moonlight," wrote Olson, "their gray hides glistened and I could see the greenish glint in their eyes." For a long and testing moment, man and wolves stood silently confronting each other. Then the wolves turned and trotted away into the night. Off in the forest, Olson heard a long howl. He was thrilled by the experience. He saw in the wolf something like himself, but unencumbered by the confusions and moral ambiguities of modern life. Something wild and noble.

Olson's reaction was unusual for the Minnesota of the 1930s, but Olson was an unusual man. He understood ecological relationships, and he had undertaken the first serious scientific study of the wolf on his own initiative. He deplored the fact that conservation meant largely "protection of herbivores at the expense of predatory forms." He saw that conservation might thus lead to the extermination of large carnivores, and hoped to suggest to the world that, "after all, lions, wolves and coyotes may be an exceedingly vital part of a primitive community, a part which once removed would disturb the delicate ecological adjustment of dependent types."

His study, published in 1938 in *The Scientific Monthly,* wasn't rigorous enough to survive the kind of peer review it would get today. It consisted of eighteen years' experience snowshoeing, hiking, and canoeing in the north woods, and of observations shared with him by trappers and timber cruisers. Olson focused especially on the wolf's food habits, which, he wrote, "determine whether or not a species is an acceptable member of any society." He declared, "The major portion of the food of the wolf during the summer months is grouse, woodmice, meadow voles, fish, marmots, snakes, insects and some vegetation." The wolf, he said, fed on deer only in the winter, when the smaller animals were in hibernation. "Close students of wildlife in the border country all agree," he wrote, "that wolves kill comparatively few deer, and then only in the late winter and early spring periods." That observation was largely based on anecdote and the sampling one winter of the stomach contents of wolves—all of which contained deer. Today, wolf biologists say wolves may eat grouse, mice, and rabbits but live mainly on large ungulates, such as deer and moose. But from examining the remains of winter kills, Olson came up with a stimulating conclusion: "The great majority of the killings are of old, diseased or crippled animals. Such purely salvage killings are assuredly not detrimental to either deer or moose, for without the constant elimination of the unfit, the breeding stock would suffer."

It was the dawning of a new era. Something dramatic was happening to our view of predators. In 1928, F. S. Bodenheimer, a German biologist studying insect populations, suggested that climate was almost wholly responsible for determining population densities, and that predation meant little to overall numbers. About the same time, the ecologist Charles Elton noted that, when populations grew large, individuals migrated, and the migrants were more susceptible to predators, because they were on unfamiliar ground and often harassed by those already in possession of the territory. In the 1930s, Paul Errington, studying the effects of mink on muskrat populations in Iowa, concluded, "Mink predation upon muskrats tends to be almost restricted to those individuals or parts of the muskrat population that may be properly referred to as pushovers." He found that muskrats would produce a surplus of young every year, and that the young would go wandering in search of breeding territories, which were limited in number. Muskrats that were too small, too irresolute, too stupid, or too slow would weaken as they traveled. His studies showed that 70 percent of the muskrats that fell prey to mink were already victims of disease or freezing.

The new view held that predators took nothing more than the expendable surplus, but it also argued the somewhat contradictory view that, without predators, prey populations would increase to such numbers that they would consume all their nutritional resources. Perhaps the most famous example came from Arizona's Kaibab Plateau. The Kaibab had been made part of the Grand Canyon National Game Preserve in 1906, but by 1920 predator control had removed its wolves and mountain lions. In the 1920s, the deer population rose from perhaps four thousand to perhaps a hundred thousand. But public sentiment would not permit shooting the deer. Within a few years, 90 percent of the forage was gone, and the deer were starving everywhere. Aldo Leopold, the famous wildlife biologist and environmental philosopher, had urged the extermination of wolves early in this century. But by 1944, he looked back on the past with regret: "I have seen every edible bush and seedling browsed, first to anaemic desuetude and then to death. I have seen every edible tree defoliated to the height of a saddle horn. . . . In the end the starved bones of the hoped-for deer herd, dead of its own too-much, bleach with the bones of dead sage." Only the wolf, he concluded, could have kept this tragedy from happening.

Durward Allen, who had studied predation in Michigan in the 1930s and 1940s, declared, "The natural function of predators is to keep big game range from being destroyed by the animals it supports." Predators might even help to keep prey numbers from falling. Allen recounted that bighorn sheep had succumbed all around the West to diseases brought to them by domestic livestock, but that the ranchers, seeing wolves feeding on their carcasses, concluded predators were to blame and redoubled their efforts to eliminate them. Allen held that the wolves, if left alone, might have slowed the spread of disease through the sheep population by culling the weak animals before they infected others. "There are strong implications that this species needs the culling of its ancient enemies, the wolf and the cougar," he declared.

Increasingly, biologists were concluding that predators served a useful function, that they kept prey populations from destroying their own food resources and removed genetic mistakes from the population, thereby keeping the species strong. Wrote Allen, "The intensive weeding out process to which a prey may be subjected, if sustained through the ages as it undoubtedly has been, could hardly fail to render the stock more vigorous and more efficient in using the protective features of its environment."

Acting upon this new view that predators didn't really limit prey posed problems. Wildlife management had developed upon human rather than biological needs. Its view of predators derived not from ecological studies, but from moralizing over the taking of game and livestock by nonhuman competitors. Allen observed that game management had evolved largely as the effort to produce harvestable animals, and that, "from the first, war on carnivores has been one with game production." Antipathy to predators was further fueled by professional writers, who "learned long ago that the atrocities committed by gore-fed carnivores are among the most merchantable material for the magazine trade." As a result, he wrote in *Our Wildlife Legacy*, "Many people think . . . the wolf doesn't live in the forest; he infests it. You don't just kill a predator; you execute him. You don't just hunt him for sport; you track him down in a crusade for moral reform."

Paul Errington cautioned, "Man may call predators robber barons or cannibals and talk of honor or lack of honor" when they talked about the relationships between hunters and hunted. But, he declared, "the moral rightness or wrongness that man sees in these relationships, after all, is only man's."

The new view of predators was based on studies of small birds and mammals. Could studies of mink and quail be generalized to larger creatures like wolves and moose? In 1939, Adolph Murie, a biologist who had already studied moose in Michigan, elk in Washington, and coyotes in Yellowstone National Park, was asked to undertake a study of the wolves of Mount McKinley National Park in Alaska. In the late 1920s, the Dall's sheep of the park were abundant, and wolves, which had been scarce, were just beginning to increase. But in 1929 and 1932, snow was deep and crusted, and sheep died in large numbers. Sheep remained scarce through 1938, and wolves were blamed. The National Park Service was by then embarrassed about having eradicated wolves in Yellowstone National Park. Park officials were uncertain whether they should pursue wolf-control programs in Mount McKinley. They wanted to know more about the relationships between wolves and caribou, moose, and Dall's sheep.

Murie spent three years in the field, observing the wolves hunt, and watched the sheep spend long hours staring at the terrain before they would cross a flat area between mountain ranges. They didn't pay much attention to wolves when the wolves were below them on a mountainside, but if a wolf were looking down on a sheep, and therefore blocking an easy escape upslope, the sheep would become nervous. Murie saw that the wolf and the sheep were locked in a single larger complex of design and behavior. At the least, the wolf caused the sheep to dwell in a rocky habitat.

Murie wanted to know whether wolves limited the number of sheep. Examining the skulls of sheep he collected around wolf dens, he found that 95 percent of them came from very young and very old sheep, and concluded the wolves were taking the weakest sheep especially. He believed that, when the sheep increased in number, they spilled out onto the low hills, where they were vulnerable to wolves. The wolves then increased *their* numbers by taking these sheep, and gradually pushed sheep back to the rocky habitat. Over long cycles of time, sheep and wolf had evolved a kind of balance. The wolves probably served to check the numbers of caribou, too. He judged, "If this check were entirely removed, the caribou might increase in numbers to such an extent that vast areas of choice lichen range would be severely damaged. Those familiar with the overuse of many big game ranges in the states can readily appreciate the importance of this consideration."

Murie's *The Wolves of Mount McKinley* was the first careful scientific study of wolf predation. Others would follow. Between 1943 and 1946, Ian McTaggart Cowan, of the University of British Columbia, studied wolves in five parks in British Columbia and Alberta. Unable to find any difference in the survival of young ungulates with or without wolves, he concluded that predators were less important than the presence or absence of suitable winter forage in determining the number of moose, caribou, or sheep. Milton Stenlund, of the Minnesota Department of Conservation, who studied wolves in the Superior National Forest from 1948 to 1952, concluded that they helped deer by reducing browsing pressure on an overburdened range.

. . .

The change in our view of predators does not seem to have been solely the result of biological observation. After World War II, people were revising their views of nature in general. In 1947, a young Canadian, shaken and made pessimistic by his experiences as a soldier in the war, decided he wanted to visit the great barrens of northern Canada. In those years, one needed a permit to enter the northern territories, and the only way young Farley Mowat could do it was to accompany an elderly American ornithologist who was undertaking a study for the Arctic Institute and needed someone to look after him. "I looked after him," says Mowat. "And I traveled."

The Keewatin area was abuzz with dire predictions of the collapse of the caribou herds, and wolves were being blamed. It was gospel in the country that wolves should be shot and poisoned. But Mowat held a contrary view. His boyhood home of Saskatoon was fiercely anti-predator. "Everybody was death on coyotes," says Mowat. "Of course the wolves were all gone by then." He was contrary even as a boy, and took a dislike to the persecution of any wild animal. At the age of fourteen, he wrote a series of pieces called "Birds of the Season" for the Saskatoon *Phoenix*. One of them defended hawks and owls, which were routinely shot by farmers and town residents. "Local fish-and-game people were furious" about the articles, he recalls.

By the time Mowat returned from fighting in Europe, he was convinced something was wrong with the way humans perceived the world around them. It is a belief he carries to this day. "I don't like saying this," he said to me in a telephone interview, "but I think we're a bad species, and the sooner we get off this planet, the better."

So, there in northern Manitoba, Mowat was bound to have a contrary perspective on the tales of wolves killing caribou. In 1948, an acquaintance, Frank Banfield, was contracted by the Dominion Wildlife Service, forerunner of the Canadian Wildlife Service, to study wolves and caribou. Mowat was allowed to go along as a student biologist. He simply made mischief for the project. "They were trying to kill wolves," he says today. "I was on the side of the wolves." So he went out on his own time and did his own study.

He talked to native hunters and trappers. Though he saw few wolves—"just glimpses"—he spent hours observing a wolf den at

Nueltin Lake. He compiled a report in which he declared that he saw the wolves "consistently engage in mouse hunting." The report noted that caribou did not occur in the area for nearly half the year, and observed that when they did occur, "they probably constitute the major food item." But he stated that while he was watching, "No wolves were seen to pull down caribou."

Mowat submitted his report to the Dominion Wildlife Service. He felt the study was too short. "There is insufficient evidence to make any estimate of the number of caribou killed by wolves," he declared. "No attempt is made to draw any conclusion from the information amassed to date. . . . I shall be continuing with wolf studies for some time and when I have sufficient data to warrant a summing up of the study, I shall do so."

In his heart, though, Mowat had reached a conclusion. He was convinced that the hunting of caribou by natives to feed their dog teams was responsible for the caribou decline. Indeed, Banfield had already calculated that wolves were taking a relatively small share of the caribou, and that the native hunters were responsible for the decline. Nearly a half-century later, Mowat recalled for me his sense of outrage: "I was sitting on a hill in northern Manitoba, and people were slaughtering wolves all around, and I just got good goddamned mad. I decided I was going to write this book." That year, he drafted *Never Cry Wolf,* a fictional account of a biologist sent by the government to the remote north to study wolves suspected of destroying the caribou herds. The biologist finds the wolves are intelligent, sociable, and confiding, and he sees them eat only lemmings and voles. He concludes that native hunters are killing the caribou to feed their dog teams, and that the wolf is blameless. "It was a deliberately contrived piece of propaganda," he admits, but he stands by the wolf lore in the book, saying it all came from his own observation and from conversations with Indians and trappers. He concedes that the wolves in the book aren't shown hunting caribou, despite his certain knowledge that they did. "I knew what they did, but I sure as hell wasn't going to make much of that in a book that was in defense of the wolf. I was going to present a heavily loaded picture in defense of the wolf. I was deliberately nonscientific."

When the book was published in 1963, biologists—even those who agreed that wolves were not responsible for the decline of caribou—didn't like it. University of Montana biologist Charles Jonkel, who was working for the Canadian Wildlife Service at the time, recalls, "We used to call him Hardly Knowit." Paul Joslin, who was studying wolves with biologist Douglas Pimlott in Ontario at the time, recalls, "When Farley Mowat's book came out, Doug Pimlott hit the roof." Pimlott and others accused Mowat of plagiarizing Murie and Banfield, and of falsely implying that Canadian government scientists believed wolves, rather than humans, were the cause of the caribou decline. Pimlott said it was "at least deplorable" that the book's publishers represented it as nonfiction. He and others felt it was no less an act of mythmaking than the stories in the hunting magazines. Banfield, in reviewing the book for *Canadian Field Naturalist,* concluded, "It is certain that not since Little Red Riding Hood has a story been written that will influence the attitudes of so many towards these animals. I hope that the readers of *Never Cry Wolf* will realize that both stories have about the same factual content."

Mowat didn't care what scientists had to say, because he felt that most of them were sold on predator control anyway. Besides, he knew then that the redefinition of the wolf's nature was as much the reflection of human psychological and spiritual need as it was the fruit of scientific investigation. "A very large part of it comes out of our needs," he says today. "There's a species-wide sense that we're in trouble, that we've done something wrong, that we're losing contact, that we're drifting into space. This has come because we've lost consciousness of the rest of nature. There's a desire to get back into contact with nature." In 1963, Mowat was coming to see it as his mission in life to spur that desire. Now he says, "It's our last best chance."

Never Cry Wolf became the most widely read book about wolves. Its translation into Russian led to a reversal of wolf-persecution policies in the Soviet Union in the 1970s. It is still widely read and quoted. A woman recently told me she had heard something about wolves saving an endangered species of rodent in Canada, and then, as we unsnarled the wool of thought about this in her mind, it became clear she was referring to the implication in *Never Cry Wolf* that wolves fed only on mice.

Nearly a half-century after Mowat wrote the first draft of *Never*

Cry Wolf, the wolf has been redefined in the mind of the public. Many people see it as a creature that kills nobly and innocently. Says Canadian biologist Tom Bergerud, who has long advocated the kinds of wolf controls that Mowat detested, "There are guys who love the wolf so much that they cannot bear to see a wolf killed. We've still got people saying we shouldn't manage because we always screw up the system. It's been the story of my life to go into a barber shop and say, 'I'm a caribou biologist,' and the guy says, 'Have you ever read *Never Cry Wolf*? We'll never get past *Never Cry Wolf*!"

Says Joslin, who credits Mowat with unshackling the wolf from the old legend, "Farley Mowat did more than anyone else to change the public's attitude. He brought the wolf out into the open, and we can never go back from there."

Do we simply invent the wolf? Mowat says, "Perhaps. Maybe I'm one of the inventors. But if we're on the right side of the invention, on the side of giving life a chance and a degree of equality . . ." His voice trails off, as if he were going to say, ". . . who can argue with that?"

If Mowat prepared the public to reinterpret the wolf, the field researchers provided the terms by which the new wolf would be defined. When Sigurd Olson judged in 1938, without much hard data, that wolves were taking the old and the weak, he was to some degree reflecting the ideas of Paul Errington. Errington had distinguished between "compensatory" and "additive" effects of predation. Predation was compensatory if the predators took individuals that would die anyway—of starvation, or winter freezing, or stress from repeated combat. Predation was "additive" if predators caused additional mortality by killing prey that would not otherwise succumb to starvation, bad weather, disease, accident, or weak genes. Errington viewed predation as generally compensatory, and a number of studies seemed to confirm this view. In one California study, for example, coyotes were removed from an experimental range to control calf losses. When the coyotes were gone, the rabbit population didn't increase; this suggested that coyotes had been taking only the exploitable surplus of rabbits. Durward Allen found that, just as Michigan hunters shot more young pheasants in the opening weeks of hunting season, the natural predators killed more young at that time of year. He wrote in 1954, "Animal populations are padded annually with a surplus that is inevitably eliminated."

Since Errington, the language of predation has grown more complex. Predator and ungulate biologists speak of limitation, regulation, and control. There has been much disagreement about what each of these terms means. "Control," for example, is defined by one authority as the maintenance of a population at some level, and controlling influences could be things like lack of water or breeding sites. Another author says control means a planned attempt by humans to manipulate population size through hunting or culling. Without agreement on their definitions, biologists have at times used all three terms interchangeably. There seems to be increasing agreement and more careful use of the terms "limit" and "regulate." Predation is a limiting factor if it causes a prey population to decline or causes a decline in the birth rate. Limiting can be either additive or compensatory. Predation is a regulating factor if it stabilizes the number of prey in a specific area over time, and if predators take proportionately more of the prey species when they are abundant than when they are uncommon. Regulating is thus a subcategory of limiting.

The essential question posed by the terms, however, is this: Can wolves by themselves cause a prey population to decline and remain at lower levels? Before the middle of the century, only Murie and Cowan had sought to answer this question. And by then, wolf habitat was too remote, and wolves themselves were too elusive, to make prolonged study seem worth the effort. To understand a predator-prey system, one would have to spend years watching wolves in a remote place with expensive logistical support. Such a study would require years of data on the age structure of the prey population and the overall number of predators and prey. It would call for evidence of the frequency with which wolves ate each prey, and thus continuous study of the contents of wolf stomachs or scats. Funding such a study, when predation was still regarded largely as a moral and economic problem, seemed very unlikely.

But in the 1950s, an accessible and contained wolf population became available for study on Isle Royale, an island national park near the north shore of Lake Superior. Until the late 1940s, wolves had been absent from the 210-square-mile island. Moose had swum to the island sometime before 1912, and, without wolves, they had grown so numerous that there were efforts to transplant them to the Michigan mainland.

Adolph Murie had studied the eruption of the moose population on Isle Royale in 1930 and had suggested introducing wolves as a way to keep the moose from overrunning the island. A similar proposal was made in 1951 by Michigan biologist A. M. Stebler, but the park superintendent opposed it because wolves were "vicious beasts" and he feared reintroduction would be bad public relations. Durward Allen says today that he had been dreaming of wolves on Isle Royale for years: "I was an undergraduate when I saw the possibility of moving some of the wolves from Michigan to that island." He confesses that when he was a student he and a close friend, Lee Smits, talked about sneaking wolves onto the island.

Smits, a Detroit conservationist, eventually did put wolves on the island. In 1952, Smits persuaded the Park Service to agree to an introduction. Four wolves born in the Detroit Zoo were brought to the island. At first they were kept in pens to acclimatize; then the pens were thrown open, in the hope that the wolves would eventually move out on their own into the wild. But the wolves were too habituated to humans, and too little skilled at moose hunting. They hung around campgrounds and cabins, scaring tourists and tearing up the nets and laundry lines of fishermen. The Park Service trapped one of them and removed it to the mainland, and shot two of the others. The fourth, Big Jim, a wolf reared at home by Smits, vanished into the wild.

Other events overtook Smits' dream. In 1948, while Farley Mowat was drafting his book, a pair of wild wolves crossed the ice bridge to Isle Royale from the Canadian mainland in winter. Unlike Smits' captive-reared wolves, they knew how to hunt, and, as wild wolves do, they remained quietly out of view. Humans occupied the island only during the summer months, when wolves generally remain out of sight in the dense underbrush, and not in the winter, when tracks are easily seen. Just months before the tame wolves were put out, a park ranger saw wolf tracks, and shortly after the release a visitor reported seeing wild wolves. Little was made of these sightings because the Park Service was committed to the introduction effort. But then, in 1953, after the release and removal of captive wolves had perhaps made park officials more attentive to wolf sightings, James Cole, the park biologist, saw a pack of four wolves.

When Durward Allen, then working for the U.S. Fish and Wildlife Service in Washington, D.C., heard that wild wolves had arrived on the island, he left his job and took a position teaching at Purdue University, anticipating that the position would allow him to study the wolves on Isle Royale. He felt a sense of mission about the prospect. "There were some things that were rife among most human beings, like the Little Red Riding Hood story," he says today. "The old-timers who lived in the woods claimed to know everything, but they knew nothing. The old-timers had taken over the sporting magazines, and the stories they told were a lot of hogwash. But people believed them. In North America, nobody was getting eaten up by wolves. Nobody took any trouble to expose wolf dynamics or what they were doing to their prey population. It was clear that the people who wanted to understand wolves were on their mettle: they had to prove everything. Before my Isle Royale days, I had concluded that the scientist is going to have to get into this and outdo the old-timers."

It was for Allen a chance to change fundamentally the way people looked at nature. "People don't think ecologically," says Allen. "This is one of our problems. They don't have an ecological background on which to hang an animal. And most of them are dominated by their religious culture," which in some places still associates wolves with evil. Allen saw in Isle Royale a chance to educate people about the importance of ecological thinking. It would take him four years to get organized—to find funding and people who could do the work.

While he taught classes at Purdue, Allen would require a graduate student to man the island research. He chose David Mech, then a recent graduate of Cornell University, to conduct the fieldwork. Mech had grown up in Syracuse, New York, in a family that camped every summer in the Adirondacks. Without knowing it, he was already engaged in a conversation with Durward Allen. Allen had written the Boy Scout wildlife merit-badge handbook which Mech had read as a boy, and through it, perhaps, Allen had summoned him years before they ever met. Mech learned trapping as a boy at a New York

State Department of Conservation summer camp in the Catskills from counselors who were wildlife-management students from Cornell. The experience gave focus to his life. "It headed me in the right direction," he recalls today. "It headed me to Cornell for a wildlife major."

As an undergraduate at Cornell, Mech worked on a study of black bears. His chore was to ear-tag the bears. It was primitive work. Bears were trapped in leghold traps; to put an ear tag on one, a gang of students had to spread-eagle the bear and hold it down while someone injected a drug which could only be administered through the abdominal wall. One person held the bear in a choker, someone else noosed a hind leg, and a third noosed another leg, until five or six of them spread the bear out and turned it over. That, and Mech's experience alone in the Adirondacks in winter trapping fisher, persuaded Allen that Mech was the right person to wrestle with wolves in the wilds of Michigan. That Mech had written popular-magazine stories about some of his trapping exploits made him more desirable. Allen saw a need to educate the public in this new view of predators, and Mech's communication skills, he felt, would be important assets.

Allen also felt he needed a student who did not jump to conclusions. At their first meeting, Allen saw a quiet caution in the younger man. Mech recalls that first meeting: "Here I was a senior in college, twenty years old, and Durward wanted to talk to me. The only other state I'd been in was Pennsylvania, and Durward was talking to me about this island somewhere out in the West, this huge island with wolves and moose. It seemed almost mystical. I hadn't a clue why he was telling me this. And I told him it sounded like a great project. I said, 'I wish you luck in it.' And he said, 'Well, I'd like you to do the study.'

"I just couldn't believe it. I never thought I'd get, in my life, to some place where I'd see a wolf track. Suddenly, my horizons were greatly expanded. I would see wolf tracks in my lifetime, and I might even see wolves."

He would indeed. Since then, Mech has studied wolves in Michigan, Minnesota, Montana, and Alaska, in Russia, India, and Italy, and on remote Ellesmere Island in the Canadian Arctic. He succeeded Douglas Pimlott as chairman of the International Union for the Conservation of Nature's wolf-specialist group and served on the recovery team for the eastern timber wolf. When the media discover wolf controversy, reporters call him. Today, Mech is considered the dean of wolf biologists. He has been called "Mr. Wolf"—both in praise and in sarcasm, for wolf biology is contentious work. Hawk-nosed and thin-lipped, he has a closely trimmed beard that is beginning to show hints of gray. His brown eyes are deep-set, and the folds over the corners tell of long hours spent looking at things. He is deliberate and watchful, with a wolflike curiosity—though he would bristle at the wolf comparison, because he would be the first to say that one should never attribute human qualities to wolves.

Mech began his work on Isle Royale in 1958. In February and March, he flew over the island, tracking and watching the wolves from the air. He found it possible to follow them thus in the snow, and even to circle and watch as they attacked and killed moose. The wolves had markedly increased in number since 1953. One February afternoon, he would later recall in *The Wolves of Isle Royale,* he was flying after a pack of sixteen wolves along the lakeshore when they stopped and pointed their noses toward an old cow moose two hundred yards away. They went inland and came out of the tree cover twenty-five yards from the moose, which fled. They chased her and caught up. One wolf grabbed a hind leg, but the moose shook it off and trotted through a clump of spruce. They caught up again, and several wolves grabbed her flanks, but she shook them off and continued. Then a wolf darted in, grabbed her by the nose, and hung on. As she stood trying to shake him off, several other wolves fastened their jaws on her flanks and rump. The wolf on her nose held its grip for at least a minute, while the cow dragged it and several other wolves a hundred yards. She shook them all off and ran into the woods, kicking and trampling wolves, to stand next to a balsam fir, so that the wolves could only attack from one side. The wolves simply lay down in the snow and waited. Every time a wolf walked close to the moose, she would threaten. But she was weakening and stiffening from her wounds. Mech had to break off the observation before nightfall. When he returned the next morning, the wolves were feeding on the moose's carcass.

If he found a kill, Mech would land and pick up bones or teeth and determine its age and condition. He discovered he could walk right up to the kill and the wolves would flee into the forest. They were never threatening. In summers, he walked the trails of Isle Royale, looking for wolf tracks and scats, and listening for howls to locate wolf dens, but he rarely saw wolves. In the fall, he counted the moose and estimated the ratio of cows to calves.

Mech found that the wolves on Isle Royale acted just as Murie said they acted at Mount McKinley: they took the young, the old, and the sick. Healthy adult moose could either escape or turn and fight, and those that fought invariably survived. Despite the arrival of wolves on Isle Royale, the moose population did not decline.

Isle Royale excited newspapers and television reporters. Mech and Allen published an account of Isle Royale's wolves in *National Geographic* magazine in 1963, and Mech wrote a number of popular articles in succeeding years. Mech's photos were widely published and they gave the impression that Isle Royale was a place where people could effortlessly watch wolves hunt. "It grabbed the public's imagination," says Mech. "Newspapers were always hounding me for interviews. The public didn't know about Olson's or Murie's or Stenlund's studies. But once it got around that this island was a closed system where predation could be studied intensively, there was a lot of interest. The study focused a great deal of public attention on wolves."

Through the articles and interviews, Mech explained that most of what was written about wolves was myth. His book *The Wolf* replaced Young and Goldman's as the standard reference on wolves for biologists and the public alike. And Allen followed a few years later with *The Wolves of Minong* (the Ojibwa name for Isle Royale), detailing the Isle Royale study and its findings. "We showed that a lot of the crap that came out in the sporting magazines wasn't true," says Allen. "Wolves don't characteristically eat people—they'll run away rather than attack a human being. We proved the wolf is a very intelligent animal that can accept human beings."

Most important, Isle Royale seemed to show that wolves did not regulate the prey. Mech had found there that moose from two to seven years of age were almost invulnerable to wolves, and that calves and elderly moose constituted nearly all the winter kills. The calves were smaller and weaker. The older moose tended to have infestations of moose tick and hydatid cysts in their lungs. Sometimes the cysts were so large and numerous that the moose probably couldn't breathe freely. In his Ph.D. thesis, published as *The Wolves of Isle Royale* by the National Park Service in 1966, Mech observed that high twinning rates among moose were generally regarded as signs of a healthy moose herd, and the Isle Royale moose herd had one of the highest twinning rates in the literature. He tentatively attributed this high rate to wolf predation, which kept the moose population within the limits of the food supply and culled out the unhealthy individuals. "The Isle Royale moose population," he concluded, "probably is one of the best 'managed' big game herds in North America."

Mech also challenged the idea that wolves were unfailing killers. He has probably seen more wolf kills than any other living human. "I've watched wolves kill everything from deer to caribou to moose to musk oxen to Arctic hare," he says, "and, by golly, they get a small percentage of the prey, pure and simple." On Isle Royale, Mech saw wolves test seventy-seven moose in hunts and succeed in killing only six of them. He believes wolves approach about twelve deer or moose for every one they actually catch.

Mech believes that this is one of his most important discoveries, because it forces us to think differently about the wolf. "The prey are always ahead of the predators," he says. "If predators were ahead of the prey in an evolutionary sense, they wouldn't hunt twelve times for every kill, and they wouldn't tend to kill fawns and old individuals, but would instead kill a cross-section of the prey population. The typical prime member of the prey population is invulnerable to the wolf. The wolf is essentially a glorious scavenger. They just don't wait for the prey to die."

Other researchers also found that wolves failed more often than they succeeded in killing prey. Ludwig Carbyn and S. M. Osenbrug studied the relationship between wolves and bison in Wood Buffalo National Park in northern Alberta from 1978 to 1981. Bison provide 80 percent of the wolf's diet there in summer, and more in winter. The two researchers watched 143 bison-wolf interactions, ranging from "close watching of bison" by wolves, to kills. If the wolves saw some sign that a bison was weak, they tested the herd with a run at them. Half of the time, bison stood their ground when wolves first charged, and the wolves gave up and moved on. Whenever the bison fled, the wolves attacked. If a fleeing bison got angry and turned to charge the wolves, the wolves would run right by that bison, continuing to pursue the rest of the herd, and Carbyn would be treated to the sight of the one brave bison passing the wolves to catch up with his own herd. Wolves would stay with a herd of bison for as long as six days, pressing the prey, waiting for something to happen. In all those encounters, there were seventeen attacks and only three kills.

Clearly, even if they are attacked, prey often escape. "We frequently observed bison with missing tails and suspected that wolves were responsible," wrote Carbyn. William Hornaday reported that in the 1880s travelers on the Great Plains saw gigantic buffalo oxen, animals that had been castrated by wolves but survived to become bison of enormous size.

Isle Royale showed that wolves—and perhaps predators in general—were not what people thought they were. Says Mech, "They have to try very hard for anything they get. And that is contrary to the views of a lot of the public, and contrary to the views of biologists long ago, and contrary to the views of ungulate biologists, who feel wolves can go around and kill any time they want. If people understood the prey's advantages, there'd be even more sympathy for the predator as being a helpless link in that system."

At the conclusion of his graduate work, Mech left Isle Royale. A series of other principal investigators—Philip Shelton, Peter Jordan, Wendel Johnson, Michael Wolfe, and Rolf Peterson—continued the study under Allen's direction. Well into the 1970s, the moose and the wolves continued to thrive on the island, a fact that helped persuade Minnesota and other states to stop paying bounties on wolves. It also helped push British Columbia, and then other Canadian provinces, to stop controlling wolves as vermin and to reclassify them as a game species. Meanwhile, with funding from the World Wildlife Fund and the New York Zoological Society, and later the U.S. Fish and Wildlife Service, Mech began another study of wolves, in northern Minnesota's Superior National Forest. Today, that and the Isle Royale studies are the two longest-running studies of a wildlife population in the Western Hemisphere, perhaps in the world.

In Minnesota, Mech would see the picture of wolf predation he helped develop on Isle Royale grow more complex. Here deer rather than moose were the prey population. Eight of the ten previous winters had been mild, and when Mech arrived the deer were thriving. But from 1966 to 1972, there were seven severe winters. Deer declined all across the Great Lakes area from 1969 to 1975. Fawn-doe ratios declined in both wolf-inhabited and wolf-free areas, and deer disappeared from the poorest habitat. But Mech found that more does were lost in areas where wolves were least persecuted by humans. He had to concede, "Wolves were one of the main causes of the deer decline in the Superior National Forest."

The differences between Minnesota and Isle Royale reflect the fact that not all wolf-prey systems are the same. Isle Royale has only wolves and moose. Sweden has only moose and humans. Gaspésie Provincial Park in Quebec has moose and bears. Bears, wolves, and humans alike prey on moose in Alaska's Kenai National Wildlife Refuge. Wolves, bears, and lynx share Denali National Park with moose, caribou, and Dall's sheep. Each system has its own wrinkles. Some, like Isle Royale, have good escape cover for prey. Cows with calves resort to small islets that fringe the island, where they are less likely to be attacked by wolves. Most wolf-prey systems once had human hunters to complicate the problem. There is still disagreement as to what role native Americans had in maintaining elk and sheep populations in Yellowstone National Park. The Kaibab Plateau deer herd may have been destined to overpopulate their range after the five hundred Indians that once hunted there were removed. The more complexity we find in these different systems, the harder it becomes to generalize from one to another.

The Minnesota deer herd's decline still did not suggest to Mech that the old view was right, or that wolves were the ultimate cause of the deer decline. Comparing the ages of deer killed by wolves with the ages of deer killed by hunters, Mech found that the wolves were still taking the young and the old, the least fit members of the deer population. He was convinced that this was "compensatory" predation: the wolves were taking deer that would succumb anyway to disease or winter cold.

Mech believed that the evidence showed that reduced nutritional resources were the cause of the decline. There was less logging going on, and so the young aspen stands which came up right after logging, and in which deer browsed, were growing less common. The severe winters severely stressed the deer, and the deep snow hindered escape. With the deer so weak and slow, the wolves committed surplus killing—the killing of more deer than they could eat.

Mech believes that the effects of one hard winter might not be seen for years. He has described what he calls "the grandmother effect": "If the grandmother is well during pregnancy, the grand-offspring stands a better chance of not being eaten by a wolf." If a doe is stressed by hard winters, her offspring might have a weaker immune system or less intelligence, and if the offspring survives, its weakness may in turn affect the third generation. "These offspring might look perfectly healthy to any observers," says Mech. "Chances are they're a little lighter-weight. But laboratory studies on rats whose parents had been nutritionally stressed show they have fewer brain cells. They're dumber; they can't learn mazes well. Their immune systems aren't as good."

Mech's contention about the victims of wolves is difficult to prove. He says: "Generally, when you come to a kill that a pack has made, there's very little left. There may not be any bones. If we're lucky, we have a jawbone, and that will give us some idea of age. If we get a leg bone, that will give us some idea of condition. But there's a real flaw in the femur fat test. If the marrow is depleted, you can say this animal was close to dead anyway. But the converse is not true. If you find a fully fat leg bone, that doesn't mean the animal was in good condition. It turns out that only 2 to 3 percent of the animal fat is in the bone marrow. If it's fat, the deer could still have been down to its last 10 percent of fat. It doesn't tell you whether the deer couldn't see very well, or its IQ was way low."

Others disagree with Mech's contention that weather and food are the chief limiting factors. François Messier, of the University of Saskatchewan, looked over Mech's data and declared that he could find no effects of snow on deer-population growth. He concluded that "in some situations wolf predation can prevent moose populations from increasing, even if environmental conditions and forage resources are suitable." A. T. (Tom) Bergerud, of the University of Victoria in British Columbia, looking at moose-population figures in Pukaskwa National Park in Ontario, concluded, "Predation limited the increase of the moose population and may have caused a decline."

Bergerud is in the forefront of Mech's critics. At sixty-three, he has long straight hair, a white mustache, gold-rimmed glasses, and a faded denim jacket. He has a cracker-barrel informality, but it is laced with a tone of controlled anger.

"My Ph.D. is in caribou and lynx," he says. "I went to Newfoundland in 1956, and I found these caribou with sores on their necks, and it was a major mortality factor. We all came out of school indoctrinated with Errington's idea that predation was compensatory mortality and that predators take the weak and the sick. I lost almost ten years in Newfoundland trying to understand why these caribou were dying of sores on their necks. I was sitting around a campfire one night feeling, 'I'm stumped. This is a wound. Somehow, something is going in there.' At the campfire, a shepherd told me, 'When lynx get my sheep, they always get 'em on the neck.' We got the skull of a caribou, and there were four tooth holes that fit the bite of a lynx. It turned out that lynx have *Pasturela multocida* bacteria in their saliva. The lynx would attack the calves, and the mother would chase the lynx off, and the calves would die later. There's a ten-year cycle in lynx and a ten-year cycle in caribou-calf crops. Lynx were definitely causing the decline."

Ever since, he has sought to show that predators can be the chief factor holding down prey populations. He put radio collars on caribou so that when they died he could find the carcasses and judge what had killed them. "The first thing we found was, the calves were being killed by bears and wolves," he says. Comparing estimates of moose populations in areas with and without wolves, he concluded, "Moose populations in North America appear to be limited by predation." Bergerud performed statistical analyses of moose and caribou populations in various areas where there were government-operated wolf-control programs in British Columbia, Alaska, and the Yukon. In lectures, he displays graphs depicting moose and caribou populations before and after a number of these hunts. Except where bears were responsible for most of the ungulate predation in the first place, he concludes that the moose and caribou populations increased after wolf controls. That, he believes, is proof that wolves had been limiting them all along. In one Alaska study, he judged that, during the year of the hardest winter, biologists found no starved caribou. He dismissed Mech's hypothesis that a hard winter may dispose calves to predation generations later as vague and impossible to prove.

Bergerud believes that biologists who view wolf predation as compensatory are misled by Isle Royale. Says Bergerud, "Isle Royale really griped me. It's in all the textbooks as a classical example. Isle Royale is really an abnormality, an artifact. The artifact is that there is no ingress of wolves." Anywhere else, he says, when there is a plentiful food supply, wolves come in from other areas.

The idea that predators do not regulate prey has come under increasing attack in Alaska and Canada. In the 1970s, biologists there were concerned about a continuing decline among caribou, and their seeming inability to rebound when habitat conditions seemed capable of supporting more of them. Theoretically, as prey populations decline, predators find less to eat and also decline, until eventually the prey population recovers. Mech's study in Minnesota showed that, as the deer population declined, the wolf population decreased by one-third. Mech believed that, as predation relaxed, the deer herd would recover. In Alaska, game officials thought they saw something different happening. Says David Kelleyhouse, director of the Alaska Department of Fish and Game's Division of Wildlife, "It was a prevalent notion in the 1950s and '60s that predation had a rapid feedback effect. If the prey went down, the predators would go down immediately. But that was wrong. They lag."

When the moose and caribou populations have declined to such a low density, says Kelleyhouse, the wolf population is still high, and the wolves no longer bypass the healthy animals to attack the weak. They are forced to hunt healthy moose and caribou, and they prove perfectly capable of killing them. "It's strictly additive mortality," says Kelleyhouse. "It's not compensatory any more." Kelleyhouse says he has seen evidence of this: "I have seen a big male wolf keep diving in on this cow moose and repeatedly getting kicked. She was a perfectly healthy cow. We shot the wolf and autopsied him. He had a broken jaw and broken ribs. Broken bones all over."

In some circumstances, as one prey population declines, wolves will switch to alternate prey. In interior Alaska, when caribou decline, wolves might switch to moose. Wolf numbers might then remain high while the caribou continue to decline, and thus, Alaskan officials believe, keep caribou or moose at unnaturally low numbers. And because many people have come to Alaska for the hunting, low moose and caribou numbers become political issues.

Bowing to hunter pressure, and backed by biologists convinced that moose and caribou declines were due to growing wolf populations, the Alaska Department of Fish and Game began killing wolves. Between 1976 and 1982, government hunters went out in helicopters and shot more than 70 percent of the wolves in certain areas. In the years following the control efforts, moose populations increased in some of the areas. William Gasaway, Kelleyhouse, and other Alaska Department of Fish and Game biologists reasoned that, if the wolves had been merely killing the weak and the sick, the weak and the sick would go on dying after the wolves had been removed. They concluded that, since the prey populations increased, the wolves must have been the limiting force. Similar hunts were initiated in the Yukon, and moose and caribou populations increased there as well. Wrote Gasaway, "Increasing evidence suggests predation by wolves and one or two species of bear is the primary factor limiting moose at densities well below carrying capacity [the maximum number of moose the habitat can support]." Declared Alaska Department of Fish and Game biologist Robert Boertje, "In the 1960s, we didn't think predation had much effect on prey numbers. Now we've gone pretty much full-circle. Where wolves are regulated, moose density is much higher, and harvest by hunters is much higher. Moose will occur in low density in Alaska unless wolf and bear populations are manipulated."

Others say that Bergerud's studies and those of the Alaska Department of Fish and Game are inconclusive, that they don't show that weather, disease, or changes in food resources or other factors didn't predispose the moose and caribou to wolf predation. Rolf Peterson criticized one study for using different census techniques on prey populations from year to year, for not taking into account differences in vegetation, and for using overly broad indicators of snowfall. Wolf controls in Alaska were not followed with long-term research to see whether prey populations remained high after the predators recovered. Moreover, since sport and subsistence hunting continued during and after the controls, the effect of human hunters clouded the picture. So far, the studies show only that one can, over the short term, increase the number of moose or caribou available for hunters by shooting wolves.

Bergerud claims victory. "Most pro-wolf people still want to make the argument that wolves are doing nice things out there, keeping us healthy," he says. "The argument is really over. I don't think many front-line biologists are saying wolves don't have an impact."

Bergerud is partly right. Says Peterson, "Nobody would argue wolves aren't important in the regulation of prey. The scientific viewpoint that predation was just a compensatory thing, that they just killed the weak and the sick, has pretty much been tossed aside." But the argument is far from over. Now biologists contend over how predation, weather, food resources, and other factors relate to one another. If wolves do regulate a caribou population, for how long do they do so? A year? Two years? Or more than that? How do logging, hard winters, dry summers, or the activities of other predators relate to the effects of wolves?

The answers to these questions—and the lack of answers—lie at the heart of nearly everything we do about wolves. If wolves limit moose and caribou, for example, should we manipulate wolf populations to boost caribou populations that are alarmingly low in parts of Canada and Alaska? Previous experience shows that we can, at least temporarily, increase some caribou populations by killing wolves. But what are the long-term consequences of such programs?

Theoretically, as prey increase, predators increase, until at some point the two reach a point of equilibrium, with predators just numerous enough to keep the prey population steady. Barring environmental changes, the number of predator and prey could remain the same for centuries. Most biologists believe there is only one such point of equilibrium. But a theory advanced by A. R. E. Sinclair, of the University of British Columbia, and Gordon Haber, an independent researcher in Alaska, proposes that there may be multiple equilibria. Thus, wolves may hold moose or caribou populations at low densities. Or the prey population may increase until there are a lot of healthy breeding adults and more than enough young and old to feed the predators. At this higher population of prey, predators are limited by other factors, such as the availability of adequate denning sites, and the inability to live in very large packs. They fight with one another. They invade one another's territories. In the stress that follows, fertility rates fall. The high prey population and the self-regulated predator population may persist indefinitely.

A study in Australia removed foxes and feral cats from an area to see whether afterward, when predators repopulated the area, they would again reduce the rabbit population. The study found that the rabbits increased to higher densities, but when the foxes returned, the rabbit population did not drop back to its earlier low density. The researchers concluded that this demonstrated the existence of multiple equilibria. If multiple equilibria exist in wolf-caribou systems, a wildlife agency might undertake one wolf-control operation and ever after allow nature to manage the system.

Wolf biologists today argue heatedly over the existence of multiple equilibria, but as yet there has been no direct test of the theory with wolves. A test could come out of a wolf-control program begun in the Yukon in 1993. The Aishihik caribou herd in the southwestern Yukon began to decline in the early 1980s. Between 1980 and 1990, the population plummeted from fourteen hundred to seven hundred individuals. Meanwhile, moose populations dropped to the lowest density recorded anywhere in boreal forests. The local native peoples, who depend upon moose and caribou for subsistence, urged the government to do something. It is not clear why the moose and caribou declined. Neither subsistence nor sport hunters appear to have exerted much pressure on the caribou population. "There is some evidence that the population was being limited by wolf predation," says Robert Hayes of the Yukon Fish and Wildlife Branch. But researchers fear that putting radio collars on caribou calves, to verify that wolves are taking them, might make them more conspicuous and therefore more vulnerable to predators.

Instead, in 1993 the Yukon Fish and Wildlife Branch began to control wolves in the area. After removing 75 to 85 percent of the wolves, it planned to prohibit hunting and to monitor the moose, caribou, and wolf populations for several years, to see whether there is a higher density of prey that would support both wolves and hunters with no future need for wolf controls.

A lot rests on the outcome of the debate over multiple equilibria. If there is only one point of equilibrium—if wolves normally regulate prey at low densities—then wildlife managers will feel they must choose between hunters, who pay their salaries, and wolves, who don't. If there are multiple equilibria, then it may be possible in the future to manage the herds for humans and wolves without shooting wolves.

We have come a long way from the view that wolves are simply a curse on the gentler forms of life, but we haven't yet arrived at our destination. What we have found is that predation is far more complex, far more varied, far more difficult to understand than we ever thought it would be.

Even if we can encompass all this complexity, it seems to me unlikely that we will ever really understand predation, for its greatest complication lies not in the woods of Minnesota or the mountains of Alaska but in our own minds. We will always have difficulty separating what killing means to wolves from what killing means to humans. And we may never agree upon what killing means to humans.

Bearhead Swaney, a member of the Flathead Tribal Council, was a big, broad-shouldered man who wore his hair in long braids and had a look of barely contained anger on his face. He loved to make white listeners squirm by accusing them of having shallow views. In 1980, at the University of Montana, Swaney lectured an audience of young whites on native-American attitudes toward wildlife. It outraged him, he said, to see biologists put radio collars and ear tags on wild animals, because this showed no respect for the spirits of animals. Then he changed his tack. "Did you ever pray before you killed something?" he roared at the audience. His listeners, mostly wildlife students, who were accomplished at neither killing nor prayer, sat in silence, paralyzed by their own sense of disengagement with life.

Swaney meant to make them feel that they stood pale and trembling outside the elemental relationships of life, and that they were immoral because they could not connect act with spirit. It was a fairly delicious act of red man's revenge.

That line between object and feeling often daunts us. During the Gulf War of 1991, for example, Americans sat in front of their television sets, gripped by the sleek precision of smart bombs which carried their own televised point of view as they dropped from the wings of jet aircraft, slid in great whistling arcs over the desert landscape, fell with eye-widening speed toward a looming bunker below, slipped soundlessly through an airshaft, and exploded inside, committing their own concussive suicide in the process. As we watched, we seemed to become machines. The war was, to our hearts and eyes, bloodless. Though one hundred thousand Iraqis died, we never saw a body, and we had no feeling whatsoever for their death or any individual suffering. Killing became an abstraction.

It seems to me that we make a similar abstraction when we talk about the roles or effects of wolf predation. Biologists ask us to overlook the individual suffering of prey species for the sake of the ecosystem and the ongoing process of evolution.

But there are people who cannot easily sublimate the suffering in the numbers. So while some of us out there in the snow are puzzling out the unfeeling facts of predation, others cannot look upon the carcass of a deer without confronting its suffering. It is fine to argue that the deer died with dignity, that this was nature's way, that predation is part of the plan, that starvation is not necessarily any better than being dragged down in the snow by a pack of wolves. But not everybody can make that abstraction. Thus the regret, the uneasy glances away from the kill on the ground, to the trees and chickadees and life.

It's the discomforting end of humankind's bargain with consciousness. We empathize with other beings and imagine we feel their pain. Either the empathy blinds us to the mechanics of life, or we shut the empathy off.

Some people have no qualms about looking upon death, or about taking the life of a deer or a wolf. Some feel the taking is a blessing that gives them food, or that taking the life of an animal is a measure of their own competence and deserving effort. Others feel the taking of any life brutalizes both the victim and the killer. Killing is something we don't agree upon. We find our own private understandings, and we shall probably never have a shared understanding of killing, never find an intersection between biology and spirit where predation makes real sense.

And because of that, we shall forever argue about wolves.

4

THE VOICE OF THE WOLF

It is late at night in Algonquin Provincial Park in Ontario. The moon is a pale glow behind the rainclouds driving in from the west. The night is full of messages. Maple leaves rustle, and the wind sighs through spruce boughs. Green frogs twang like loose banjo strings. Spring peepers shrill like police whistles. Mosquitoes whine. John Theberge steps quietly out of his battered blue truck. He is tall and rangy, with a salt-and-pepper beard and bushy black eyebrows. Behind his thick eyeglasses, he has a faraway look, as if he is dreaming. His wife, Mary, a petite, sharp-featured, energetic woman, slips out the passenger side. Both are dressed for the field, socks pulled up out of their boots and over the cuffs of their trousers to keep the blackflies out. They are careful not to slam the doors. The lights are turned off.

Theberge stands quietly, adjusting his ears to the small noises of the night. Then he tilts his head back and howls. It is a loud, straightforward unmusical howl, in slightly descending notes—not deep and melodramatic, not a movie sound-track version of a wolf, but more like the long steamwhistle shift call at a steel mill. He doesn't bother to cup his hands to direct the sound. He has been doing this so long that he knows it is unnecessary.

From off in the darkness comes the yipping, and then the crooning, of wolf pups howling back. Next a deep, throaty adult voice joins in, rising slowly and then descending, howling repeatedly, its refrain something between a moan and a song. It is the sound Stanley Young held was "so frightfully piercing as to go through your heart and soul." The Theberges are transfixed by the performance.

The voice of the wolf is the only aspect of the animal most people have any experience with. They have heard wolf howls, if not in the forests of Minnesota or British Columbia, in the sound tracks of television movies. Because this sound lodges in an older, less tutored part of the mind, it is the most emotional point of contact we have with the creature. If wolves clicked or burped or brayed or wheezed, they would be, in our minds, a far different animal.

The Theberges spend their days setting traps in order to radio-collar wolves. They spend their nights howling. The wolves' responses give some idea of whether they're breeding and how many are out there without collars. "Usually when you howl," says Theberge, "you get the pups answering first. Then the adults answer." Often they get no howls back, but one wolf howled back a hundred times before tiring of the game. Sometimes a wolf creeps up to them in the darkness, hidden in the bushes. "It will whimper, as if wanting to come along," explains Theberge, "but it doesn't have enough incentive to come out of the bush."

Says Theberge, "Mary and I have undoubtedly howled at wolves more than anybody." Indeed, he has been howling at wolves and listening to their replies for more than thirty years. But he still can't say what the return howls mean. "I don't know if that means they think we're insane or it's a note of pleasure." What the howl means is a question he has been pursuing in one way or another for most of his life.

Theberge began in 1959, when he was just a high-school student, working for the Canadian biologist Douglas Pimlott, who had begun that year to study wolves in Algonquin Provincial Park. Pimlott had seen that hunters and farmers were pushing the wolf toward extinction all over its range, and he felt wolves belonged in the ecosystem. "It might be said that the wolf was one of the last natural resources to be included in the great modern movement toward conservation," Pimlott wrote. He wanted to see Ontario's bounty on wolves eliminated, and he knew that the development of an argument that would convince hunters, livestock owners, and the politicians who listened to them would require the study of wolves.

No one had studied wolves in a forest environment like Ontario's. Murie had done his study in the open landscape of Alaska; Ian McTaggart Cowan had done his in the Canadian Rockies, where wolves were more visible. Here, in Algonquin Provincial Park, the forest of oak, maple, birch, poplar, fir, pine, and spruce has a thick understory. Pimlott had started with little idea of how to see enough of wolves to learn anything in this leafy environment. He hadn't the luxury of radio collars, which hadn't yet come into wide use. He did have a Labrador retriever, which he let run loose in the woods. "It actually did come back with two wolves one night," says Theberge. "But it didn't work out as a research tool."

Pimlott wanted to know how many wolves were in the park, and decided that getting wolves to howl would help locate them. He had found that captive wolves would reply to recordings of their own howls, so he mounted huge, trumpet-shaped speakers on the hood of an old government truck and proposed to drive the truck down logging roads in the park, broadcasting recorded howls. His first recording was of three timber wolves, three coyotes, and a coyote-dog, all howling together, which he thought sounded like a mixture of adult and juvenile wolves. He would crank up the big speakers, play the recording, and wait to see if other wolves responded.

On one of their first tries, Theberge recalls, they drove out into the woods. "I was about to play the tape of wolves howling. Doug said, 'I'm going to go to the top of this hill to see what happens. Give me five minutes and turn on the tape and blast it out.'" Theberge waited the requisite five minutes and turned on the tape. Immediately, wolves began to howl all around him. It was an unexpected and unnerving reply. And almost as immediately, there came a crashing in the bushes, and Pimlott, breathless and frightened, came running to the safety of the truck. "He didn't know what we had stimulated," says Theberge. "None of us knew what we were doing in trying to trigger the howls ourselves. We wondered if we were triggering aggression."

Having only studies by Murie and Cowan to go by, Pimlott and Theberge knew almost nothing about wolves, and the weight of centuries of folklore hung upon their imaginations. "We all believed that wolves were safe," says Theberge, "but we had a few exciting encounters." Wolves would sometimes run toward the researchers when they howled them up. Once, three wolves came toward them at a gallop, not howling back. "If you didn't know what was going on, you'd be scared," he says. "The whole experience is open to misinterpretation. I knew that no one had been attacked by a wolf, but I thought, 'Who's the most likely to be the first?' "

Paul Joslin was a student of Cowan's at the University of British Columbia when he joined Pimlott's Algonquin Park study. "I didn't know anything about wolves at that stage," he recalls. Today, Joslin is director of research and education for Wolf Haven International, a wolf rescue-and-education center in Tenino, Washington. Beneath his white hair and beard, he is placid and soft-spoken, and he has a broad, outgoing curiosity. "Doug wanted to use howling as a way to census wolves," he says. Joslin would pack the big speakers on his back and spend the whole night walking miles into the woods along a railroad right of way, stopping now and then to play Pimlott's recorded howls and listen for responses.

Like Pimlott, Joslin was unsure how the wolves would respond to him. Finding himself nervous about being out alone in the woods, he thought he might be experiencing an ancient fear of predators, something ingrained in the human psyche by millennia of life in the woods. "I found in those early weeks that there wasn't a sound that you didn't right away identify as friend or foe." A beaver slapping its tail on the water in the gloom of night a few yards away would just about put him into the air. "If you had a slap right beside you, you would suppress the impulse to leave the ground before you consciously heard the sound. And then it would hit you: 'It's a beaver.'

"I had the odd time where I accidentally got too close, and I howled, and the pack was right next door to me, and they came to check me out. They sounded like dogs coming. They came crashing through the bushes. I could hear them shuffling around me. I couldn't see anything. And the moment they were downwind of me, they were gone.

"But one night, I was howling back and forth with some wolves down a valley, then I hear *crash, crash, crash!*" A wolf was coming toward him through the bushes. "This wolf stopped in the brush and started to bark and growl. It kept moving back and forth, growling and barking. I put my back against a tree, then moved to another tree and put my back to it. I went from tree to tree, all the way up the hill. It carried on for twenty-some minutes, by which time I'd already gotten to the top of this small mountain and back to my trail.

"It's one thing if I have a moose encounter. Then you go around the problem. But here the animal I'm working on has threatened me. It's like falling off a horse: you've got to get back on, you've got to deal with it. I went back the next day with the only weapon I had, which was an ax. I didn't find any wolves, but there was a freshly killed deer there. That was enough.

"I talked to Doug and said, 'I need one of three things: an assistant, a pistol just in case as a backup, or a dog as a companion.' Doug came up with a high-school student. He was built like an ox. That solved the problem." Or at least it made Joslin comfortable enough to see that there wasn't really a problem. Wolves never attacked him, the student, Pimlott, or Theberge. He and Theberge would find that, even if they approached wolves feeding on a kill, the wolves would run off before they got within two hundred yards of them.

At first Joslin got few wolves to howl back at him when he used the speakers. On a whim, he tried howling himself. "After some time, I finally found a wolf pack. They howled back. For a time I tried a combination of the recorded howling and my howling to get them to respond. At the end of the summer, I had 15-percent better results howling myself than using the recording." He believes his own howls got more replies because the wolves took the recorded howls for the boasts of a strange pack suddenly turning up in their territory and were intimidated. Also, the recorded howling broadcast the sound farther, and a pack miles away might have responded, beyond the reach of Joslin's hearing, and exhausted its will to reply before Joslin got within hearing range. Once wolves have howled, they will typically remain silent for twenty or thirty minutes. Perhaps they are listening for other packs, or they are musically sated—no one knows why.

When Joslin heard howls, he would record the compass bearings from which the sound came and plot them out on a map, then, in daylight, go in to look at the site and perhaps find dens. Sometimes he would find something unexpected. Once he found a dead wolf near a den in a rotted-out tree. When he took her into camp and necropsied her, he found she had more than ten broken ribs. He went back to the den. Looking it over carefully, he found bear hairs on the opening of the den.

So little was known about wolves that discoveries came quickly. They found that there were two peaks in wolf howling—one in the winter, when they are courting, and one in the summer, when the pups are out of the den. Joslin found that, after they left the den, wolves would move to a meadow in the forest, where adults would leave the pups while they hunted. They would keep to that meadow a few days or weeks, then move to another. "I thought, 'What am I going to call these things?' I looked at Murie. He said they rendezvoused together. I said, 'Fine, we'll call it the rendezvous site.'" As radio collars came into use, Theberge began trapping and radio-collaring the wolves, and the researchers could spend their nights riding the logging roads, radio receivers in hand, trolling the airwaves for wolves.

Pimlott, Theberge, and Joslin were growing comfortable with wolves. The ones they caught in traps or snares seldom offered any resistance, displaying instead "a total fear response," says Theberge. "We've hardly ever had one lunge at us. In some, the parasympathetic nervous system kicks in and they're quiet, basically fatigued." He and Mary once followed a radio-collar signal to a wolf caught in a snare. "It was alive," he says, "but it was cartwheeling and strangling itself. We didn't have any drugs or equipment. We went up to it and it just went into shock." Theberge put a snowshoe over it to restrain it while Mary released it from the snare. The wolf lay in the snow passively until they left, then scampered off. "We could do a lot of this work without drugging the animals," says Theberge. "But holding them down stresses them, so it's better to have them out."

In the early days in the dense Algonquin woods, howling proved to be the best way to locate wolves. Since the team heard wolves more than they saw them, they grew especially interested in the meanings of wolf vocalizations. They found that wolves make a complex and varied array of sounds—howls, barks, growls, whines, squeaks, and combinations of these noises. Traditional peoples and modern writers alike have held that these vocalizations constitute a language that conveys specific information. Jason Badridze thinks his Georgian wolves have distinctive howls to inform other members of the pack when they have killed something. He believes wolves can even convey numbers vocally. Farley Mowat wrote in *Never Cry Wolf* that an Eskimo told him that wolves passed on news of moving caribou herds from one pack to another. People who keep captive wolves frequently say the wolves know who is coming to visit and convey this to other wolves many minutes before the guest arrives.

Whines and whimpers are soft, high-pitched sounds that humans generally regard as plaintive or begging, whimpers usually being described as short bursts of sound, and whines as longer, more drawn-out vocalizations. Both can start as squeaks at the upper ends of the human ability to hear. Dr. Michael Fox, who has done extensive studies of wolf behavior at Washington University and the St. Louis Zoo, reported "undulating long whines" by an adult as it brought pups out of a den. Pups whine in pain, but they also whine or whimper to adults during greeting ceremonies after the adults return to the rendezvous site or the den. Joslin was once howling near a den when a pup howled back at him and then rushed toward him, whining.

Adults also whine when greeting or submitting to a higher-ranking adult. A captive wolf will go from growl to whine and back again if a human handler tries to take its food away. It will lay its ears back and wag its tail and whine and attempt to lick, then bare its teeth and flare its ears and growl, then lay the ears back again and whine and wag its tail submissively.

Barks are short bursts of sound lasting less than a tenth of a second. Humans interpret them as alarms or warnings, something like our shouts of "Fire!" or "Scram!" Joslin feels that single barks are alarms and seried barks are threats. A growl ending in a bark by an adult may send pups running into the den. Often a sharp bark ends a howling session. When researchers have approached den sites or rendezvous sites too closely, they have heard one or two sharp barks and a drawn-out bark in a series of lower pitches. Joslin once heard such barking repeated for twenty-seven minutes, with growling between the barks, and felt it was a warning. John Fentress, of Dalhousie University, could make a captive adult bark by entering the cage and howling while it was howling. When Murie heard wolves bark, he could usually see them, and they seemed to him to mean to be seen. Once he attempted to get close to a den and adults came toward him barking. Barking may have other meanings as well. In captive wolves, barking has been observed in animals trying to solicit play. Pimlott heard two wild wolves howl and bark as they approached each other, then go off into the woods howling together.

Growls are deep-throated sounds which humans universally interpret as threatening. A wolf may growl at another wolf when its tail is high, its legs are stiffened, and its hackles are raised, and follow with an attack. Growling is usually heard among wolves of higher status. Very subordinate wolves seldom growl; typically, they whine in submission. In a dominance fight at the Folsom City Zoo, Lupine, the lowest-ranking wolf, was attacked by all three of the other wolves. While they mauled her, they growled. But she neither growled nor whined; she curled her lips and showed her teeth and snapped her jaws loudly and repeatedly. Growling is the least recorded of the wolf vocalizations, probably because it is usually directed at fellow wolves. Observers watching den sites are usually too far from the wolves to hear a growl if it is uttered, for growls do not carry long distances. Rabid wolves attacking humans have been said to growl and snarl.

The howl is the form of expression that most fascinated Theberge. To human ears, the howl is the wolf's masterpiece. A drawn-out, continuous sound, sometimes lasting more than ten seconds, in which the wolf's voice rises and falls melodically, the howl is produced by vibration of the animal's vocal cords. Just as a violin string vibrates at several different frequencies—at halves or thirds or quarters of its entire length—when it is plucked, wolf vocal cords vibrate at more than one frequency, and thus produce complex tones. Theberge's studies showed that, whereas human ears hear only one tone, the howl actually consists of a fundamental tone and a number of harmonic tones, usually only one, two, or three, but sometimes as many as a dozen.

Howling is most frequent early in the morning and late in the evening, but occasionally wolves howl in daylight. Though hunters' stories in the popular press often depict wolves howling while they hunt, Mech, Peterson, Haber, and Theberge, who have between them more than a century of wolf study, all say they have never heard a wolf howl while hunting. Howling increases in midwinter, when breeding season comes on and the social interaction of a pack is more intense and active. After breeding takes place, the wolves are quieter. Wolves howl least when pups are very young, perhaps to keep other predators from discovering their dens. But by July and August, the pups begin to howl, and it becomes relatively easy to find a rendezvous site. Captive pups will howl from as young as one month of age.

A howling session can last many minutes. Usually one wolf starts it and howls once or twice before the others join in. The others give long, low howls and work up to a series of shorter, higher-pitched howls. It is not uncommon, when wolves howl in the woods, for other species to respond: coyotes may yip and howl in what seems like an answer, owls hoot, loons call, even sleeping songbirds pull their heads from under their wing feathers and chirp.

When a pack howls, the interbraiding of voices and shifting harmonics gives the impression of many singers. General Ulysses Grant recalled in his *Memoirs* riding with a guide and hearing wolves howl. The guide asked Grant how many wolves he thought he heard, and Grant, wanting to appear wise in woodlore, concluded it would be best to underestimate. He said, "Oh, about twenty." When they actually came upon the wolves, there were only two.

Howling fascinated Theberge. He went on to the University of Guelph, returning each summer to work with Pimlott, and ultimately he wrote his master's thesis at the University of Toronto on the howling of wolves. He built a pen in the park, put a captive wolf inside it, and waited to hear the wolf howl. "I couldn't get it to open its mouth," he recalls. Though he played it tape recordings of wolf howls, and stood outside the compound at night and howled, the wolf would not respond to him. Then Mary came to visit him. She howled once at the wolf and it howled back.

Why did the wolf respond to one and not the other? They set about trying to understand. "We started to study the harmony of what was different in our voices. I can get up to middle C, and Mary can get down to middle C. The wolf would not answer me but it would answer Mary. However, it insisted on a live performance and wouldn't respond to a tape recording. After a while, we understood that every wolf had its own set of harmonics that shifts, predictably, as its voice slides up or down the scale. They have a far better ability to discriminate between harmonics than humans, which means they can certainly identify who's howling. Which raises the question, why do they howl back at humans?"

And why do they sometimes not howl back? Joslin recalls, "I had one occasion in Algonquin where a wolf came up and saw me when I was howling, and I had a deuce of a time. It took two or three weeks before that pack would howl back. There's usually one or two wolves that will lead the howl. I suspect it was that one that saw me."

Theberge and Joslin would talk much about why wolves howled or didn't howl. Theberge tried a police siren; it worked. Wolves howled to loggers' chainsaws. They got so habituated to Theberge's own overtures that they would sing out when he accidentally slammed the door of the truck as he was getting out to howl.

There were a variety of explanations offered for howling. Murie often saw wolves howl after hunting had separated them, and usually found the wolves reunited shortly after those howls. He guessed howling helped them to get back together. Theberge's captive howled when he left camp, and the researcher thought it likely that the wolf was expressing a desire for companionship. Erich Klinghammer, who studied the howling of captive wolves at Wolf Park in Indiana, found that older and higher-ranking animals did more solo howling and then were joined in chorus more often than younger wolves, females, or lower-ranking males. He concluded that solo howls attracted lone wolves, especially females.

But clearly there were other meanings. During dominance fights or other great excitement, Klinghammer found that females not directly involved in the fight were "running about and howling repeatedly." And when animals became alarmed near the den, individuals might bark and howl. Mech and Fred Harrington, a professor of psychology at Mount St. Vincent University in Nova Scotia, got more wolves to return their howls near kills and considered howling "important in territorial maintenance."

Theberge thought howling might help wolves identify each other as individuals. Howard McCarley, of Austin College in Texas, had found that he could identify individual red wolves by the way they started and ended their howls. Mech was coming to the same conclusion about gray wolves in Minnesota. Theberge wanted to know if each howl had an individual signature. He and J. Bruce Falls analyzed the harmonic structure and pitch of howls of captives from Algonquin Park. They found, "Each individual had a tendency toward a certain type of beginning, ending, pitch range, and pitch change throughout the howl."

Others had similar thoughts. Z. J. Tooze, John Fentress, and Fred Harrington recorded the howls of captive wolves temporarily isolated from their packmates and made spectrograms of the recording. Analyzing fourteen different variables on the spectrogram, they found that each howl had its own frequency or its own way of varying the frequency. In other words, each wolf had its own distinctive voice.

The findings suggested that howling may have evolved, at least in part, to keep wolves from engaging in unnecessary aggression. Harrington believed aggression is expressed by howling at lower frequencies; bigger animals produce sounds of lower pitch and harsher tonal quality, and a deep voice is apparently as intimidating to wolves as it is to humans. Harrington found that wolves that responded to his howl by moving toward him howled back with deeper voices than those that stayed put or moved away.

Howling probably serves notice to other packs or dispersing individuals that this pack is here and the territory is claimed. Howling has often been observed to help packs avoid one another. When one pack howls, another, near the boundary of the territory, may move away from it. In Minnesota, when one pack howls, the neighboring pack may respond, and a third pack over the ridges may join. Robert Stephenson, of the Alaska Department of Fish and Game, says that in the open terrain of Alaska's North Slope humans have heard wolves howling ten miles away. In Minnesota, electronic microphones picked up the replies of wolves responding to the howls of researchers 6.8 miles away. "Thus on calm quiet nights," wrote Mech, "a single howling session could advertise a pack's presence over an area of 50 square miles to 140 square miles or more."

Theberge thought there might be two layers of meaning in wolf howls. One level was "universal," in that all wolves might understand, for example, the proclamation of territorial rights in a howl. The other was "individual," in that wolves of long acquaintance might read nuances in each other's howls that indicated individuals' moods. He found that spontaneous howls contained higher notes than howls he had elicited, which suggested to him that subtle differences in howls might convey distinct meanings. And researchers who worked in a particular area for long periods said they recognized different howling styles in different packs. Jason Badridze believes that each of the packs he has worked with in the Republic of Georgia has its own dialect. Mary Theberge recalls that the Foys Lake pack in Algonquin Provincial Park had a beautiful, trailing call she found distinctive.

For John Theberge, the wolf's howl seemed much more than mere reflex. "We also determined that there was some emotional content in the howl," he says. "When this wolf in the pen was agitated, he would howl differently. For example, we went blueberry picking one day and left him alone, and he started to howl differently than when we were in the cabin with him. We started to analyze these howls, and we found there were significant differences in emotional content. Howling has significant communication value. It identifies the wolf and tells something about its emotional state.

"We think we can discriminate a mourning howl. It is a haunting howl. It drops in pitch twice and goes into minor keys. It's quite emotional. At least, you can read the emotion into it." Once, Theberge visited a man who had just killed a wolf near the park. The wolf that had been killed was a dominant female with pups left behind in the woods. Theberge went outside with his radio receiver and located other radio-collared wolves. "One," he said, "was giving the mourning howl over and over." When Pimlott euthanized a captive wolf that had been injured in its pen, Theberge heard this howl from a compatriot wolf. Badridze says, "There is a howl of loneliness. It's very distinctive for everybody. Everybody can just feel it. It must be a universal language."

The howl seems to many people to express the essence of the wolf. For those who think of the wolf as evil, the howl is frightening. Stanley Young described it as "perhaps the most dismal sound ever heard by human ear." He compared it to a dozen railroad whistles braided together, hooting until one after another has faded off, leaving a last, long, heart-wrenching wail. We still hear such sounds in the sound tracks of movies to suggest the lurking of supernatural powers. On the other hand, those who spent time watching wolves and came to see them as noble creatures began to hear joy and reassurance in their howls. Lois Crisler heard such howls.

In 1953, Crisler and her husband, Herb, a cinematographer, went to Alaska. He had shot footage of elk in Washington's Olympic Peninsula for use in a Walt Disney *True Life Adventure,* and now he wished to film the great caribou migrations in Alaska. The Crislers planned to spend two years in the Arctic, filming wildlife. They wanted to film wolves hunting caribou, but found that wild wolves wouldn't let Herb get within camera range. So they acquired wolf pups dug out of a den by Eskimos at Anaktuvuk Pass, which they hoped to rear to maturity and use to get the footage Herb needed. The wolves, however, took on additional purposes. Alone in a camp in the Brooks Range of Alaska, Lois Crisler looked long and hard at the wolves. She became enthralled by their dignity, their intelligence, their gentleness, their remoteness, and their mystery. And she began to write about them. Her sensitive portrait of the wolves, published in 1956 as the book *Arctic Wild,* would influence a generation of wolf researchers.

Crisler observed that people thought in a stereotype about wolves. "Everybody 'knows all about' wolves," she wrote—everybody thought them to be vicious, conniving, sneaking, and cruel. "And people who have never seen a wolf will defend their myth-wolf pattern with betraying fury." She saw in that perception a reflection of the human observer, whom she described as "a naked, nervous, angry species."

She found no suggestion that wolves were combative or cruel. In fact, she felt wolves were exemplary. "Wolves have what it takes to live together in peace," she wrote. "For one thing, they communicate lavishly. By gestures . . . and by sounds, from the big social howls to the conversational whimpers."

She found that they cared for one another. "They feel concern for an animal in trouble even when they cannot do anything for it," she observed. "A dog got his nose full of porcupine quills on our walk one day. All the way home the wolf Alatna hovered anxious-eyed around his face, whimpering when the dog cried in trying to tramp the quills out." When a newly acquired dog cried through the night, a wolf stayed near him, whimpering when he cried. A young dog wandered off from their daily walk, and "the wolf with us ran to me, cried up to my face, then standing beside me looked searchingly round, call-howling again and again. When the dog sauntered into view, the wolf bounded to him and kissed him, overjoyed."

The wolf's howl was to Crisler the sound of a noble conviviality. "A howl is not mere noise, it is a happy social occasion. Wolves love a howl. When it is started, they instantly seek contact with one another, troop together, fur to fur. . . . Some wolves love a sing more than others do and will run from any distance, panting and bright eyed, to join in, uttering as they near, fervent little wows, jaws wide, hardly able to wait to sing."

It was the beginning of a change in the way humans heard the voice of the wolf. Mech and Pimlott had not published their studies, and Mowat had not yet published *Never Cry Wolf.* But as people like Crisler, Pimlott, and Theberge got to know wolves, the howl meant, not the approach of darkness and evil, but something more reassuring. Crisler saw her glimpse into "wolfness" as a glimpse into "wildness." She declared that wilderness was emblematic of wolves, not because they were ferocious but because they were independent. Pimlott and Theberge used the same terms. Theberge would write his own book, *Wolves and Wilderness,* in which he would declare, "Today the howl reminds us that our past is deeprooted in wildness. . . . It epitomizes the wilderness we have fought so successfully to conquer and now must fight to save." He saw in wildness and wilderness the source of our own better nature. If wolves were considerate and expressive, communicative, reluctant to kill members of their own kind, and ready to display joy and exuberance for life, so might be humans who threw off the shackles imposed by citified life. Wildness and wilderness, in other words, had healing qualities.

Underneath this new view of wildness as healing lay a kind of misanthropy. Crisler described humans as a "seething, hating species." It was the era of the atomic bomb and the cold war, which brought a new awareness that coldly rational humans could at a moment's notice snuff out a considerable portion of the life on earth.

Two world wars had made us question the nature of our species, and we looked for new evidence that humankind was not an evolutionary blunder, bent ultimately on destroying itself and the earth in the process. Hoping to find the answers in our origins, we became intensely interested in the ways other species lived, and in the ways evolution shaped human behavior.

Anthropologist Ashley Montagu, for example, wrote, "Animals in the state of nature do not make war upon their own kind; they have no Attilas or Hitlers. They seldom exhibit the kind of savagery that civilized men exhibit toward one another." It was precisely because humans denied their own nature that they devastated much of the world, and wilderness offered a chance for redemption. Declared Montagu, "Man may yet restore himself to health if he will learn to understand himself in relation to the world of nature in which he evolved."

The idea that humans needed to rediscover their nature precipitated an intense debate over wilderness. By 1960, the United States and Canada were considering laws to enable them to preserve wilderness areas. The debate was argued, on one level, about saving ecosystems for plants and wildlife, but on a deeper level it was about human character. The "old" view held that wilderness was the source of ignorance and violent impulse that humans tried to overcome. Robert Wernick, a writer who viewed human material culture as the species' triumphant achievement, declared that wilderness "is precisely what man has been fighting against since he began his painful, awkward climb to civilization. It is the dark, the formless, the terrible, the old chaos which our fathers pushed back, which surrounds us yet. . . . It lurks in our own hearts, where it breeds wars and oppressions and crimes." Replied Montagu, "Civilized man, especially in the western world, has projected an image of his own violent self upon the screen of nature." Author Wallace Stegner declared, "Wilderness is something that has helped to form our character." Stegner believed "we are a wild species," and suggested, "One means of sanity is to retain a hold on the natural world, to remain, insofar as we can, good animals."

Are we at heart savage and predatory and not to be trusted? Or is wildness the best in us, the instinct for freedom and motion, adventure and care, for close engagement with each other and with the world? We continue to debate the nature of wildness on all sorts of levels. We debate it when we ask why we are violent and when it is permissible to be violent. Characters in television dramas who kill one another are part of the debate. So, too, is argument about abortion or gun control, or whether prisons should punish or rehabilitate—or about the morality of hunting. And we debate wildness when we talk about wolves. If we see nature as treacherous or insane, we shall see in the wolf a reflection of that nature. And if we think of wildness as the source of kindness and joy, we are apt to see our better nature reflected in the gaze of the wolf.

The voice of the wolf has increasingly symbolized the new view that wildness is a source of good. All over the range of the wolf, humans now go out into the night and howl, as if seeking affirmation of their faith in wildness. Theberge did much to popularize the practice, and his experience of howling up wolves has become the most widely shared such experience in the world.

When the Minnesota Science Museum's traveling exhibit, "Wolves and Humans," was installed at the American Museum of Natural History in New York City, the museum called Theberge and asked if he had recordings of wild wolves they could use for a cassette tape and record to give to members as a premium for joining. Theberge had written articles on wolf howling for *Natural History,* the magazine of the American Museum. He happily provided some recordings, because they would help educate the public about wolves and create more interest in their conservation.

When the museum issued the recording, *The Language and Music of Wolves,* there was a press conference, and Mech and Theberge came to speak. They appeared on television talk shows and *The New York Times* ran a front-page article on wolf howling. Someone enlisted Robert Redford to read a script on the record, and Theberge helped write the script. Columbia Records bought the recording, which was sold in record stores and is still widely used in television news stories and documentaries. Unfortunately, Theberge had signed a release that precluded him from earning royalties—something he rues today, when he considers all the radio collars and flying time that royalties might have provided for his project.

Pimlott, Theberge, and Joslin are also responsible for the most popular howling event of all. In the summer of 1963, Algonquin

Park officials asked them if they would hold a campfire talk in which they took visitors out to hear wolf howls. They expected a few dozen campers to show up. Recalls Joslin, "When the time came, it was a shock. There were six hundred people. The Royal Canadian Mounted Police were all upset, because they hadn't been informed, and here was all this traffic on the road. We heard some wolf howls, and you could have heard a pin drop. All this horde of people standing under the stars, silent." Ever since, wolf howls have been held on occasional Thursday nights in August at Algonquin Provincial Park—only in August because then the wolves are likely to be at rendezvous sites along Highway 60, and only on Thursdays because a weekend howl would flood the park with vehicles. In 1990, sixteen hundred people attended each of two howls. Since 1963, more than sixty-five thousand people have attended. Says Algonquin Park interpretive naturalist Mike Runtz, "In the calmness afterwards, there's always applause, thunderous applause, from fifteen hundred people. And as the people drive off, they yell, 'Thank you, Thank you!' at the howlers."

In recent years, wolf howling has become a tourist activity at Algonquin Provincial Park, Prince Albert National Park in Saskatchewan, Riding Mountain National Park in Manitoba, and Jasper National Park in Alberta. It is done also at the International Wolf Center, a project fostered by David Mech near Ely, Minnesota.

Wolf howling sort of creeps up on people. Wolf researchers even start howling at other people as a kind of good-humored gesture of hope. Theberge once howled for the Canadian Minister of the Environment on a busy Toronto streetcorner. Mech had a group of wolf researchers troop into the Minnesota Legislature's chambers when the legislators were about to vote on appropriation of funds for the International Wolf Center, and called upon the researchers to give a howl. Recalls Theberge, "It was pretty tacky, but we did it. And the legislators howled back."

Once we start hearing a nobler voice in the howl of the wolf, we are likely to start arguing for wolves and wilderness. Pimlott and Theberge sought to do just that in Algonquin Provincial Park. Pimlott felt the wolves in the park were unfairly persecuted. His study showed that the deer population was, if anything, alarmingly high—fifteen deer per square mile. In 1966, perhaps finding Pimlott's views threatening, the Ontario Department of Lands and Forest, which felt it had a responsibility to produce moose and deer for the benefit of hunters, stopped funding the study. Pimlott took a position at the University of Toronto, where he could express his views more freely. He organized the Canadian Nature Federation, the Canadian Arctic Resources Committee, and the International Union for the Conservation of Nature's Wolf Specialist Group. Until his death in 1978, he was one of Canada's most eloquent and insistent voices for wilderness and wildlife conservation.

Theberge followed in Pimlott's footsteps. He says today, "I started as a young person interested in wolves and ended up a person interested in having places to put them." Theberge has become a leading opponent of wolf-control programs in Canada and Alaska. He sits on the International Union for the Conservation of Nature's Wolf Specialist Group, and he has tried to get the group to add to its International Wolf Manifesto a provision saying it is unethical to kill wolves simply to find out the effects of predation. "I will oppose killing of wolves on ethical grounds," says Theberge. His declaration has not been without cost. "A Canadian biologist said to me, 'You're a disgrace!' There are wildlife managers now who won't even speak to me."

Today, he continues the work Pimlott started, studying the wolves of Algonquin Provincial Park. But this project has itself become an enormous challenge. Local residents hunt the deer and moose heavily themselves, and when they see the deer decline, as they do periodically, the hunters are quick to blame the wolves. Many of the older local people do not hear anything redemptive in the voice of the wolf. One day in 1992, after flying to radio-locate wolves, Theberge landed at the Bonnechere, Ontario, airstrip. Two men drove up in a pickup truck and snarled out the window at him, "We don't like wolves and we don't like you. We get to hunt deer two weeks a year and the wolves hunt 'em fifty-two."

Logging inside the park had, since the 1880s, opened up clearings into which grew stands of poplar, aspen, red maple, dogwood, and raspberry, which increased the deer population. But when the United States imposed tariffs on Canadian softwoods in the 1970s, logging subsided, and deer habitat in the park declined. In winters, the deer moved increasingly into wintering yards of cedar and red osier dogwood near the communities outside the park. The year 1992 was a poor one for acorns in the park, and the park's deer massed in wintering yards in nearby townships. Almost all the wolves in the park followed the deer out. Local hunters, seeing more wolf-killed deer in the yards, concluded that wolves were laying waste to the deer. Said Ralph Bice, a lifelong trapper and a guide in the park, "Nobody's been to the woods like I have but admires wolves. But they need to be controlled." The mayor of Round Lake said she feared her children would be attacked by wolves.

The winter of 1992–93 proved to be a severe test for the wolves of Algonquin Park. Local residents declared they had a plague of wolves and set out snares. They gathered thirty snowmobiles and staged a wolf drive. Over the winter, 55 percent of Theberge's radio-collared wolves died. Says Theberge, "We were radio-tracking to people's houses and knocking on the doors and asking for our collars back." Theberge followed the signals of radio collars into people's yards and found his wolves dead and skinned. The carnage was great. One pack of twelve was reduced to two females. A pack of eight was reduced to a single female. "This is an obscene level of killing," said Theberge. With help from the World Wildlife Fund, Canada, he got the Provincial Ministry of Parks to establish a ban on snaring wolves in three townships outside the park—which made many of the local people even angrier.

One winter afternoon, John and Mary Theberge went to retrieve the carcasses of several wolves a man had killed near Round Lake. While they were sitting at the kitchen table talking to the man and his wife, wolves started to howl outside. Theberge would be the first to admit that he doesn't know what the howls really mean, but what he hears in the voice of the wolf makes him fight for the survival of the creature. Almost reflexively, he went outside to listen.

The wolf hunter and his wife remained at the kitchen table. She shivered and told Mary, "Oh, I hate that sound. You just can't go outside the house!"

5

LEADER OF THE PACK

The North Central Forest Experiment Station field laboratory is an old ranger station on the Kawishiwi River in Minnesota's Superior National Forest. It is a log cabin with a big living room in front of a wide stone fireplace, a small kitchen, a couple of bedrooms, and a screened-in sleeping porch, set in a forest of balsam fir and Norway pine.

Inside, David Mech is on the telephone with someone who says there is a freshly wolf-killed deer in the snow outside Ladd Williams' outhouse. Mech has been following a pair of wolves in the neighborhood and is glad to have news of them and a leg bone from the carcass with which to judge the health of the victim. He hangs up and addresses himself to the morning's main task, which is to find Wolf 171, a young female wearing an expensive, high-tech radio collar. It is time to run some blood tests on the animal and to retrieve the information stored in the radio collar's small computer.

Wolf 171 is nine months old and already traveling alone. Mech has been interested in what determines whether a wolf stays for life with a pack or becomes a disperser, an animal that drops out of the pack at a young age and wanders. Dispersing wolves may travel enormous distances. A female wolf left Mech's Minnesota study area and was shot by a farmer in Saskatchewan, five hundred miles away. A wolf from northeastern Alaska went almost to the west coast, a distance of 450 miles. A wolf from Alaska's Nelchina Basin turned up in the eastern part of the Brooks Range, more than four hundred miles away. In Kenai National Wildlife Refuge, Rolf Peterson found that males tended to disperse more than females. Most of the dispersers that travel more than two hundred miles seem to be males.

It is a risky business for the dispersers. They may starve without the shared hunting experience and concentrated killing power of the pack, or be attacked, injured, or even killed by other wolves as they wander. Launched suddenly onto unfamiliar territories, they are immensely more vulnerable to hunters.

Young and Goldman held that a pack was a pair of wolves, their pups, and their offspring from previous years. Murie believed the two packs he studied in Alaska were such families. But dispersers show that wolf society is more complex. Some dispersers find mates on unclaimed territories and start their own packs; others join existing packs. Mech and Diane Boyd have seen wolves successfully move from one pack to another. Tom Meier, of the National Park Service, reports that, in 1992, 35 percent of the radio-collared wolves in Mech's Alaska study area were living in a pack other than the one they had been born into. Meier more often saw unrelated wolves adopted by other packs than he saw dispersers form new packs. Adults may change packs more than once in life. Possibly neighboring wolf packs have familiar associations, and some wolves find friendships back and forth across territorial lines.

Dispersal has obvious benefits to the species—it is a way of mixing genes in the wolf population, of reducing inbreeding and sharing traits over broad geographic ranges. If packs kept unrelated wolves from joining them, wolves would ultimately be breeding only with their own family members. Inbreeding leads to the loss of such genetically determined traits as immunity to disease, and also to reduced fertility: sperm in inbred animals typically shows much higher rates of deformity and lower rates of motility.

If a disperser is to share its genes, however, it must become the dominant wolf in a pack. And because of this, dispersers highlight the conflicts in wolf society between community and individuality. Wolves are social beings, enjoying warm, companionable, and highly emotional lives within the pack. But wolves are also individuals, occupying different roles in a pack and competing, sometimes violently, for social standing and the right to pass on their genes. It is in a wolf's interest to accommodate and coordinate with other wolves, yet also to contend with the same wolves. This dual nature draws the interest of humans almost as powerfully as the act of killing, because it mirrors our own contending social natures. Humans and wolves both evolved as group hunters. In consequence, they have surprisingly similar social lives: Both have dominance hierarchies; both care deeply for their young and their families and can display precise and exacting coordination in complex tasks; both take pleasure in companionship, children, touch, song, smiles, and play. Yet both can be aggressive and violent. Like wolves, we are sometimes altruistic and cooperative, sometimes ruthless and domineering.

Since a dispersing wolf may either form a new pack or join an existing one, it is not clear whether a disperser is a leader or the loser in some lupine social contest. Understanding what makes a wolf disperse may help us understand what makes one wolf a leader and another wolf a clown. Is the best-adapted wolf the swiftest and strongest, the most aggressive, the most perceptive, the most convivial, or the most caring?

What makes such questions especially difficult is that, again, we ask the same questions of ourselves. What is the core of our natures? If we are like wolves, are we cooperative and nurturing, or are we individualistic and competitive? Do we live for ourselves, or for the pack? If we do a little of both, how do we balance one against the other?

Mech has just arrived at the field laboratory after spending two weeks of working at his office in St. Paul. Mike Nelson, the on-site director of the research program, brings him up to date on recent wolf sightings and tells him that bad weather has grounded the planes that usually go up to locate radio-collared wolves. Joel Norton and Eric Seabloom, two volunteer technicians who do the tracking and radio-collaring in the field, are putting on extra pairs of socks and sweaters, readying themselves for a day in the snow. Volunteers like Norton and Seabloom are the dispersers of wildlife management, recent graduates who, in order to catch on in the profession, must find a project like this to give their time to, sometimes for years, before they go on to graduate school. The field is full of aspirants, but the funding is always uncertain, so they work for experience and hope to win the sponsorship of an established wildlife biologist.

Mech moves from one room to another, reading letters and telephone messages, catching up on the current location of wolves and radio-collared deer in the study area. He looks quietly at things, but he looks long and carefully. "Curiosity runs in my family," Mech says. His father grew up on a farm and had a sixth-grade education, but he was eminently curious. "He was a methodical and expert fisherman," says Mech. Heart surgeons, psychiatrists, business executives would ask to go fishing with him. "He'd take them out to pick their brains. He'd show 'em how to fish, and then he'd talk their ears off, asking them questions."

Even as a boy, Mech indulged his curiosity by trapping. "As a fourteen-year-old fur trapper, putting scent posts up to catch foxes, I was really curious about why is that fox interested in this urine? What does that mean to the life of the fox?" Today, when life gets too full of snarl and shrillness, Mech goes off to trap mink.

The habit, he believes, reflects the predatory nature of humankind. He talks of "my long predatory instincts," and adds, "I don't take any pleasure in hurting anything, but I also don't think we as humans shouldn't kill things. Because we, like wolves, are predators."

We have long compared ourselves to wolves. Much of the comparison has focused on the wolf's social nature. Jack London's fictional sled dogs harked back to their wolf ancestry, and when one's true nature asserted itself, the "dominant primordial beast" emerged. Buck, the lead dog in London's *Call of the Wild,* killed his way to dominance and then used his leadership to instill efficiency in the dog team. In the same vein, Adolf Hitler, whose first name comes from a Teutonic word for "fortunate wolf," referred to himself as "a wolf . . . destined to burst in upon the herd of seducers of the people," and often used the pseudonym "Wolf." As Führer, he was called by his admirers "Uncle Wolf," and the implication of the name was that his dominance provided an efficient and companionable order. It is a long way from such mythical views of wolves to the scientific understanding Mech hopes to achieve.

At the heart of our understanding of wolf society is the idea of the alpha wolf. In the 1930s, Norwegian biologist Thorlief Schjelderub-Ebbe developed the concept of a pecking order while studying domestic fowl. He observed that each bird in a flock tended to peck down on a bird of lesser status. By the 1940s, the concept of a dominance hierarchy, an ordering of status from most dominant to most submissive, was being applied to other species. In the 1930s and 1940s, Rudolph Schenkel studied captive wolves at the Basel Zoo. He called the highest-ranking male and female "alphas." In theory, wolves have such dominance orders because, though they must hunt in groups, there is such a risk of overrunning the food supply that only one pair in the pack may breed. In general, among wolves, only the alphas breed, whereas, among other species, dominance hierarchies don't reflect breeding patterns. A female chimpanzee, for example, will mate with a succession of males, one right after another, without regard for rank. Humans, too, breed regardless of status, and in some societies producing a lot of children is the chief means of gaining standing. Even among wolves, it is not always the alphas who reproduce. Mech observed that the breeding female in an Ellesmere Island pack was a subordinate female. And in one of Murie's Mount McKinley packs, the dominant male was not the breeding male. Such matings pose the possibility that dominance is not just an expression of the sharpest bite or the strongest will.

Dominant wolves have prerogatives. Any wolf in possession of food is likely to have a zone of one to two feet around it that no other wolf will enter. But dominant wolves have wider zones of personal space. Mech observed that, after the Ellesmere pack fed on a musk ox, the alpha pair took possession of the carcass. At the kill of a musk-ox calf, he saw a subordinate male propel itself toward the dominant male, with ears and lips back and body low in emphatic submission. The subordinate male crouched below the alpha and pawed at his face, like a puppy. The dominant male snapped at the subordinate and it fell to the ground. Mech wondered whether the gesture expressed a kind of homage.

Dominant wolves may have powers we do not clearly perceive.

Eric Zimen studied a captive pack in Germany in which two females bore pups several years in succession, but the pups of the subordinate female always died. Though he could not say conclusively why the pups died, he had the impression that the subordinate mother and her young were heavily stressed, perhaps by psychological pressure from the alpha female. Zimen suggested that the alpha female kept subordinate females from coming into breeding condition by some similar psychological means. He found that all of his subordinate females had vaginal bleeding at estrus. They ovulated and had the same hormone levels as the alpha female, and some even produced milk, but they did not become pregnant. Klinghammer found that, with the alpha female present, subordinate wolves were in estrus only twenty to thirty days; if he removed the alpha, the subordinates' estrus periods lasted forty-seven to sixty-five days. Similarly, Mech found that, whereas wolves normally don't breed until their second year, pups reared apart from adult wolves would come into season and even breed at nine to ten months of age.

Rank is often reflected in personality. Dominant wolves are confident, sober, outgoing, and assured. Low-ranking wolves are nervous, shy, and sometimes withdrawn. If a dominant wolf loses its rank, its personality may change. Zimen's captive subdominants were friendly with strange wolves and even sought them out, something they could only do when away from the dominant members of the pack. One of Zimen's captives would get out and go to the village, where it played with dogs and children. But when it became alpha male, it ceased to do that, and attacked people and strange dogs through the fence. It even attacked Zimen. Once it lost its dominant status, it reverted to its old congenial personality.

In practice, the idea of the alpha has been stretched to mean "leader." The alpha in a school of fish is whichever fish happens to be swimming in front. When the school turns, a different fish may be the alpha. With wolves, it is often presumed that the alpha directs hunting and the movement of the pack. Some observers believe that the alpha keeps the pack together as a society; but whether the alpha does so by being aggressively intolerant of disorder or by fostering a sense of companionability is not known.

People inevitably say of Mech that he is the "alpha" of wolf researchers. Indeed, he is the researcher with the broadest and longest experience watching wolves in the field and arguing for them in committees, in state legislatures, and in the halls of the U.S. Congress. When reporters want to know about wolves, they call Mech—and they call often, for wolves are an enduring source of controversy. Mech's efforts to bring what he has learned about wolves into the discussion of what we should do about them has kept him at the center of that controversy for decades. Minnesota had a bounty on wolves until 1965. Mech testified in the legislature against continuing the bounty, arguing that only wolves known to have attacked livestock ought to be killed. When the legislature did not reauthorize the bounty, farmers and hunters alike were outraged. Though the wolf was declared an endangered species by the federal government in 1967, legal protections didn't start until 1974. Even then the Minnesota Department of Natural Resources continued to trap and kill wolves, and did so until the federal government warned that it was violating the Endangered Species Act. Minnesota has never liked having the federal government preempt its control over a species, and today still wishes to see the wolf delisted and a sport-hunting season opened on it. In 1978, the federal government downgraded Minnesota's wolves from endangered to threatened, to try to reduce conflict. But in 1983 and 1984, when the state tried to allow the sport hunting of wolves, conservationists sued and barred the hunt. There is still a lot of anger over wolves in Minnesota, and Mech, who has repeatedly come to their defense, has borne a lot of the anger.

He takes it from both sides. Defenders of wolves also criticize him for his willingness to permit wolves that prey on livestock to be killed. "In a pluralistic world," he says, "I believe we have to manage most of our wildlife. We can't have bison running through wheatfields. We have to manage bison when they're in areas where they cause damage. And we have to manage wolves."

He sees that the constituency for wildlife has changed. "Since Rachel Carson and Earth Day," says Mech, "there's a whole new breed of people who've become interested in wildlife, maybe more from reading or television. Many of these folks didn't grow up hunting and fishing and trapping. A lot of these folks turn more to animal welfare and animal rights and wildlife rehabilitation, which from a biological attitude makes very little sense. They think that every wild animal out there is like a pet. It's a very emotional approach to life, and it leads to such absurdities as two people independently asking me why the government doesn't go out and round up all the wild wolves in Minnesota and give them physical exams and euthanize the ones that aren't fit and feed the wolves so they won't have to go through the gruesome thing of killing. The people who get interested in that phenomenon are very important to conservation, but, alas, it's for the wrong reason. We have people worrying about every individual muskrat while people are out there draining the marsh. If we can save the marsh, we can have muskrats forever."

It is in part because Mech has stood between these conflicting forces that he is recognized by nonscientists as the leading authority on wolves. But if he is a leader, he does not—at first glance—seem very wolflike about it. He is not aggressive, and it is hardly in his nature to speak ill of someone else. Careful with his words, he is apt, when speculating about why wolves do something, to use two or three qualifiers in a sentence, to say "maybe" or "almost," or to apologize for suggesting a mere analogy. Perhaps this is all a reflection of his watchfulness: he does not speak for what he does not know. And when he is in a room full of biologists, he is likely to seek consensus. He chairs the International Union for the Conservation of Nature's Wolf Specialist Group, which advises the IUCN and comes up with action plans for wolf conservation all over the world. "Most governments will listen to us," he says. But the meetings of the group are strewn with controversy. Should the group oppose the use of poison? Should it condone aerial gunning of wolves as a method of research? If Italy's last three hundred wolves prove to have interbred with dogs, should the group support their protection on ecological grounds? Such questions may be argued heatedly. Mech is apt to end the discussion before it gets too contentious, trying to save tempers and working relationships until the issue develops a semblance of civility. Says Mech, "We try to work by consensus. I think it's a better way to go if you can do it. Why embroil yourself in controversy if you don't have to?"

As he says this, it is hard not to think about what some people say about alpha wolves being not the meanest and most aggressive, but the ones that are best able to bring harmony to the pack.

. . .

Mech is currently conducting research in Alaska's Denali National Park, on remote Ellesmere Island in the Canadian Arctic, and in the Superior National Forest of northern Minnesota. He delegates much of the Alaska and Minnesota work to graduate assistants who stay in the field, but prefers to do the Ellesmere Island work himself, because Ellesmere is open terrain, and in midsummer the sun does not set, and he can watch the wolves twenty-four hours a day. Since 1966, however, he has devoted the bulk of his time to research in the Superior National Forest.

The Minnesota study is wide-ranging. The big question has always been what effects the wolves have on the white-tailed deer population. Wolf trapping goes on from June to October, deer trapping from January to April, with radio-tracking all year round. An extremely diverse collection of scientific publications have come out of the twenty-six years of study, from papers on scent-marking and papers on howling to papers on home-range size. Mech is constantly coming up with ideas for scientific articles he feels he should write. "I know far more about wolves than I've ever published," he says. "The major work from this project has yet to be written." He writes much of his research by dictating into a miniature tape recorder during the five-hour drive between the lab and his office in St. Paul. He is working on a book on predation that will explain what he has learned about wolf-prey systems.

One of the things Mech is interested in is what makes some wolves disperse. "What I'm trying to get at is, why do some subordinate wolves stay with the pack for years and years, and others take off? What is the triggering mechanism that makes them go? Are they kicked out, or do they go on their own?" And that is why we are about to go out to find and capture Wolf 171.

We put on three pairs of socks and thickly insulated snow boots. Mech pulls a Gore-Tex shell over his fleece pants and I zip up a goose-down parka. We get into his Fish and Wildlife Service Chevy Corsica, and he follows Norton and Seabloom, who are driving a pickup truck and hauling a trailer carrying two snowmobiles. The sky is low. It is ten degrees out, and snowing. Mech drives forty miles an hour over the snow-paved roads of the forest, steady as a rock at the wheel.

As he drives, Mech mentions that, in thirty-five years of study, except for Ellesmere Island, he has had only fifteen encounters with wolves without the aid of radio collars or airplanes. "All were fairly brief," he says, most of them windshield sightings as he drove along a forest road. One wolf crossed a highway in front of him in Alaska, and he almost hit it. He points out a spot on this road where he saw a pack cross in front of him. He recounts another in Ontario, and others in Minnesota. He remembers every one. So rare were the encounters, he says, that "each one was memorable. Each one was an event. If one ran out in the road in front of us right now, it would be extremely exciting."

The point is, says Mech, "it's difficult to find these animals from the ground." So you use either airplanes or radio collars, and preferably you use both. The National Park Service forbade the use of radio collars on Isle Royale because visitors would object to scientists' meddling with the lives of wild wolves in a national park. But in Minnesota, the United States Forest Service had no such objection. Today, radio collars allow Mech to follow the movements of individual packs and to find and retrieve their kills. He can also collar deer to get a better idea of how many of them fall prey to wolves. "Primarily, everything we've done since 1968 is based on radio-tracking," he says. "We've radioed five hundred wolves in this area. One wolf we followed for eleven years and three months."

The first collars were simple radio transmitters beeping out an unvarying ping-ping-ping, twenty-four hours a day, an electronic case of hiccups that kept up until the batteries died. They allowed researchers to plot the location of wolves on maps and to find kills and dens. To allow researchers to follow individual wolves a long time, the batteries had to be replaced, and to do that, the animals had to step into traps. Many wolves would not do that twice in a lifetime.

"Twenty years ago," Mech says, "I had the idea to make a collar through which you could drug an animal. For ten years, I just thought about it. Then I put a student on it." The 3M Company hired the graduate student and put close to $1 million into development of the collar. After five years of work, they came up with a collar that would dart a wolf in the field. The collars must be handmade, and they cost about $1,750 apiece, but they can do an enormous range of things.

The collar has both a standard transmitter, to emit beeps, and a receiver, to receive commands from the researchers. It also has a computer, so it can interpret the commands. The researchers can turn it on and off. When it is on, a mercury switch on the collar senses tilts of more than ten degrees. Each tilt is recorded on a counter, and the record is stored in the computer. One pattern will indicate that the wolf is up and running, another that it is sleeping, another that it has died. The collar thus keeps a record of the wolf's most recent thirty-six hours of activity, and the researchers can instruct the collar to confide its observations into a laptop computer in the field. If the collar fails, it has a mechanism that allows the technicians to instruct it to fall to the forest floor and continue to broadcast signals until they find and retrieve it.

The collar has two gunpowdered-charged syringes inside metal sleeves strong enough to survive wolf bites. The syringes are armed with the drugs Telazol and Rompun in a solution of glycol to keep them from freezing. By triggering the darts, the researchers can immobilize a wolf from a half-mile away, and then follow the radio signal to capture the animal or take blood samples. The cylinders stand out on the wolves' necks like small antlers. It wouldn't be at all unlikely for a hiker or a hunter in the woods to glimpse one of these animals and mistake it for a young deer. And if wolves insult one another, the ones with the collars must take some hard kidding about looking like badly dressed herbivores.

Wolf 171 is one of four wolves and two deer wearing this collar in the study area. Says Mech, "With the deer, we've directed the collar to listen for high activity counts in twenty-four hours. We know that deer are chased by wolves. I have absolutely no idea whether that is once a day, once a month, once a year, or once in a lifetime." With the collar, they can get some idea when the deer has been chased. "Once we recorded not only the chase, but the wolf killing the deer," says Mech. "It probably was one of the collared wolf packs in that area that killed the deer." So it is possible to monitor both the victim and the killer simultaneously, an unveiling of woodland secrets hitherto denied to an observer on the ground.

The next step with this collar will be to have it locate the wearer precisely without having to fly or triangulate, to work something like the ETF loran navigation system used by boat operators. It will report by radio signal the wolf's latitude and longitude within 154 meters every fifteen minutes. Currently, researchers get one or two locations a week by flying, and that only when weather permits. With the locating collar, a computer could map the wolf's movements, and the researchers would always know where to start looking for an animal that had stopped broadcasting.

Mech is always looking for ways to spy on wolves. He would like to use satellite collars, which transmit exact locations to orbiting satellites several times a day, but, he says, "It's extremely expensive, and not that practical." He has tried implanted transmitters—cylinder-encased radios, slightly larger than a lipstick, surgically placed inside the abdominal cavity of a wolf. "We have not used them routinely. With a radio under the skin, you don't get a signal that's very powerful—you can't put a very big battery in a subcutaneous transmitter. And our wolves travel far and wide.

"I'm constantly looking for things that will improve our data-collecting ability," says Mech. "Technology is the only thing that has made wolf studies possible. Without that technology, we'd be back doing studies like Sig Olson, slogging around in snowshoes at one mile per hour."

On Road 424, we stop and get the radio out of a metal case in the bed of the pickup truck. Seabloom turns it on and broadcasts an instruction to Wolf 171's collar to turn itself on, and Norton, standing on the roof of the cab, sweeps the horizon with a hand-held antenna, searching for an answering signal. We get nothing. There is just the stillness of spruce and pine, a latticework of bare birch and aspen branches, and the silence of snow. We drive another half-mile, stop, get out, broadcast, and listen. In this way, we troll for wolf, traveling down miles of snow-covered logging road. In fair weather, they would do this in an airplane and have the location in a half-hour or less. In the falling snow, it is two hours before we get a signal back. Seabloom instructs the collar. It replies with twenty double beeps, as if to say, "Hello, hello." Then there is a steady signal. The wolf is probably less than a mile away.

In the cold, it takes us half an hour more to get the snowmobiles down into the snow and our gear into a sled to be towed behind one of them. When at last the gear is ready, Mech offers the back of the sled, where one can stand like a dog musher and feel more engaged with the winter woods, to anyone else. No one claims the honor, so Mech takes it, and we set off down a snow-covered logging road that winds into the forest. After a half-mile, we stop to retake bearings. The signal is faint and intermittent. We go another quarter-mile. The signal is a slow, rhythmic pinging, indicating that the wolf is lying down, probably asleep, within a half-mile of us. The cawing of two ravens off in the trees may indicate a kill there; the wolf may have fed and moved away to sleep. We get off the snowmobiles and Seabloom and Norton chat the computer through another set of commands. Norton instructs the collar to report on the wolf's activity. The collar answers with the slow pinging of a sleeping wolf. Another command arms the dart.

Norton sends a command that fires the dart. We hear a sudden rapid series of beeps as the wolf is startled by the eruption inside her collar and the sting of the needle. She gets up and lurches off. For a minute, there are rapid beeps. Then they slow. Within three minutes, the wolf has collapsed and the slow, rhythmic beep resumes.

What must she have felt from the jab of the dart until the drug took hold and scattered her consciousness? Did her legs grow heavy and the air thicken in a riot of gravity? Did the trees jelly and streak as she lunged to escape? Somewhere in the woods, she lies motionless, eyes staring, tongue lolling.

We put on snowshoes. With Seabloom carrying the antenna and the receiver, we follow the radio signals into the woods. The snow is deep; it is still falling in thick flakes. As we brush by trees, branches crack and snow falls down the backs of our necks. We must climb over logs and go around densely clustered spruces. At one point, Mech stumbles noisily. I wonder what the wolf is hearing, how clumsy she must think humans are, how noisy and awkward of movement.

In the hilly terrain, it takes a half-hour to find Wolf 171. She has plowed into the ground sixty feet from where she lay when the dart struck. Her tracks show she has come from a bed in a cluster of spruces at the bottom of a hollow and has crossed a hillside, keeping to cover, perhaps flattening herself out a little to go under a deadfall. The drug overtook her as she entered a small forest clearing. Her legs are folded under her, as if she had started to rise and then forgotten what she was about to do. Her nose is wedged into the snow. An attending raven croaks perplexedly from the trees.

Norton and Seabloom remove the collar carefully because the needle is still in the wolf's neck. After extracting the dart, they spread out a heat-reflecting space blanket, lift the wolf onto it, and drag her into the shelter of a cluster of balsam firs. Norton carefully dusts newfallen snow from the back of the wolf, and pulls the edges of the space blanket over her to help keep her warm—one effect of the drugs is that, despite the blanket, the wolf's temperature will drop from a normal 101 to below 96. As she sprawls on the blanket, her eyes still glow bright yellow. She yawns and blinks, unable to shake off the chemical restraint.

Wolf 171 was collared in September at the age of four months, in hopes of understanding how hormonal changes relate to dispersal. She was darted in November for tests, and again in December because the collar needed new batteries. Now, for the third time, they take a blood sample. This is difficult, for the wolf's veins do not stand out, and in the ten-degree cold, the researchers' hands are not agile. After they get a baseline sample, they inject a hormone into the wolf. The hormone will stimulate the wolf's pituitary, which in turn will prompt her to release sex hormones. Every ten minutes for the next hour, more blood will be drawn, and each of the seven samples will ultimately be screened for sex hormones. "We're interested in the degree of maturation in relationship to when they disperse," Mech explains. When I ask if the hypothesis is that animals that mature earlier are more likely to disperse, Mech says, with characteristic lupine caution, "Something like that, but I'm not saying that. The hypothesis is that there is some relationship. We're not sure what it is."

Mech's curiosity is covered with cautions; he wants to attack the problem only where it is vulnerable. He is trying to see into the mechanisms of wolf society. What makes a wolf disperse? Is it kicked out, or does it have some lupine equivalent of gumption? Some authorities suggest that dispersers are genetically different from residents, but no one has yet demonstrated a wolf gene for wandering. Murie believed that dispersers might be subordinate wolves that got less to eat, and lack of food put pressure on them to go off to hunt by themselves. Michael Fox held that lone wolves were wolves genetically destined to become alphas, wolves with a swaggering pride which left them unable to hold an inferior position and compelled them to leave the pack. It is hard not to frame the question in human terms. Is the disperser a winner or a loser in the great lottery of wolf life? It's the question we pose of the alpha, too. Is the alpha the biggest, baddest wolf, the one most capable of aggression? Or is it the most intelligent, or the most caring, or the most civic-minded? Are alphas despots? Or are they diplomats?

Either character may be found within the pack. Wolves display a spirit of friendliness and cooperation. Old or wounded wolves that don't share in the killing are allowed to eat what other wolves kill. Cooperation begins with the pups. Adult wolves seem to love pups. In zoos and breeding facilities, it has often been noticed that when a wolf gives birth all the wolves in the facility, even those in distant cages, show great excitement. Adult wolves in adjacent cages will try to entice pups to crawl under fences to play with them. And pups can roughhouse with adults almost without constraint; they are seldom punished for biting or clawing. Adults have even been seen to hold up a bone or a piece of deer hide to distract pups who are overly rambunctious.

Adult wolves continue to demonstrate the companionability they learned as pups. Whenever the pack gets up from a rest or leaves a kill and is about to travel, there is likely to be a rally or a play session. A wolf will arch its back and twitch its tail, put its front legs flat to the ground, and open its mouth with tongue dangling, making a play face. If it gains a playmate's attention, it may turn and race away. The companion may give chase. They play games of keep-away with bits of stick and bone. A racing wolf turns and pounces on a pursuing wolf, and the two tussle in mock battle, biting at each other's neck ruffs, slamming hips into each other's sides. They may slide down snowbanks with expressions of glee. And the session may end with a greeting ceremony, in which all gather around the dominant male, their ears back and eyes narrowed in submission and pleasure, nuzzling his face, licking his jaws.

It is thought that this kind of active and friendly intimacy helps to develop and maintain the close coordination wolves need for hunting. Pups and adults alike play at hunting behavior, stalking and ambushing, pursuing and snapping. Wolves need to have a sense of each other's quickness and aggressiveness, watchfulness and impulsiveness. When they are pursuing a moose or deer, that very sense gives the pack coordination and may save individuals from injury. A wolf that has played these games often is more likely to sense when an erstwhile playmate is darting in for a bite or dashing away from the moose's hooves.

Companionability also assists in the rearing of the young. Adults will bring meat from a kill even to pups that are another pair's offspring. If a lactating mother is killed, sometimes other females in the pack will lactate and feed her pups. And pups learn adult behavior by playing with adult wolves in the pack.

There is a competitive side to the games of chase and keep-away. If things get out of hand, a wolf may close its jaws over the neck of a pup or a subordinate adult and pin it to the ground in a posture of submission. Submission does not seem to instill fear or resentment in wolves. Dr. Michael Fox measured the resting heart rate of wolves in a captive pack. Heart rate declined when a wolf watched with interest a falling leaf or sunlight glinting on water, or engaged in submissive greeting or friendly contact. Indeed, Fox found that being licked or groomed by a companion lowered a wolf's heart rate by half. Heart rate also declined when a wolf was seized by its muzzle and pinned to the ground by a superior; Fox likened this to the tonic immobility of a frightened fawn or frog.

A hierarchy of strong to weak is constantly evolving and changing in the pack. Wolves compete for status from an early age. Fox found wolf pups routinely working out dominance hierarchies at eight weeks; Mech saw four-week-old pups fighting for status. Wolves have a complex language of gesture and posture for dominance behavior. Dominant wolves may need only to stare at a subordinate to freeze it in its tracks. In humans and wolves alike, a direct stare is a threat. Some wolves look right through you as if you were not there. That is a dominant stare; it expresses the looker's loftiness. A subordinate wolf, like a submissive human, looks down and away, or makes a lot of quick, glancing eye contact—looking, then looking away, licking with the gaze rather than skewering with the stare. Within the gaze itself, there is more content: changes in pupil size accompany changes in emotion and help to communicate fear or playfulness, pleasure or pain.

If a stare is not enough, a dominant may lunge at a subordinate, growling with bared teeth, erect ears and tail, and bristling hackles. An inferior wolf tucks its tail under its body and curls its lips, showing all its teeth. If the dominance display overpowers it, it may flop on its side or back, head stretched out, hind legs raised, and genitals exposed. Then the superior animal is likely to stand over the supine individual and sniff at its genital area.

Usually, bites within the pack are inhibited and do not draw blood. A dominant wolf may discipline a subordinate wolf merely by placing its mouth around the subordinate's muzzle. Says Randall Lockwood, who studied wolves at the St. Louis Zoo and in the wilds of Alaska, "They use no more force than is necessary to get what they want out of a situation. I've seen hundreds of wolf fights. I've seen blood drawn only once, and that was a wolf that bit its own tongue." Lockwood concludes, "The wolf pack is based on affection, appeasement, and solicitation."

Konrad Lorenz, who pioneered the study of aggression in animals, observed these bloodless gestures of triumph and submission and read in them a mechanism for maintaining peace. The vanquished offered his neck to the victor, and, wrote Lorenz, "a dog or wolf that offers its neck to its adversary in this way will never be bitten seriously. Should a dog or wolf unrestrainedly and unaccountably bite the neck of his pack-mates and actually execute the movement of shaking them to death, then his species would certainly be exterminated within a short space of time." Later writers believed that these submission signals were inviolate, and they extolled wolves as models of conduct. Michael Fox, for example, declared, "Severe injuries [from wolf fights] are rare. As soon as a wolf gives a surrender signal and shows submission toward the other contestant, the latter will immediately stop fighting. Wolves do show chivalry."

But not always. Terry Jenkins, curator of the Folsom City Zoo, recalls a captive pack whose alpha female was very aggressive with all the females in the pack. "Whichever female was on the bottom, she always had bites on her." At two years of age, one of her daughters fought back, and two of the daughter's litter mates, a female and a male, joined in. "They really beat her up. It was obvious they were trying to teach her a lesson," says Jenkins. The conflict went on for days. "The mother would go into the den for a while, then come out and call them down to her, so they could fight more. She really wanted to get this settled." Then, for days, they left her alone in the den, and there seemed to be no communication between the mother and her daughters. Finally, the two daughters went into the den, grabbed the mother, dragged her out, and mauled her. Even her mate now joined in the attack. "It was evident from that point that they were going to kill her," says Jenkins. Zookeepers were able to get the female out, but she died that night. Klinghammer has also seen a captive female kill her own mother. An aggressively domineering female was mobbed and displaced from the top of her pack at the Julian Science Center in California. Paul Kenis, who keeps the wolves at the center, says he separated her to keep her from getting killed, but wolves attacked her through the fence, and she died after a wound suffered in one of these cross-fence battles became infected. In all three examples, the wolves may have been made more murderous by captivity. In the wild, scapegoat wolves that are fiercely attacked by higher-ranking pack members tend to become dispersers. Mech says he has never seen such murder within a wild pack, and knows of no example in the scientific literature.

Wolves are routinely murderous, however, when they encounter each other across territory boundaries. Dispersing wolves trying to join a new pack may be repeatedly attacked, and even killed. In 1974, four Isle Royale wolves broke off from the East Pack and moved into territory occupied by the West Pack. West Pack wolves killed two of them and badly wounded a third. Gordon Haber, who studies wolves in Denali National Park, tells of wolves leaving their territories and going on raids into neighboring territories, to all appearances bent on murdering their neighbors. Says Haber, "They actually go out of their territories looking, searching, for other packs. It's a primitive form of warfare. One pack will go into the neighboring territory and then go back, and the other pack will follow it, and they will go back and forth a half-dozen times like that." If they meet and one pack senses a strong advantage, it will attack the other. "It's an extreme form of aggression," says Haber. Rolf Peterson reported that a trespassing pack on Isle Royale killed a female pup he supposed to be a member of the resident pack, and marked the corpse heavily with urine before they returned to their own territory. Tom Meier reports that in Denali National Park between 1985 and 1990, thirty-seven wolves were known to have died, and nineteen of them were killed by other wolves, most of them from neighboring packs.

Terry Jenkins was confounded by an event she observed. The old male in the Folsom City Zoo pack had a degenerating spine and was losing the use of his legs. Zookeepers decided to euthanize him. Because Jenkins wanted the rest of the pack to understand that he had died, she let them come over to investigate the corpse. The big male sniffed at the dead wolf and whined like a puppy, then walked a few steps away and regurgitated as if the dead wolf were a begging pup. Next, the subordinate female went over and sniffed at the corpse, started jabbing at it, and then grabbed it, and dragged it to the center of the pen. She became excited. The other two wolves came over, and all started biting the dead wolf. "I thought they were going to tear him apart," says Jenkins. "It was touch and go to get him away from them. It was a very emotional situation for all of them."

Competition and strife are part of pack life. The dominance order is constantly changing as dispersing wolves enter and leave packs and as individual wolf personalities change. Wolves mature, and wolves age. Young subordinates suddenly become bold and strong; older leaders weaken and become yielding. There is constant testing, constant pressing of one's advantages.

But because we see so little of wild wolves, our explanations for all this complex behavior still rely on guesses. Much of our guessing falls back on comparisons between humans and animals. Are wolves really "generous" when they share food? Are they "mean" when they fight? Part of our problem with wolves is that we are so bewilderingly like them that we can't always see where one species leaves off and the other begins.

It is almost impossible, even for the most cautious researchers, to separate human qualities from wolf qualities. Mech, in speculating about what some wolves were doing crossing a road, suggested they "thought" something; then the careful researcher and peacemaker within him caught hold of the statement. "It is so natural to express yourself that way," he says, and he confesses that he is not always able to resist the temptation. "Where it is easiest to fall into the trap of humanizing the animal is when I'm living with them up in the Arctic. You're seeing them up close and minute to minute. The pack of wolves I was living with killed a musk ox. I was right next to that musk ox when it was killed, and I never felt sorry for it." On the other hand, he says, "The pack attacked another wolf and I saw the other wolf just after they had wounded it. For an instant, I felt sorry for it. That's not a very appropriate feeling [for a scientist]. But that wolf was a very bedraggled wolf; that poor wolf was just so thoroughly chewed. It died right after I left it."

Who knows what even a careful scientist will freight onto the behavior of wolves? Says Mech, "I even dream about seeing wolves, because they're such a part of your life. I have dreamed about traveling a road I've never been before and seeing them traveling up on the hill, going somewhere."

Wolves have been more extensively studied than just about any other wild species. The more we know, however, the more we don't know. We may feel tempted to follow them into the woods. But what are we following: wolves or humans? Sometimes, out on that trail, the likeness we share overtakes us.

It was hard, for example, to escape that sense of likeness at the 1992 International Wolf Symposium in Edmonton, Alberta. As it convened, wolf biologists, state and federal and Canadian wildlife officials, wolf fanciers, graduate students looking for jobs, and animal-rights advocates assembled in the commons room of a University of Alberta dormitory for an opening reception. Mech was there, as were Rolf Peterson from Isle Royale; Tom Bergerud from British Columbia; Bob Ream from Montana; Ludwig Carbyn from Alberta; Robert Hayes from the Yukon; John Theberge from Ontario; Bob Stephenson and Warren Ballard from Alaska; Todd Fuller of the University of Massachusetts; Erich Klinghammer from Wolf Park; Steve Fritts, the Northern Rocky Mountain wolf-recovery coordinator for Fish and Wildlife Service; Luigi Boitani of the University of Rome; Gao Zhong Xin of China; Nikita Ovsyanikov of Russia; Yadvendraved Jhala of India; Julio Carrera of Mexico; plus graduate assistants and recent Ph.D.'s still looking to make a name. It was as complete a gathering of wolf biologists as had ever been assembled. As they entered and looked about the room and greeted each other, sometimes stiffly, sometimes warmly, it was clear that there were divisions and ranks. Scientists have their own dominance orders: one generation of young aspirants is always seeking to unseat its predecessors. Wayne Brewster, a biologist at Yellowstone National Park, looked around the room and said, "The alphas are all here and their tails are all high."

Unlike primate biology, which has well-known women at the forefront, wolf biology is essentially a male hierarchy. It got started in an era in which game departments assumed that men were strong and independent and women were not, and department officials wouldn't have dreamed of sending a woman out into the woods alone. By 1992, there were female field technicians tracking, trapping, and radio-collaring wolves in Algonquin Provincial Park, in northern Minnesota, on Isle Royale, and in northern Montana. But there were few women on the symposium program: only Diane Boyd and Cheryl Asa, who does her research at the St. Louis Zoo, would present papers.

The fact that wolf biology is still largely a man's field may color the way wolf biologists deal with one another. They argue fiercely about things—fighting about the effects of wolves on prey populations, disagreeing over whether there ought to be programs to shoot wolves in order to increase hunters' take of deer and moose and caribou, skirmishing over theories, over one another's publications, over who is the ranking authority. When you get a roomful of wolf biologists together, you are apt to see a lot of behavior that reminds you of a wolf pack. You are apt to hear scathing criticisms. At Edmonton, one heard a lot.

"There are only a half-dozen real wolf biologists," says one of the biologists, "based on field study and publications." He goes on to name himself first on the list.

"Bergerud is way out in right field," says another. "He sees the world through predator-colored glasses."

Says a third: "Guys study it and love the animal and they don't want you saying anything contrary to 'It's a white, shining wolf.' Mech, Peterson, Theberge, and Haber just love that animal."

. . .

Mech's thirty-five years of wolf study have been guided by his broad and roving curiosity. To him everything about wolves is worth studying, but to many of the younger scholars there is something amateur and Victorian in studying wolves just to get to know them better. To these younger scientists, who have come of age in a time of scarce funding, science should aim at solving immediate problems. In part, their attitude is a measure of the extent to which scientific inquiry is defined by the sources of funding. Most predator research today is aimed at reducing damage to livestock or increasing hunter harvest. Mech had the good fortune to start work when so little was known about wolves that a blank notebook was worth having. Younger researchers envy Mech the freedom of his inquiry. He recalls a younger scientist who visited him in Minnesota. When Mech dropped him off at the Minneapolis Airport, the visitor's parting words were, "I don't know how you do it. If I can't do an experiment in two weeks and get a publication out of it, it bothers me."

Mech has also had to struggle, through the years, to fund his research. "Wolves live twelve to fifteen years," he says. "Deer are the same. When you're looking at population changes in long-lived species that are elusive to begin with, it takes a long-lived study to get results. But budgeting is done on an annual basis in the government. When I started the Minnesota study, it was a one-year project. It's still going. But even the agency can't go on a year-to-year basis. Our budget year begins in October. I still don't know what our budget is going to be this year." Nevertheless, he has gotten grants from the New York Zoological Society, the World Wildlife Fund, the National Geographic Society, the National Park Service, the U.S. Fish and Wildlife Service, and private donors. That kind of success seeds envy in the hearts of younger aspirants. So does Mech's contact with the media and with public officials. He has amassed prizes that other biologists covet. And all this is clearly part of the reason many of his fellows see him as the alpha. They look for ways to challenge him.

Says one: "There hasn't been an objective study of wolves in the last ten years. The question of raised-leg urination might be interesting, but no game agency is going to fund it. They want to know what the effects of predation are on an ungulate population. You ask a question you want to get an answer to."

Says another: "There's a little bit of Mech's makeup to try to control things. But when you take it as a whole, what a lifetime of work. He changed people's thinking about the wolf. He's a high-energy person, bubbling over with ideas."

Says a third: "Natural-history studies are not acceptable any more. You have to have some explanation of what is going on."

Says another: "Dave Mech has made more difference than all the biologists put together. The question is, how many Meches can there be?"

On the second day of the conference, Mech had just completed a talk on the future of wolf research, during which he had referred to a paper by Dr. François Messier, of the University of Saskatchewan, who had been studying wolf-moose dynamics in Canada. In his paper, Messier had contested Mech's theory about the relationship between winter snow depth and deer populations. Mech said in the talk that he had revised his own paper and that it should overcome the objections Messier had made.

As Mech stepped down from the stage, Messier boiled up to the front of the room. Round and bearded, intense and heavy-jawed, he moved quickly, like a badger. Mech, though caught by surprise, was eminently calm, and his voice did not rise at all. Messier's jaw moved rapidly; his dark bushy eyebrows alternately flared and furrowed. "I believe what you are doing, but you put out a comment that is not appropriate!" he said, and declared that it was not fair for Mech to rebut his work when Messier could not reply. Mech, however, looked upon his remarks not as a rebuttal but as a revision of his own work. "I would never do that to you," he said. "I would never rebut you without giving you an opportunity to rebut." Mech's gaze was unwavering, his mouth relaxed, and his voice soft and steady. Messier was dogged in his attack, rapidly going over key points in his objection to Mech's argument: "The first winter there was no effect. The second winter there was no effect. . . ." His eyes glanced at Mech and veered off again, and his teeth flashed.

If we had been watching wolves here, we'd have to have said the postures showed Mech to be the dominant, resisting the challenge by not barking back.

. . .

Although writers often ascribe leadership of a pack to the dominant male, dominance is not necessarily leadership. Ernest Thompson Seton held that when a wolf attains great size and cunning "he attracts a numerous following." Durward Allen wrote of the alpha male's "privileges of leadership." Eric Zimen held: "My own theory is that . . . it is the alpha female, the mother of the cubs, that is the principal figure governing the life of the pack." Erich Klinghammer's view may be more to the point. He says, "We tend to think in categories that don't mean too much to wolves." Mech points out that leadership is a human concept, and we aren't sure how to define it among wolves.

Leadership is a complex enough issue among humans. We know even less about it with regard to wolves. The alphas do seem to have some leadership role. Peterson found that, 70 percent of the time, it is the dominant male or dominant female that runs at the head of a wolf pack. Allen believed that, before setting out for a hunt or for travel, when wolves came together for a rally, nose to nose, with wagging tails, the alpha male was the center of the ceremony. Lois Crisler observed that, among her captive wolves, the female seemed to lead, especially presiding over mood. "If she was gay," Crisler wrote, "their eyes turned gay; they trooped and tossed with her in gaiety. If she 'talked' and she was the sole one that truly, wolfishly did—they uttered their own indescribable ows and wows."

None of the expected indicators of leadership correlate with dominant status, however. Pack howling is initiated by one wolf, but not always by the pack leader. Fox believed that hunting success was synonymous with dominant rank; yet it is not necessarily the alpha which kills. When experimenters in Germany released rabbits into a wolf enclosure, they found that the lowest-ranking wolf caught and killed more rabbits than any other wolf. In Alaska, Robert Stephenson could find no correlation between social rank and hunting success. Zimen wrote, "No member decides alone when the pack is to move or exercises sole power of command in any of the other activities that are vital to the cohesion of the pack. The autocratic leading wolf does not exist."

What makes an individual dominant? In sheep and antelopes, size and age are the most important determinants of rank. In Old World monkeys such as Japanese macaques, rank may be derived from one's mother's standing. Among squirrel monkeys, dominance correlates with levels of adrenal hormones. It may be that among wolves the alpha is the dominant not by virtue of an aggressive nature or a sharper bite, but by virtue of an ability to hold the pack together, to give it comfort and coordination and belonging. Zimen noted that subordinates seemed less able to form subgroups; the strongest bonds of friendship seemed to be between the highest-ranking individuals. Noting that the alpha male is the center of the greeting and rallying ceremonies, he declared, "No wolf is more interested in a friendly, cooperative atmosphere" than the alpha male. "He is the experienced initiator, the watchdog and protector against external dangers, the friendly and tolerant center of the pack." He felt that the alpha male preferred peace because peace favored the successful rearing of his cubs and put off the day of his overthrow as dominant male. Self-interest may decree that one of the alpha male's roles is simply to reduce tension.

If the alpha isn't the biggest, baddest wolf, or even necessarily the wolf that breeds, one may ask, what is the evolutionary advantage of the hierarchy? Sociobiologists point out that evolution proceeds on more than one level—not just on the individual but on the group as well. Thus, ants and bees have evolved sterile castes of workers that do not reproduce. Hierarchies produce stable social systems and so are advantageous to a group if not to the individual. And, in a roundabout way, altruism helps even the subdominant wolf to pass on its genes. Since most fellow pack members are siblings, increasing the fitness of one's fellows by cooperating in defense of territory or helping to feed the young improves the chances of passing on the genes that both dominant and subordinate have inherited. At the same time, in societies with strong dominance hierarchies, newcomers are a threat to the status—and therefore to the perpetuation of the genes—of every member of the group, and every member of the group is apt to attack a newcomer.

We are still laboring under the assumption that with wolves, if not with people, it is toughness, hard will, and readiness to contest things that determine rank. The cinematic view says life's prizes go to the contestant with the strongest impulses. The evidence, however, doesn't always show that's the case. Some races go to the swift, some to the strong, some to the sly, some to the sociable. It may be that each wolf is different and each pack is different. It may be that there is no fixed rule. And if the absence of simple explanations of wolf society discomforts us, it is probably because we are of more than one mind about our own.

Back in northern Minnesota, Wolf 171 blinks a yellow eye. Seabloom is having a hard time finding a vein from which to draw more blood. Mech shows him. Their hands are blue with cold, but they do not put on their gloves between blood samples, because there is too much to do. Mech pinches the wolf's leg and directs Seabloom to align the needle parallel to the bone. Seabloom finally finds the vein and draws blood. "Thank you," he says to Mech. "Wolf Topography," Seabloom says to no one in particular, offering the words to the trees satirically, as an imaginary course title or a cartoon caption.

It takes nearly four hours to complete the work. The wolf is bled, weighed, and measured. The collar's stored thirty-six hours of data is dumped into a portable computer. The darts are replaced, and the collar is put back on. By the time we are ready to inject a drug that will pull the wolf out of sedation, everyone's hands are numb. It is still snowing. The wolf is injected. She licks her nose and blinks. In a few minutes, she raises her head, pulls her legs under her, and tries to wobble to her feet. We put on our snowshoes and move away. Mech moves out quickly, wanting to leave the wolf with as little experience of humans as possible—not just to avoid spooking her, but presumably to avoid habituating her. As I pull on my bindings, I look back across the clearing at her. She drags herself away, looking at me uneasily, the cold yellow fire glowing in her eyes proclaiming still that there is a dimension to wolves we haven't fathomed.

6

THINKING LIKE A WOLF

In his famous story of the outlaw wolf Lobo, Ernest Thompson Seton held that the wolf walked on the backs of sheep at night to attack the goats around which the flock clustered, not because he was hungry for goats, but so that the sheep would scatter and the wolves could pick them off at their leisure over the following weeks. Seton described his own meticulous care in putting out poison to kill the wolf. He melted cheese and kidney fat in a china dish, cut it with a bone knife "to avoid the taint of metal," and injected strychnine and cyanide into chunks of the bait. All the while, he wore gloves "steeped in the hot blood of a heifer." Carrying the baits in a rawhide bag which was bathed in blood and always suspended from a rope, he dropped baits over a ten-mile circuit; the next day, he returned to find that the wolf had gathered up baits, piled them in the trail, and defecated on them. The wolf methodically sprung the traps he buried in the trail, too. "The old king was too cunning for me," confessed Seton.

Such stories crowd our view of wolves, but their proclamation that wolves are sagacious is largely an interpretation born of storytelling. The artful tracker reconstructs a hunt from tracks, and often fills the gaps between what he knows with speculations about how a wolf thought.

We really know very little about what a wolf plans or thinks. However, those who spend time observing wolves see plenty of evidence that the mind of the wolf is complex, purposeful, and full of feeling.

For example, in 1970, the Alaska Department of Fish and Game sent Robert Stephenson to Alaska's North Slope to find out what was happening to its wolves. Hunters had been shooting them from airplanes, and the wolf population appeared to be in decline. Eskimos said dens that were usually active were empty, and hunters said they weren't seeing many wolves.

How might Stephenson tell if wolves were any more or less common than they had been ten or twenty years before? There had been no rigorous surveys of the wolf population. It was still early in the days of radio telemetry, and the equipment could not withstand the extreme arctic conditions. Stephenson began by talking to the local Nunamiut Eskimos, who still hunted for much of their food on open tundra.

He found that the Eskimos watched wolves. From the 1940s to the 1960s, Alaska paid bounties on wolves, and the bounties brought the Nunamiut enough money to buy guns and cartridges with which to hunt for meat. They would hunt wolves in summer. It was treeless country, one of the rare places that allowed prolonged observation of wolves. Says Stephenson, "They would camp up high, watching adult wolves through brass ship's telescopes to see where they went and try to find the dens. They had spent thousands and thousands of hours watching wolves in open country. They knew about seventy dens. Some of them became real students of wolf behavior. I realized that with a modest budget this was a place to learn about wolves."

So, over the course of three years, Stephenson sat on mountainsides with these Nunamiut wolf-watchers, a campstove, and a spotting scope. In the summer, when the sun never set, he says, "We'd stay up all night watching wolves coming and going. Sometimes you didn't see any wolves. Once we watched two wolves hunt small mammals for eighteen hours."

The Eskimos were full of lore. "They'd say, 'That wolf is going to lay down,' and it would lay down. 'That's a male. That's a female. That's an old one: see how its hair is shedding differently?' We'd talk about old stories. We'd travel in the winter, too, tracking wolves. We'd trap wolves with them. They knew all kinds of idiosyncratic things, like different kinds of howls."

Stephenson's study fell somewhere between experimental science and anthropology. He wasn't manipulating his subject or generating complex statistics, but he saw things other observers had not seen. He watched a wolf sneak up on a bald eagle and, at the last minute, make noise to scare the eagle. He believes that wolf was "just fooling around." One day, looking down on a valley with clumps of willow in it, he saw a wolf tease a grizzly. "We'd see this wolf fooling around at the edge of the willows," he says, "and then a grizzly came out of the willows and chased the wolf." The wolf disappeared into the willows, and so did the bear. When the wolf came out, the bear came out chasing it. "It looked like the wolf was having a good time. The wolf had this smiling face, like he was really enjoying the hell out of it."

Wolves sometimes kill bears. They have been known to eat grizzlies. In 1992, a pack of wolves in Alaska killed a sow grizzly while her cubs escaped. Bears kill wolves now and then, too. But one day, says Stephenson, "we saw a wolf traveling with a bear—just walking along, thirty feet apart." His Eskimo hosts said they'd never seen that before.

Nearly everything humans see of wolves is fragmented and incomplete, truncated by wolves' shyness and hidden by their furtiveness. Stephenson, however, was seeing long sequences of complex and puzzling behavior, and much of it raised questions about how wolves think. Stephenson quickly agreed with the Nunamiut: "When you talk to them, they'll say it's a real smart animal."

We equate thought with purposeful behavior. Until midcentury, we supposed only humans could foresee the consequences of their actions. "The cat runs after the mouse," wrote psychologist William James in 1890, "not because he has any notion either of life or of death, or of self, or of preservation. . . . He acts . . . simply because he cannot help it; being so framed that when that particular running thing called a mouse appears in his field of vision he *must* pursue. His nervous system is to a great extent a preorganized bundle of such reactions." Indeed, experiments in the 1940s showed that one could predictably cause cats to sleep or fight by passing small electrical currents to specific parts of their brains. Thus, animal intelligence was thought to be a matter of instinctive responses to specific stimuli, and each animal was thought to be more or less identical to all others of its kind.

In the middle of the twentieth century, our views of animal intelligence began to change. In 1948, Rudolph Schenkel published a study of wolf body language and expressive postures he had conducted at the Basel Zoo. He concluded that, because the wolf could change the intensity, direction, and quality of its personal behavior patterns, it had a higher level of intelligence than the more stereotypically behaving birds, reptiles, and fishes.

By the 1960s, scientists were seeing in wolves complicated purposive behavior. Norma Ames, who kept wolves in New Mexico, tells a story that suggests insight and purposeful action. "I had two female wolves that didn't get along and kept them apart with their families in different enclosures," separated by a chainlink wire fence. One day a friend brought her some bones from an elk he had shot and cleaned, and she gave a bone to one of the females. "The wolf who had the bone," says Ames, "took it and put it a few inches away from and parallel to the fence. And she stepped back and waited and waited. And, sure enough, the other wolf put her nose through the fence, and the waiting wolf attacked her. That took deliberate thought to do." Biologist John Weaver saw similar evidence of planning among wolves as they hunted in Jasper National Park. In one hunt, a deer was bedded in a blowdown. Four wolves came upon it before it sensed them, and they split up and surrounded it. "It was like they had this deer in a corral," says Weaver. Another pack caught scent of a bedded fawn. Three of the five climbed the back side of a ridge, while the other two came along the base and cut off the deer's escape. "You can see where the deer got up out of its bed and made two jumps and *pow!*" Dr. Michael Fox, who had been observing wolves in the St. Louis Zoo, felt compelled to report to the *Journal of Mammology* something he observed in 1969. Wolves at the St. Louis

Zoo were kept in a double pen with a sliding door between the two enclosures. The sliding door was closed by pulling a rope inside the inner pen. When a keeper came in to hose down the outer pen, the male wolf would herd all the pups into the inner pen, jump up, grab the pulley rope, and close the door. When the hose was turned off, the wolf would open the sliding door with its paw and muzzle, and let the young out. Fox considered this to be an example of "high order behavior in wolves."

That same year, Harry Frank, a professor of psychology at the University of Michigan–Flint acquired a female wolf pup from a Chicago zoo. Frank already owned a three-quarter wolf-dog hybrid and a malamute. With all three animals in his household, Frank began to see that "there was some fundamental difference in the way they processed information. The hybrid was intermediate in some ways. The first time we brought a wolf home, the wolf went into submission, and the hybrid knew exactly what it was. The dog didn't. When the malamute didn't respond, the wolf snapped at him. He grabbed her by the back of the neck and threw her against the side of the house. She could not understand an animal that was not communicating his status, and she had no use for an animal that was going through all these shenanigans."

The wolf had an insatiable curiosity and an insistent playfulness. She got into everything. "She could turn on the faucet in the kitchen just to play. She'd spin the lazy susan on the kitchen table just to make things fly. She'd play hockey with a flattened coffee can on the iced driveway—and play it with the wolf-dog hybrid. She'd invent games.

"The hybrid, though, could communicate fully with humans." When Frank was building things inside the animals' enclosure, the wolf would nip at his backside or tug his pant leg and dart off as if inviting Frank to play. When he failed to give chase, she would grab a tool or a box of nails and run off with it. "All I had to do was give the wolf a dominating look and give the hybrid a look as if to say, 'Okay, it's your job. Go over and take care of this problem.' He'd go over and grab the wolf by the neck and neck-pin her."

Above all, the wolf seemed smarter. "Our kenneling facility and outdoor compound were separated by a door that required two distinct operations to unlatch," says Frank. "First the handle had to be pushed toward the door, and then it had to be rotated. Our malamute watched us perform this task several times a day for six years and never did learn to do it himself; our wolf-malamute hybrid was able to unlatch the door himself after watching us for only two weeks; and our older female wolf learned the task after watching the hybrid once. Furthermore, she did not use the same technique: the hybrid used his muzzle, and the wolf used her paws.

"We have a lot of other sorts of evidence of the cognitive process," says Frank. "Little wolf puppies, instead of manipulating an object, look at it, study it, and then do the right thing. You have to hypothesize that the animal did its trial and error internally. Never did we see the dog puppies do the same thing." If dogs solved the puzzles Frank posed them, they did so after extensive pawing, tugging, and prodding of the puzzle objects—after going through a process of trial and error.

On the other hand, one of Frank's students had tried for six months to teach the wolf to heel and sit on command, but the wolf never acquired the behavior. The hybrid watched Frank put the malamute through his obedience routines for six days and, without any training himself, obeyed the commands as reliably as the dog.

Frank was intrigued by the differences between wolves and dogs. He had read John Paul Scott and John L. Fuller's 1965 study, *Genetics and the Social Behavior of Dogs,* which established standards for assessing the mental qualities of dogs. The tests consisted of mazes, mechanical puzzles, barriers around which dogs had to navigate to reach dishes of food, and other trials. Scott and Fuller had speculated about what wolves might be capable of, but never tested them. Frank thought it was worth trying. Says Frank, "We thought we should find that dogs are very trainable but not very good at solving complex problems, and that wolves should not be very trainable but very good at solving complex problems."

To test his hypothesis, Frank and his coworkers acquired four wolf pups from a Minnesota game park and reared them in close contact with people. At the age of six weeks, he subjected them to a series of tests. The first of these was a test of insight, in which a pup and a bowl of food were separated by a plywood barrier with a wire-mesh window, so the pup could see and smell the food. The test was designed to see how long it took a pup to discover that it couldn't get the food directly, but had, instead, to run around the barrier. If Frank saw a wolf suddenly change its strategy by reversing direction and going straight to the food, he took that to be evidence that the wolf had restructured the problem in its mind—or, in other words, displayed insight. He compared his results with Scott and Fuller's results with dogs in the same test, and found wolves were far more adept at getting to the food. They less often reversed direction in a single trial or stopped to sniff along the way. Frank concluded from a wide range of tests that wolves were capable of "abstractness, flexibility, complexity, foresight, mental representation, and insight into rudimentary means-ends relationships."

Dogs, however, did better on tests that required instruction. Frank designed tests in which dogs and wolves were graded on how they reacted to flashing lights or whistles that had no cause-and-effect connection with the desired response. "When a light flashed, they'd be required to turn right, and they couldn't. The dogs could. When we started working with the dogs, they did it right. They were cuing on the behavior of the lab assistants." But wolves didn't seem to read the lab assistants' clues. Says Frank, "Wolves, unlike dogs, have great difficulty associating a to-be-learned behavior with a wholly arbitrary one."

Frank concluded, "Dogs did better than wolves on all the training tasks, and wolves did better than dogs on all the problem tasks." The explanation seemed clear: "Intelligence is the capacity to adapt to changes in one's environment. What defines intelligence depends on the environment one is in. There really is a qualitative difference between the natural environment and the human environment." A wolf can see the mechanical relationship—say, of a fallen tree resting precariously against a boulder in the forest, and so will stay away from it. "For the dog, the principal environmental feature is the human being. In a human environment, the means-ends relationships are invisible. What could be more magical to a dog than an elevator? You get on and push a button, and it goes up. There is no way to visualize what actions will lead to what consequences. Also, in the man-made environment, those things that control the environment are not within a dog's reach: they are intended for animals with human height and fingers. It's much more advantageous to develop a very keen understanding of human behavior and to communicate wishes to a human, because the human is the most important feature of the environment, and we give a lot of visual and auditory cues." Evolution hasn't honed the dog's problem-solving skills, just its people-reading skills.

Domestication, Frank explains, selects for infantile qualities such as docility and dependence, and turns animals into infantile forms of their wild ancestors, since the infantile form is more trainable than the adult form. "Much of the character of the wolf capacity to process information seems to be a feature of the adult," he says. You can see the difference dramatically if you mix wolves and dogs. "Ben Ginsburg, of the University of Chicago, said, when you introduce a dog to a bunch of wolves, you might expect the wolves to kill it. But in fact they treat it as a juvenile. If you want a real thumbnail sketch of the difference between wolves and dogs, it is that the wolf is the adult form. That is it. The dog is a juvenile wolf. The wolf demands dignity and respect; the dog you treat as a child."

There is much evidence that domestication physically changes the brains of animals. Darwin noticed that the brains of domestic rabbits are smaller than those of wild rabbits. A German scholar in the 1920s held that, on the average, domestic forms had brains 30 percent smaller than those of their wild ancestors. Brain-size reduction has been shown in rats, mice, rabbits, pigs, sheep, llamas, and domestic cats. German researchers found the brains of wolves to be as much as 29 percent bigger than the brains of dogs. Brain size, though, is not in itself a very reliable indicator of intelligence, and, at least in mammals, tends to correlate closely with body size. Some researchers believe that, if one scales the size of the brain to the size of the body, the shrinkage associated with domestication becomes insignificant. But there are structural differences between the brains of wild and domestic forms. In 1973, German researchers compared the brains of poodle dogs with those of wolves and found that the wolf brains were not just larger, but larger in particular regions. The wolf brains were 40 percent larger in the hippocampal region, which guides and regulates emotional reactions, aggression, and motivation, a finding that is consistent with the fact that domestication selects for gentleness and tractability. Some areas of the brain may also have a greater density of nerve cells—a difference that has been confirmed in comparisons of the brains of wild and domestic forms of cat.

What might these structural differences mean to the relative intelligence of wolves and dogs? Many people believe wolves have powers dogs don't have, including faculties we generally regard as extrasensory.

Elizabeth Andrews, who worked with captive wolves at a Washington State wolf park, speaks of a "strange intelligence in the wolves." She lived an hour's drive away from the wolf park, and arrived at a different time each day, but whenever she arrived the owner was standing at the door, expecting her. He told her that the wolves had begun to howl fifteen minutes before she arrived, and they would howl in such a way that he knew it was she, and not her husband, though they drove the same car.

Dr. Michael Fox tells of a captive wolf released two hundred miles from its home in Alaska that made a beeline almost two hundred miles back to its cage. He believes that dogs also have such abilities. There was a veterinarian's dog, Fox says, that would stand waiting at the window for his master to come home exactly fifteen minutes before his arrival, even though the veterinarian returned at different hours on different days. The veterinarian's wife concluded that the dog got up and started waiting at exactly the moment the veterinarian closed his clinic and started walking home. Fox calls the phenomenon "psi"—for "psychic"—"trailing." He confesses, "There's no logical or mechanistic explanation for it," but adds, "One of my theories is that animals that are capable of doing this are attuned to the pathosphere, a feeling realm of awareness."

No one has yet demonstrated these psychic abilities empirically. It may be that we want the wolves to lead us into new dimensions in intelligence, and that the stories tell more about our wishes than they do about the abilities of wolves. Or it may be that a wolf's senses are acute enough to account for the wolf's apparent clairvoyance. The 1973 German study also found that wolf brains were larger in areas that dealt with sense impressions. It is in the area of sensory perception that wolf abilities are really remarkable.

All wolf watchers eventually remark about how observant and curious wolves are. Lois Crisler watched a female wolf wade back and forth in a brook for ten minutes, "looking down as if it were a live thing," lifting her paw and touching the water. When the creeks froze and Crisler's young wolves saw their first ice, they became so absorbed in it that they spent the morning at an overflow pond by a brook, "looking down at the ice, rearing back and drumming on it." David Mech gives another example: "In 1991, on Ellesmere Island, I'd been watching wolves for weeks on end and had a certain type of big heavy parka and a wool hat on. Suddenly, the weather turned warm—the temperature went up to fifty degrees. I changed my clothes, and I was in the same position on the four-wheeler. You'd think the fact that I had a different jacket or a different hat on wouldn't make a difference. But they were very shy and wary. They wouldn't come up within ten feet of me." When he put his old parka and hat on, the wolves immediately resumed their confident familiarity.

Sometimes, the wolves' curiosity posed problems for Mech. "They're so curious you don't want them coming around your tent, because they'll chew up the tent. Chewing is curiosity in wolves. They only have their paws or their mouths to work with, and can't really explore the upper corner of the tent with their paws, so they explore it with their mouths." One day, Mech looked back at his campsite to see that the wolves were at the tent, and one of them had his head inside the drawstring window. "He had his head in the tent, and he was jerking on the sleeping bag, pulling it out. He looked just like he was gutting a moose."

Mech sees simple reasons for their watchfulness. "They are keyed in to judging subtle differences in prey. If they attack something too strong and healthy, they could get killed. If they don't attack the ones that are vulnerable, they could starve." So there is an evolutionary premium on being able to read and analyze small details quickly.

The sense humans rely most upon is vision. We know only a little about how visually adept wolves are; we have little idea how their brains organize what they see. Laboratory studies have shown that wolves can distinguish red, yellow, blue, and green. The visual receptors of a dog's eye are 95 percent rods, an adaptation to night vision. Dogs lack the fovea, the area of the retina which in the human eye is densely packed with receptor cells, so dogs and wolves may not see the very sharp outlines and fine details that humans see.

Still, wolves rely heavily upon what they see. They communicate with elaborate and complex visual signals, reading each other's eyes, ears, and mouths, and their faces are designed to emphasize emotional content. Lips are black and often contrast with white or light colors on their muzzles, which make their mouths more expressive. Their bright-yellow eyes are set off by black eyelids, and frequently light patches around the eye areas draw attention to the eyes. The yellow eye color contrasts sharply with the black pupil, and changes in the size of the pupils indicate changes in mood. Wolves also read important messages in each other's body postures. A threatening wolf puts its ears up, purses its mouth, or snarls, perhaps even showing teeth; it stands tall, so that its body seems larger and more powerful; its tail rises. A submissive wolf cowers, or shrinks, to seem smaller; its ears go back; it grins submissively, tail tucked between its legs.

Wolves process what they see with remarkable quickness and precision. Standing outside an enclosure near Ely, Minnesota, a young man edges closer to two captive wolves. Jedediah, the male wolf in the compound, comes over and stands sideways, showing his full size and power, and growls. He paces along the fence, challenging all the visitors, and turns so that his side is flat against the fence. His head and tail up, he makes eye contact with each person on the other side of the fence as he passes. And he growls a deep, throaty, serious threat. He fixes his gaze hard on the young man, who has gotten down to eye level and has stupidly sought to challenge him. The wolf growls fiercely, stands taller—on the very tips of his toes—and challenges more stiffly. The wolf's owner patiently explains, "Eye contact is reserved for the alpha." The visitor doesn't take the hint, and keeps on his witless challenge, but then, without really knowing why, grows uncomfortable and steps back a fraction of an inch. Suddenly, the wolf brisks off to patrol the rest of the fence. His perception of the victory is so instantaneous and uncelebrated that it is hardly noticeable.

The wolf trots over to his owner, stands on his hind legs, and licks her mouth. His ears go back. His tail goes down. He rolls over submissively. She explains that he is testing her, just as he tested the visitor. He is looking for any sign of weakness, and he does this every time she comes into the pen, so she stays out if she's feeling tired or weak. It's not just that the wolf can overpower her, but that he reads her much faster and more accurately than she ever will read him.

Wolves read the postures of their prey with similar quickness and fluency. They bluff charges to get an animal to reveal its vulnerability, and if they see that a moose is strong and healthy, they move on.

Rolf Peterson tells of a wolf who was skilled at reading moose on Isle Royale. "A female alpha in the East Pack was in charge of the pack for eleven years. She went through four mates, and may have been in on the kills of five hundred moose, which means she probably tested more than ten thousand moose. Once the pack of eleven were walking down the lake ice. There were two moose browsing on the shore. She took one look at them, and paid no more attention." While she recognized immediately that the moose were invulnerable, four youngsters didn't, and they took off running after the moose. "She just sat down on the ice, waited for all of them to have their fling, and when they came panting back, she got up and led off at the front of the line."

Wolves probably read the landscape with the same fluency. Dispersing wolves tend to turn up in exactly the same locations where people have seen wolves before, years and even decades apart. Dispersing from Russia into Finland, wolves have used the very routes wolves used fifty years before. Wolves moving south from Canada into Montana turn up in exactly the same spots wolves were seen decades ago. John Weaver, after several years studying the wolves of Jasper National Park in Canada, suggests that wolves have inborn search images that enable them to recognize habitat productive of moose, deer, or elk. They might be responding to the scent and shape of trees, the aroma of soil fungi, the texture of rock and dirt underfoot, the faint odors of mice and squirrels on the grass.

The wolf's deeper engagement with the world must owe something to its capacity to read its setting through several senses at once. Some people have suggested that much of the wolf's uncanny perceptiveness is due to an extremely keen sense of hearing. Roy McBride, who trapped wolves for Mexican ranchers in the 1960s and '70s, recalls two captive Mexican wolves on a Chihuahua ranch where he was trapping. After observing that the captive wolves were fed only tortillas and beans, he began to bring them coyotes, bobcats, and javelinas caught unintentionally in his traps. The wolves began to distinguish McBride from the host of cowboys and fence builders that came and went on the ranch. Says McBride, "The cowboys told me that, long before I would arrive in the afternoon, the wolves would begin to pace excitedly, and after perhaps twenty minutes the ranch dogs would also detect my approach and begin to bark."

McBride left the ranch and returned six months later; long before his arrival, the wolves were up and pacing and wagging their tails. McBride attributes their awareness to their sense of smell, but it is just as possible that they heard his vehicle. No one has yet tested the auditory powers of wolves, but they probably hear better than dogs—and dogs can hear sounds at forty kilohertz, about double the upper range of human hearing. Coyotes have been shown to hear sounds above eighty kilohertz; wolves could have equal or greater abilities.

The wolf's sense of smell may be its most acute sense. The sense of smell in other creatures has always been a difficult thing for humans to appraise, in part because ours is feeble compared with that of other animals, and in part because olfaction is rooted in the deeper core of the brain, which also presides over emotions, and not in the neocortex, which governs both vision and reason. Odors can arouse fear or nostalgia or courage. As rational creatures, we defer to our eyes as the supreme judges of truth, and often belittle or deny what our noses tell us. The wolf, however, has a sense of smell that challenges our imaginations. By presenting dogs with odorants in ever-smaller dilutions and testing their responses, researchers have shown that dogs are at least ten thousand times more sensitive than humans. Paul Joslin says that preliminary experiments conducted with dogs at Wolf Haven show dogs may perceive odorants in one-hundred-thousandth the concentration at which humans can smell them. They can also recognize much finer nuances of scent than humans can discern. Dogs can distinguish by smell between human twins, and detect the odor of six-week-old human fingerprints on glass slides.

Though wolves haven't been tested, their olfactory abilities may be even more acute. Trappers so respect wolf noses that they boil their traps in oak leaves, and soak their gloves in calf's blood. They stand on pieces of cowhide while setting a trap, or try to set them from horseback, and jealously guard their personal formulas of fox glands and coyote gallbladders in the scent baits they use. None of this may be of much avail, however, for all the while they are shedding dead skin cells and exhaling organic chemicals in their breath that probably cling to the ground and become the olfactory equivalent of billboards to passing wolves.

Wolves live far more by their noses than by their eyes. Much of what they communicate among themselves is expressed or interpreted in odors. Young and Goldman observed that, on their runways, "wolves have what are commonly referred to as scent posts or places where they come to urinate or defecate." Farley Mowat's fictional wolves put up a fence of scent marks which neighboring packs seemed not to cross, and his fictional biologist urinated all around his camp to keep wolves out. David Mech and Roger Peters put these speculations about scent-marking to scientific scrutiny in northern Minnesota between 1971 and 1974 by following the tracks of wolves to see exactly where and how they scent-marked.

Studies by Peters at the Brookfield and Como Park zoos had already shown that wolves urinate in two ways: females and subordinate males squat; dominant males stand, raise a hind leg, and squirt a small amount, perhaps a sixth of an ounce, of urine onto a stump or a rock or a clump of grass or some other object elevated above the surface of the ground. In one zoo study, 60 percent of raised-leg urinations were accompanied by snarling, growling, or biting, whereas none of the squat-posture urination was. The act apparently carried an emotional content.

Mech and Peters skied and snowshoed 240 kilometers of winter wolf tracks in northern Minnesota and recorded the rate of raised-leg urination. They found raised-leg urination increased markedly at breeding season, and that squat urination by females was often covered by male raised-leg urination. They noted seventy instances of nose-shaped indentations in the snow where wolves stopped to sniff, and found under thirty of them raised-leg urinations. Wolves clearly paid attention to the scents other wolves left behind.

The two researchers also looked at scent-marking in summer. They found that scats were deposited most heavily on trails leading to central points of rendezvous sites. Scats may bear odors of hormonal secretions imparted by the anal sacs, on either side of the anal opening. Mech and Peters had long observed wolves veering off their paths to sniff at scats.

Dominant males scratch the ground stiffly with right fore and left hind legs, then left fore and right hind legs. Various authors had seen dogs do this and concluded either that they were burying their waste or spreading it around to make it more prominent, but wolves neither bury nor spread their waste. The scratching is apparently intended to leave both visual and olfactory signs; it is probable that glands in the feet impart scents to the scratches.

Mech and Peters found a much higher rate of scratches, urinations, and scats at the junctions of roads, trails, and other commonly followed wolf paths, and far more raised-leg urination occurred on these paths than in the bush. Far more scent marks were found at the edges than at the centers of wolf territories, from which Mech and Peters concluded that wolves continually marked their territorial boundaries. If other packs crossed such tracks and the residents encountered their sign, the residents generally increased their rate of marking. Wolves leave scent marks every two or three minutes as they travel. "The entire territory," concluded Mech, "is studded with olfactory hot spots." A wolf can always tell whether or not it is on its own territory. Wherever it goes, scent marks may tell it which wolf passed recently, what that wolf ate, when it ate, and possibly what mood it was in.

Dispersing wolves tend to follow the boundaries of other packs' territories, but scent boundaries aren't absolutely inviolable. Mech and Peters found that, though wolves were generally reluctant to spend much time inside a neighbor's territory, scent marks didn't always repel them. In one instance, tracks showed that members of one pack chased a deer into a neighboring pack's territory and seriously wounded it there. When the deer fled deeper into the neighboring territory, however, the intruding pack refused to follow it. They scent-marked heavily and returned home. The neighboring pack encountered the deer the next day and ate it.

What might wolves be reading in all that scent-marking? Might they be reading, not just declarations of ownership, but expressions of a pack's willingness to fight for territory, signs of readiness to mate, and measures of size, strength, aggressiveness, and health? Might these messages include boasts, confessions, jokes, and insults? Wolves obviously make much of what they discern in other individuals' leavings. Paul Joslin tells of a wolf whose sense of olfactory propriety was sorely tested. While working in Algonquin Park, he says, he approached a captive wolf he had approached many times before, but this time he was carrying a fresh scat which he had collected in the wild. "I showed it to the wolf inside. He took one sniff and bolted up the hill. He would never have anything to do with me again." Even the next summer, when Joslin came back from a year at school to work in Algonquin, the wolf would not let him approach. "Whatever was communicated by that scent," says Joslin, "was something he didn't want to have anything to do with."

And what may its keen sense of smell tell the wolf about the nonlupine world? Mech was able to judge, in fifty-one hunts he trailed or observed, whether the wolves sighted or scented moose. In forty-two of them, the wolves seemed to be scenting. In one, Mech concluded the wolves had smelled a cow and twin calves from 1.5 miles away.

What a wolf does with its fine perception is bound to be different from what a human does. Wolves live deeply immersed in nature, but humans have removed themselves from it. There is debate as to whether humans are themselves domesticated animals. There is some evidence that Neanderthal brains from thirty thousand years ago were as much as one-sixth larger than modern brains. Whether or not our brains have grown or diminished, there may be differences in specific parts of the brain that have come about as we gave up hunting and gathering and lived farther and farther from nature. It may be argued, for example, that we have lost perceptive powers, as urban noise deafened us and urban smoke and sweat favored people whose sense of smell was not too sensitive. City life may favor conservative strategies of caution and camouflage over acute perception.

Wolf and human intelligences may focus on quite different things. The human mind focuses on objects. Human lives are transfixed by things that are inanimate, cars or television sets or clothes that give the illusion of imparting life to those who possess them. We confuse convenience with liveliness. Wolves are likely to think differently. The objects around them are alive, in motion, independent of wolf will. Their relationships with the world around them are likely to be more full of twists and turns, more alive and dynamic. Such a world requires a constant and free-flowing curiosity, a watchfulness, a capacity for sharp perception. It may be more fruitful for us to appraise wolf thinking in terms of perception than in terms of calculation.

. . .

We must consider wolf intelligence on yet another level—that of emotion. An individual's mind is an interplay between thought and feeling. The stronger feelings—which we call emotions—are mechanisms perfected through evolution that lead animals to intensify and focus behavior when it is advantageous to ignore other things. For example, emotions push a hunter to pursue and kill another animal despite the danger of sharp hooves or treacherous terrain. Emotions also impel parents to play with and care for their young when they might otherwise choose to hunt or sleep. At the same time, emotions can contradict thought or competing emotions, and thereby make an animal behave in unexpected ways. Emotions thus provide flexibility, so that life is not simply a matter of responding blindly to set stimuli. They give a species more kinds of response to its environment, and thereby make the species more adaptable to changes in that environment.

Many things about wolves suggest an emotional life that is rich, complex, and, like our own, full of contradictions. Wolves clearly have feelings, and observers frequently describe them as joyful or sad or sullen or playful. However, appraisal of animal emotions has so far eluded science, because animals cannot explain what they feel in human terms. When it comes to precise description of animal emotions, we humans are simply guessing.

The way the wolves Stephenson saw played tricks on eagles or bears seems to me very suggestive of the wolf's rich emotional life and the flexibility it imparts to wolf behavior. Consider, for example, the way wolves relate to ravens. Over much of the range of wolves, if you want to find a wolf, the first thing you do is look for ravens. These birds often follow wolf packs or fly ahead of a hunting pack, wait in the trees for it to catch up, then fly farther on. Stephenson found that ravens were attracted by wolf howls, and when he howled to locate wolves, ravens would appear. Mech frequently saw ravens tracking the packs on Isle Royale, flying directly over their footprints in the snow. They perched in trees, waiting for wolves to finish feeding on a carcass, and as soon as the wolves left, they darted in to feed themselves. When the wolves moved on, the ravens dropped down and picked at their scats, swallowing the edible portions of incompletely digested meat and bone and hide.

Ravens will come from great distances to feed on a wolf kill.

Christophe Promberger, of Wildbiologische Gesellschaft, a wildlife-conservation group in Munich, Germany, conducted a study for the Yukon Division of Wildlife, which wanted to know how dependent ravens are on wolf kills in winter. He placed a dead moose in the snow and watched to see what the ravens did. They came, but stood around the carcass and did not eat. For two days, Promberger drove back and forth between the study site and Whitehorse, but the ravens continued to stand around the carcass without eating. He was about to give up when he saw another dead moose alongside a road. There were two wolves eating on one side of it, and a raven eating on the other. It occurred to him that perhaps ravens were unable to open a carcass—that they needed wolves to put their dinner on a platter. He went back to his bait kill, cut it open with an ax, and threw blood and hair and bone around in the snow. The ravens dined. Promberger continued his study, and eventually he found that ravens take an enormous share of wolf kills. In one day, ravens took almost ninety pounds of meat off one of his simulated kills. He concluded that wolf kills may be the principal source of food for ravens in winter.

After the wolves have eaten and are lying around in the snow sleeping off the torpor of a big meal, ravens will stand among them. Don Murray, one of the Isle Royale pilots, saw a raven alight for an instant on the back of a wolf. Mech and others have seen ravens play with wolves. When wolves were resting on lake ice, ravens would dive at them or walk up and peck their tails. If a wolf lunged at a raven, it jumped aside or flew a few feet away. Wolves then would stalk ravens, which would jump out of the way at the last minute, fly a few feet, and stand looking back at the wolf. It seemed to Mech like a game.

Other animals also feed on wolf kills. Eagles, wolverines, lynxes, and foxes may take half the meat off a wolf kill. To save as much as they can, wolves dig holes and cache meat, and they kill their competitors. There are records of wolves killing otters and mink. They are extremely intolerant of foxes, which they seem to kill wherever they can. But why, then, don't they as a matter of course try to kill ravens?

It has been suggested that they don't because the ravens provide the wolves with an important service. "There is this theory that ravens lead wolves to a kill," says Promberger. Mech saw ravens attend an attack on a moose on Isle Royale, swirling around the moose and wolves as the wolves brought the moose to bay. When the moose had been wounded, one raven sat in a tree cawing, as if urging the combat on. "I'm sure ravens track wolves in the snow," says Stephenson. "Wolves tell ravens things. Ravens tell wolves things."

But there is no fixed rule to the relationship. On Isle Royale, Rolf Peterson and Don Murray once saw a wolf catch and eat a raven. At the Julian Science Center in California, ravens caw and soar over the wolf pen, but every now and then Paul Kenis has found bits of wing and flutters of black feathers where a wolf has leapt up and caught a careless bird.

It is not necessarily axiomatic that playing and killing are emotional acts, but it is hard to look at wolves engaged in such acts and deny the feeling inherent in the behavior. And the fact that some wolves play with ravens while others kill them suggests that wolf emotions are varied and that wolf behavior is flexible.

There are several possible ways to view this flexibility. One explanation is that wolf societies, like human societies, consist of different personalities. Stephenson says the Eskimos taught him to see that all wolves are individuals. The things he has seen out there in the wilds of Alaska don't fit into scientific reports because they sometimes fall outside the realm of generalization, and, he says, "We want to generalize pretty quick with an animal." He believes that, as research becomes more airborne and more office-bound, we generalize more and more, and we lose the vast range of wolf experience; in fact, there are soft wolves and hard wolves, kind wolves and malicious wolves, soldiers and nurses, philosophers and bullies. Perhaps the poor raven that is killed by a wolf has had the misfortune to run into a barbarian, whereas the boon companion has run into a Rotarian.

Another way to view this complexity is to recognize that even an individual may have conflicting emotions. Wolves and humans alike demonstrate that a creature that can love is also a creature that can hate. Both animals cover a wide range of character and behavior. Both are of many minds.

7

THE SILENT WOODS

The cry of a loon trails off the water of Lake Superior. It is muffled by a summer fog clinging to the ridge tops of Isle Royale. The island is wet and dripping, as if all the low greenery meant to delay the runoff of rainwater, to hold it on leaf and stem and petal in round pregnant drops, then in the soup of duff and twigs and sand on the trail, to miser it from the unimaginably vast sea of fresh water a few feet away. Along the trails, the balsam firs have a spicy, Christmas aroma. Lichens cover the trunks of birches and balsam firs. Shy yellow violets and blushing pink pyrola blossoms color the forest floor. It is a place that could be lively with the sounds of animals, but it is not.

"It's quiet," says Rolf Peterson. "You don't hear howling much. The howling has dwindled over the last decade."

Indeed it has. In 1980, there were fifty wolves on this 210-square-mile island national park. On late-summer nights, visitors could hear them howl and sense something at once thrilling and familiar. Isle Royale's were the best-known wolves in the world. Twenty-two years of careful and continual observation had gone into studying them, and they had been the subject of thousands of news stories and magazine articles. But since then, something has happened to the wolves. From the peak in 1980, their population plummeted to twelve in 1988, and they have remained at that level for five years.

"Howling in summer is prompted by having lots of pups around that like to howl," says Peterson. "And having large packs around. With just a male and a female constituting a pack, there's not much reason to howl. It's quiet. And every wolf track is worth noting these days."

Since 1970, Peterson and his wife, Candy, and, later, their two sons have spent their summers in a seventy-five-year-old fisherman's cabin on one of the outer islands. It is a snug hidey-hole of hand-hewn logs nestled in the forest at the edge of the water. The rest of the year, they are in Houghton, Michigan, where he is a professor of wildlife ecology at Michigan Technological University. He comes out to count the wolves for several weeks every winter, when the trees are leafless and tracks are easily seen.

Peterson's blue eyes and fair complexion suggest a Viking ancestry, and perhaps an inherited penchant for seeing the ends of things, of unexplored rivers and unexplained biological cycles. The son of a Minneapolis architect, he attended a YMCA camp in Canada, which taught him woodcraft and canoeing. He majored in zoology at the University of Minnesota at Duluth and spent his summers in canoe expeditions. In 1968, at the age of nineteen, he and five friends descended the Dubawnt River to the barren grounds of the Northwest Territories, passing near where Farley Mowat had watched wolves twenty years before. "I think we were only the fourth group of white people to go down the river, and the first group not to try to live off the land," says Peterson. "It was really rough." It was so cold that their feet would go numb by day and warm up only at night, and all members of the party suffered vascular damage. For thirty-five days, they saw no other humans. "We never saw wolves. All we saw of caribou, except for one band, was skeletons." The great herds of barren-ground caribou were gone, and so perhaps were the wolves that had lived off of them. The sheer adventure of the trip confirmed in Peterson a desire to stay in the North. By the time he was a senior in college, "it was a toss-up between limnology and wolves."

Peterson became aware of wolves when he visited Isle Royale as a young backpacker in 1967. Two years later, thinking the Isle Royale study had been discontinued, Peterson wrote to Douglas Pimlott asking for a position studying wolves in the Canadian Arctic, but Pimlott had nothing for him. Peterson then saw Durward Allen on a television program, and thus discovered that the Isle Royale study had *not* ended. He wrote Allen asking to be hired. Impressed by Peterson's summer expeditions, Allen took him on as a graduate student. Eventually, Peterson succeeded Allen as director of the project, and moved its headquarters from Purdue to Michigan Technological University at Houghton, which offered a greater likelihood of future funding.

Peterson had never seen a wild wolf before he came to the island. He had to visit Chicago's Brookfield Zoo before he came out to Isle Royale, to begin to get to know the animal. And his first sighting of a wild wolf was a confusing blur. He was with pilot Don Murray, who had been flying for thirteen years on Isle Royale and knew the wolves well. Murray found a kill and circled the plane over it. "He pointed out the wolves," says Peterson. "I'm not sure I really saw them."

Isle Royale's study was blessed in that it had only two pilots for most of its history, lending a regularity to the observations on which estimates of moose-population trends are based. The pilots urged Peterson to gain his own sighting skills. Murray would not announce there was a wolf below; he would circle and circle until Peterson saw what he wanted him to see. In time, Peterson learned to identify wolves from the air by looking at their tail markings. With gyro-balanced binoculars, he could keep a wolf in view while the plane was turning in tight circles and rough air, and could thus count wolves and observe their behavior in winter.

In spring, he would go out on foot to search for den sites and to try to confirm the existence of pups. It was hard work, because the Park Service forbade the researchers to use radio collars, lest park visitors complain that the wolves were being exploited or overmanaged. Peterson would search for them by tracking and howling, and sightings were seldom lengthy events. "Basically," says Peterson, "it's wolf sees people, wolf runs." He recalls, for example, watching a rendezvous site from a nearby ridge for ten days as a pack with seven pups played and howled and returned from hunts with food for the young. Then, one day when Peterson was watching the study site waiting for the wolves to appear, the male and female arrived—but there were no pups. "They just sat down and howled, the longest, most mournful howl. There was an answering from the other wolves off in the forest. And then they were gone. I think they knew we were there."

In summers, when the wolves hide in the forest undergrowth, he would walk the game trails of Isle Royale, collecting moose bones to determine what the wolves were eating and how that related to moose-population cycles. He now has what must be the world's largest collection of moose antlers, skulls, and assorted other bones, including bones collected by Adolph Murie on the island in the 1930s. The bones contain a coded history of the last sixty years of the island's moose and, when arranged chronologically, tell of times of overbrowsing, and deformed bones and antler, and times of plenty, when moose were healthy and strong.

For years, the news from Isle Royale was that nature looked after itself, and smiled both on wolves and moose. Declared Allen in *The Wolves of Minong,* his detailed history of Isle Royale up to 1976, "These 18 years have been biologically productive." But Allen warned, "Happy times do not last forever. The wolves probably have reached the maximum density they can tolerate, and they will be reduced." And in the 1980s, the wolf population suddenly declined. In 1981 and 1982, fifty-two wolves died—some killed by other wolves, some by starvation, some, Peterson suspects, by canine parvovirus, an often-fatal disease carried by domestic dogs. By 1982, only fourteen wolves survived on the island. And in the ensuing years, there was no rebound. Pups were born, but few survived, and for every pup that lived, an adult died. After thirty years of watching the wolves increase, the researchers were stunned by this crash.

In 1988, the Park Service decided that the need to understand the collapse outweighed their desire to keep wolves free from human interference, and it allowed Peterson to begin using radio collars. These days, Peterson follows radio signals to find the wolves. From May through June, he tries to locate wolf kills in the woods. "We want to know about where the wolves are and if they're alive," says Peterson. "We're mainly interested in whether they reproduce."

He looks intensively for dens. If he finds one, Peterson insists on being the first person to go in and howl and look at the rendezvous site. The most important research task is finding out whether the wolves are having pups, and, if they are, what is happening to them. Pups are afraid of airplanes, and hide from them so effectively that Peterson can't make observations from the air. So he hikes in. "It's the first time they're seen, and I have to be out there," he says. He is always careful not to get too familiar with the wolves. "We try to conceal the human form from wolves. Only rarely have we chased wolves off a kill to get a bone." This may be overprudence, he allows: "I'd be surprised if wolves don't know even that we're in the plane. It could be they have a complete understanding of our whole operation."

Twenty years ago, the Isle Royale wolves consumed every moment the researchers could spare. "In the 1970s, we were knee deep in wolves," says Peterson. "There were up to three litters of pups going simultaneously. We had to find those pups on the ground without the aid of electronics, and get them to howl so we could count them."

Now there are so few wolves that "it's impossible to find them without radio collars. The packs are too small, and they don't move around much." Two wolves feed on a kill for five or six days, whereas a big pack would feed one or two days and move on. Besides, two wolves don't leave many tracks, so they're harder to find from the air.

"The activity hasn't kept us very busy," says Peterson, "because there's nothing going on out there."

What is happening to Isle Royale's wolves? There seems little likelihood that humans are poaching the wolves in winter, when there are no humans on the island. Poachers would have to use airplanes, and, "If someone was doing that to wolves here," says Peterson, "their behavior towards us in the airplane would change." One winter in the 1970s, the wolves suddenly became afraid of the airplane. "It took them all year to get used to the plane," says Peterson. Mech came up with an aerial photograph of a pack of fifteen wolves. Though the photographers represented it as having been taken in Minnesota, Mech knew of no such large packs in the area described by the photographer. The pack resembled one of Isle Royale's. Apparently, someone had harassed the wolves in an airplane that winter, to hunt or take pictures. But Peterson has seen no such shyness of airplanes since the population crash.

Before 1988, when no wolves were radio-collared, the researchers could not find and retrieve the carcass when a wolf died. Since 1988, only one dead wolf has been recovered for autopsy, and it provided no convincing explanations, so the researchers can only guess what is going on. Says Peterson, "It's either disease or it's food or it's genetics."

The question of disease is hard to answer.

Each year, seventeen thousand people visit the park, and at least half of them are fishermen or others who come in their own boats. Although rules prohibit them, the boaters bring dogs, some of which may carry canine parvovirus or distemper. But none of the remaining wolves on the island have been found suffering from either disease, and with so few wolves surviving, researchers are unlikely to locate the remains of a victim of one of these diseases even if they are present.

Another possible cause of mortality is Lyme disease. A University of Wisconsin lab has held that moose ticks could carry the disease, but the claim is disputed by a University of Minnesota lab. As yet, there is no clear demonstration that either wild or captive wolves have come down with the disease. Explains Peterson, "The University of Wisconsin lab cultured moose ticks. Only one out of about three dozen cultured positive for Lyme disease. We do not have the deer tick, which is the common host for the disease, here. We don't know whether the spirochete that causes Lyme disease is really here or not. Until wolves somewhere in the world come down with Lyme disease, I think there will continue to be a lot of skepticism."

Peterson hopes the decline was caused by a scarcity of food. The moose population crashed in the 1970s, reaching a low of five to six hundred individuals in 1977, shortly before the wolf population crashed. In the 1980s, the moose population expanded again, hitting seventeen hundred in 1988, but the wolves did not recover with it. The moose population then dropped once more—not because of predation by wolves, but because of infestation by winter ticks. Moose are normally hosts to ticks. With mild winters, the ticks drop off the moose earlier in spring and go through their reproductive cycles earlier, because the soil, in which eggs develop, is not frozen.

That means more generations of ticks, and greater rates of infestation. The greater infestations weaken the moose by drawing more blood and by causing hair loss, which increases the moose's heat loss in winter and places it under greater nutritional stress. Says Peterson, "In 1989, winter ticks were controlling moose."

The moose survived the ticks, however, and their numbers increased once more. In 1992, numbering sixteen hundred, they were about to meet a more sinister adversary. "This year," said Peterson, "it's starvation. The numbers of moose are as high as we have ever seen here, but the forest is really in bad condition. On the west end of the island, there is a lot of balsam fir, the moose's principal winter browse. But it has all been eaten away to heights moose can reach. Balsam fir only live about a century, and then they die. On the western end of the island, the old firs are starting to die, as they would naturally, but no new firs are coming up to replace them: moose have nipped the new growth in the bud. On the eastern side, there are regenerating trees, but the larger trees are browsed from the ground to the upper reach of a moose's mouth. Moose calves are in poorer and poorer condition. They are generally small and more vulnerable to wolves." There are fewer twins.

Most people predicted that, as the moose recovered from their earlier decline in the 1970s, wolf numbers would also increase, on the theory that more moose meant more food for wolves. But that hasn't happened. Peterson's theory is that the moose population is dominated by vigorous young adult moose that were born in the 1980s. Few moose between the ages of two and eight are taken by wolves. Numerically, more calves succumb to wolves, and most of the prey biomass comes from old moose. If food is the limiting factor, Peterson believes that as the present cohort of moose age, the wolf population will bounce back. "If we look at the number of old moose and the number of wolves in the past," he says, "there's a remarkable correlation, and that's reasonable, because that's what wolves eat. In the next few years, there's going to be an amazing accumulation of old moose. If the wolves are food-regulated, they ought to turn it around." By the end of 1993, however, the wolves hadn't turned it around.

The third—and, some think, most likely—reason for the wolf decline is genetic. If the wolves don't come out of the decline, the reason may be that they are too inbred to survive. Inbreeding is potentially destructive to any species, because it increases the likelihood of expression of recessive and nonadaptive genes and because, continued for generations, it makes individuals more and more alike. Nature punishes uniformity—identical individuals are identically vulnerable to disease or changes in food resources or new competitors. Variation among individuals is a way of storing resources for use against such changes.

Cheetahs in Africa suffered a major population decline as human settlement expanded in the nineteenth and twentieth centuries. The reduced population is so inbred that it has less than a tenth of the genetic variety of humans or domestic cats; cheetah genetic resources have been compared to what one would find after ten to twenty generations of brother-sister matings. Immune systems within a species are normally diverse, but skin grafts from one cheetah to another are not rejected—an indication that they have identical, or nearly identical, immune systems. The consequences of such uniformity can be grim. In some zoos, 70 percent of cheetah cubs die without reaching maturity. When an epidemic of feline infectious peritonitis ran through a captive population in an Oregon wildlife park, it was hardly noticed in the lions, but it infected all the park's forty-two cheetahs and proved fatal to twenty-five of them. Inbreeding usually reduces a species' reproductive success, too. Abnormalities in sperm cells begin to proliferate. Studies have found 70 percent of cheetah sperm to be abnormally formed, compared with only 29 percent of domestic-cat sperm.

This danger is also faced by many other wildlife populations, as humankind cuts their habitat into small disconnected biological islands. Lions, for example, once occurred more or less continuously across Africa and Asia. By the end of the nineteenth century, the Asiatic lion had been hunted down to perhaps a last twenty-five individuals. All their descendants live in a small protected habitat in India, where no other lion genes enter their prides. Ngorongoro Crater in Tanzania has an isolated population of lions that was reduced to about fifteen individuals by a biting-fly infestation. They now number about 110 individuals, but the genetic consequences of the bottleneck are still with them: they have only a third of the genetic variety of the lions of Serengeti, a few miles away.

Both the Indian and Ngorongoro lions have high rates of abnormal sperm and low testosterone concentrations. In the United States, black-footed ferrets have as little genetic variety as cheetahs and are known to be very susceptible to canine distemper. Bighorn sheep in the United States are thought to be much more prone to pneumonia and respiratory infections after population reductions.

Wolf reproductive strategies allow for a certain amount of inbreeding. Among wolf packs, as discussed earlier, typically only one pair reproduces, and those may be closely related individuals. Says Peterson, "In Kenai, Alaska, we had a grandfather and a granddaughter mate. They didn't know they were related, but we did. They did fine. Their pups survived. And when the female died, the grandfather sired another litter with her sister." In natural settings, such inbreeding does not go on for long. Eventually, the dominant wolves will cease to breed, and a disperser will come along to bring something fresh to the pack's gene pool.

But there is a boundary around Isle Royale. The ice bridge crossed by a pair of wolves in 1948 does not form in most years, and to get to it wolves must cross a region now densely populated by humans. There has been no immigration of new breeding stock from the Canadian mainland since 1948, and the wolves on the island are therefore thought to all descend from a single pair. For thirty-two years, the Isle Royale wolves increased in number, spreading out into three or four packs to occupy different territories on the island. But all the while, they may have been losing genes. As a single pair breeds, it passes some genes on to one offspring, others to another. Different offspring carry different genes. And since not all offspring breed, some genes are lost.

Inbreeding can correct itself if new genes enter the population from the outside or if enough of the offspring breed to pass on most of the genes. "Here," says Peterson, "we had probably two founders and never a large population. Here, given the number of breeders, we would estimate we would lose 15 percent of the genetic variability with every generation. Maybe what held this population together so long was that the original breeders lasted so long. The number of new breeders that came on line were kept to a minimum." But by 1980, perhaps the loss of genes began to tell. "Certainly," says Peterson, "the genetics are a problem."

It's not easy to confirm that inbreeding is the cause of the decline.

Says Peterson, "We could go one step further and look at male sperm. To do it would mean you'd have to do it during the breeding season. In February, you'd have to dart them from helicopters, and after that, these wolves would regard you differently forever. You get basically one crack at the animal. It would change the study techniques." And even if the sperm samples showed no abnormalities, that would not eliminate the inbreeding hypothesis. "It may not be in the males," says Peterson. "It may be in the females." Perhaps they are miscarrying in their dens, or they are infertile. There are too many possibilities, and too few wolves with which to test them. Peterson says, "It's premature to act on the basis of what we know about wolf genes right now."

If inbreeding is not the cause of the population crash—if the cause is the relative invulnerability of a population of moose that consists of adults too old to be easy prey but too young to be weakened by ticks and tapeworms and other infirmities that eventually make them vulnerable, or if disease has caused a temporary setback—the wolf population could rebound. Theoretically, an increase up to a hundred animals would make the population viable again. In the absence of a compelling case for one explanation or another, says Peterson, "the present course is just to let things go, whichever way they will go."

If the Isle Royale wolves do go extinct, says Peterson, "I'm guessing it will take a long time—we might be looking at ten or twenty years. From my personal standpoint, that's bad, because I'd like to see the thing resolved sooner. And it would be bad for the moose."

What is going on at Isle Royale may be a preview of what is to come for wolves everywhere. Two thousand years ago, wolves ranged over most of Eurasia and North America. They inhabited tundra, taiga, temperate forest, prairies, and mediterranean and subdesert environments. They exploited deer, elk, moose, bighorn sheep, bison, musk oxen, and caribou in North America, and red deer, moose, reindeer, chamois, and saiga antelope in Eurasia. In this century, scientists have recognized thirty-six separate subspecies worldwide, ranging from the small *Canis lupus arabs* of Saudi Arabia to the large *Canis lupus pambasileus* of interior Alaska. But with the growth of human populations and the replacement of wild ungulates with domesticated cattle, sheep, and goats, wolves have been steadily pushed into a smaller and smaller range.

Devoted as its inhabitants were to the keeping of sheep, the British Isles led the way. In the tenth century, King Edgar tried to eliminate wolves from Wales by imposing a tax of three hundred wolf skins on the Welsh King Ludwall, and granting amnesty to rebels who brought him a hundred wolf heads. The last English wolf had been killed by 1500. Wolves persisted in Scotland for another 250 years. In the reign of Mary, Queen of Scots, it was the duty of all men to participate in at least three wolf hunts a year. Scottish wolves were probably affected less by organized hunts than by the removal of the forest to make sheep pastures. Deprived of its natural prey, the wolf was forced to live near intolerant humans. The last one was captured in 1743. In Ireland, the last wolf was killed in 1770.

Denmark killed its last wolf in 1772, but France and Germany had remote mountain forests in which wolves survived much longer. From the time of Charlemagne, the French appointed military officers to lead organized wolf hunts, but wolves persisted in France until about 1940. In 1992, a wolf is thought to have appeared in the French Alps, possibly having walked in from Italy. The Germans hunted wolves even more assiduously than the French. In 1817, more than a thousand wolves were killed in Prussia alone. By the mid-nineteenth century, wolves were rare, and by the middle of the twentieth, they were presumed extinct in Germany. Germans had so little experience of wolves that in 1976, when eight wolves escaped from a research project in Bavaria, the neighboring people panicked, and troops were called out. In six weeks, seven of the escapees were shot. The remaining female was at large for two years; when she was finally shot, in July 1978, she had a litter of pups, proof that there was at least one wild wolf still out there.

In Spain, wolves survived through 1970 in Galicia and Castile-León. However, two children were purported to be victims of a wolf attack in 1974 near Orense, and that led to renewed persection. About a hundred wolves were killed within a few months. Wolves still survive elsewhere in Spain and in Portugal, however, and over the decade of the 1980s, their range seemed to be increasing.

Wolves once ranged down the spine of Italy and inhabited Sicily. By the beginning of the twentieth century, they had vanished from most of the country. In 1973, there were perhaps no more than a hundred surviving in the Apennines. Restricted to mountains and to farmlands around Rome, they were protected in 1977. They have since increased to between 300 and 350, but through most of their range there is no native prey, and they live off of garbage, discards from butcher shops, and occasional forays into livestock pastures. They sometimes kill and eat dogs, and that makes them the target of hunters, who put out poison for them. And the wolves interbreed with domestic dogs. It is feared that the entire Italian population may now consist of wolf-dog hybrids.

Wolves were hunted out of the Scandinavian countries, and survive there today only because they disperse from the countries of the former Soviet Union. There are thought to be only about a dozen each in Norway and Sweden, and perhaps a hundred in Finland. They came back into the Scandinavian countries only after clear-cutting in Soviet Karelia opened the habitat to moose; when the moose population increased, so did the wolf population, whose dispersers sometimes headed west. Despite their rarity, it is still legal to hunt wolves in Finland.

The countries of the former Soviet Union once had enormous populations of wolves. During the world wars, wolf controls slackened and wolf populations grew. In the late 1940s, the entire Soviet wolf population was estimated at 200,000. Through the decade of the 1950s, intensive hunting, poisoning, and aerial gunning took 40,000 to 50,000 of them a year. The Soviets were headed toward the kind of eradication that took place in the United States, but in the 1970s, owing to a surge of conservation spirit, control efforts waned. In the middle of the decade, wolf numbers increased, following which there were renewed efforts to control wolves; in 1979, more than 32,000 wolves were killed. Soviet researcher Dmitri Bibikov feared that the population might drop below 50,000, and that wolves might soon disappear from half their former range. A 1990 population estimate, however, put the number at about 70,000. Wolf populations in the former Soviet Union today seem healthiest in the Caucasus, Central Asia, and Siberia, where wolves are not very accessible to aerial hunting. With the disintegration of the

Union of Soviet Socialist Republics, it may be decades before an accurate tally of wolves can be made in the region.

In Eastern Europe, wolves have tended to survive in the border areas, thanks in part to the statesmen's habit of putting borders where there are the fewest people and then keeping new settlers away from them. Wolves could repopulate Germany from the border areas of Poland and the Czech Republic. Poland has healthy wolf populations in the Carpathians; near Biescady National Park, hunters in recent years have harvested wolves at a rate of one wolf per ten square kilometers. Czechoslovakia was thought in 1992 to have about 35 wolves. The border regions of the Carpathians are the stronghold of wolves in the Balkan States. In Rumania, where wolves are still shot for the bounty, there continues to be a sizable population. In Yugoslavia, there were an estimated 2,000 before the civil war began, with the densest populations in Serbia and Macedonia.

Estimates of Greece's wolf population range from 300 to 3,000. In the 1960s, hunters and shepherds were killing an average of 700 wolves per year, and there was a state bounty on wolves until 1980. To this day, the government sanctions organized hunting and the use of poison. The most important limit on Greece's wolf populations might be the wild game they feed upon, which is disappearing as human populations expand.

Wolves can still be found in most parts of Turkey. In 1973, more than 1,000 wolf skins were sold in shops in Istanbul. But as the Turkish population grows, there is, as in Greece, less and less wild game left for wolves to prey upon. In Iran, there are at least 1,000 wolves. They are found all over that country, but the healthiest populations are in the northern mountains. There are wolves in Iraq, Afghanistan, and Pakistan, but their populations are not well studied. India has an estimated 1,500 wolves, most of them in the mountains of the north and west.

There are 110 to 150 wolves in Israel, and most of them have to feed, in the absence of wild game, on livestock and garbage. Wolves have been protected in Israel since 1954, but farmers nevertheless poison them when one gets into livestock or chickens. There are an estimated 30 wolves in Egypt, 10 in Lebanon, 200 in Jordan, and between 200 and 500 in Syria.

There are wolves in northern China, but no survey has ever been done to estimate their numbers. Mongolia, believed to have at least 10,000 wolves, also has the highest ratio of livestock to people of any nation; that poses conflicts between humans and wolves. In the 1980s, Mongolia exported nearly as many wolf skins as the top two exporters, Canada and the Soviet Union.

Canada is likely to be the best hope for the species. An estimated 50,000 wolves survive there. They never existed on Prince Edward Island or the Queen Charlotte Islands, and were extirpated from New Brunswick, Nova Scotia, and portions of southern Ontario and Quebec by the turn of the century, and the last Newfoundland wolf was killed in 1911. But across the north and west of the country, despite repeated organized government wolf-control programs aimed at increasing moose and caribou populations for the benefit of hunters, there are wolves.

There are thought to be 4,500 wolves in the Yukon, 10,000 in the Northwest Territories, 8,000 in British Columbia, 4,300 in Alberta, 4,000 to 6,000 in Saskatchewan, and 8,000 to 9,000 in Ontario. Trapping takes a smaller and smaller portion of the population each year, as prices for wolf pelts decline: 7,000 were trapped in 1982, but only 2,000 in 1990. Hunters take approximately 2,000 wolves a year in all of Canada. There are year-round hunting seasons on wolves in the Northwest Territories, Ontario, and British Columbia, and limited seasons in Quebec, Manitoba, Alberta, and the Yukon. Though in some areas wolves may be overharvested, the annual hunting-and-trapping take ranges from 4 percent to 12 percent of any province's collective population. Despite hunting and trapping, populations in British Columbia and southern Alberta, which have the highest harvest rates, and in Ontario and Quebec all seem to be increasing, rather than decreasing.

Wolves are legally protected on only 2.7 percent of Canada's land area, but even that figure is misleadingly high: Canadian national parks established since 1978 are open to native hunting. John Theberge calculates that wolves are actually protected on 1.2 percent of Canada's land area, a total of about a dozen places, which average only five hundred square kilometers apiece. "In those areas," says Theberge, "there may be only a few packs whose range doesn't overlap park boundaries. When you add it up, that's nothing." Moreover, existing protections are not ironclad. In 1994, Thelon Game Sanctuary in the Northwest Territories was off limits to hunting, but it had been proposed for national-park status. If it became a park, natives would be permitted to hunt wolves. "Thelon," said Theberge, "is over half the protected range of wolves in Canada."

The United States engaged in the most organized eradication effort of any nation. Bounties were paid for wolves, and poisons put out for them. Meanwhile, wolf habitat filled with people and emptied of game. By the twentieth century, the wolf was extinct in most of the states east of the Mississippi. By 1970, populations remained only on Isle Royale and in Minnesota and Alaska, with occasional migrants from Canada seen in Montana and Idaho. Several of the subspecies are no longer to be found anywhere in the wild. *Canis lupus monstrabilis,* the wolf of West Texas, and *mogollonensis,* the wolf of the southwestern mountains, are extinct, and if *nubilus,* the buffalo wolf of the Great Plains, survives at all, it does so only in captivity. *Canis lupus baileyi,* the Mexican wolf, has not been seen in the wild in the United States since 1970.

To stop the decline, the U.S. Department of the Interior has invoked the Endangered Species Act. The eastern timber wolf and the red wolf were both listed as endangered in 1967. The Northern Rocky Mountain subspecies, *Canis lupus irremotus,* was listed as endangered in 1973, the Mexican wolf in 1976. In 1978, the Fish and Wildlife Service stopped classifying separate subspecies and extended protection to all gray wolves in the contiguous forty-eight states. The protections thus afforded under the Endangered Species Act provide stiff penalties for anyone who kills or harasses a wild wolf in the United States outside of Alaska, and generally require permits from the U.S. Fish and Wildlife Service for those who wish to keep wolves in captivity. The protections have had the effect of stimulating a modest recovery of wolves in the lower forty-eight states.

About seventeen hundred wolves survive in northern Minnesota, where their numbers have been maintained by immigrants from Canada, and where they are classified as "threatened." With protection, wolves have begun to reestablish themselves in northern Wisconsin and northern Michigan. In 1994, there were thirty to thirty-five wolves in Michigan's upper peninsula, and at least fifty wolves in fourteen packs in Wisconsin. And wolves have been turning up in increasing numbers in Montana, Idaho, and Washington, as changing hunting and forestry practices and warmer winters in British Columbia and Alberta increased survival rates there. In Alaska, there are believed to be some seven thousand wolves, despite an annual hunting-and-trapping take of more than a thousand. The wolf population is believed by the Alaska Department of Fish and Game to be stable or slightly increasing.

Before humans made war on wolves, the geographic reach of a wolf's genes must have been impressively long. The pattern today, however, is ubiquitous and clear. Everywhere human habitation spreads, wolf populations decline. Richard Thiel, working in Wisconsin, found that, wherever there was more than .58 kilometer of road per square kilometer of habitat, wolves failed to survive. It wasn't just the presence of roads, but of humans, who shot, snared, trapped, or ran over the wolves. David Mech, Steven Fritts, Glenn Radde, and William Paul did a similar comparison of road density and wolf populations in Minnesota and got comparable results.

The grave threat is that eventually there will be broad areas without wolves that will serve as barriers to the dispersal of wolves and the sharing of genes. Even in Canada and Russia, where the largest expanses of healthy habitat remain, there is some threat that ultimately wolf populations will grow isolated from one another, broken into biological islands, and winnowed down to replicas of Isle Royale. The last wolf won't succumb to a bullet. It will weaken from the slow but inexorable loss of genes. It will die of uniformity.

If the wolves vanish from Isle Royale, Peterson unhesitatingly believes they should be reintroduced. Not everyone agrees. Says Peterson, "People say they got here on their own, they went extinct on their own, let 'em come back on their own. I've had a lot of trouble with the word 'natural,' as in 'Let them come back naturally,' lately. Their chances are affected by the fact that there are a hundred thousand people on the north shore of Lake Superior."

This is a difficult issue for the Park Service. For several decades, Isle Royale was the only United States national park outside Alaska to have wolves, and Isle Royale has become known especially for its predators. The Park Service has viewed the fact that the wolves came on their own as a key part of the story, and they have kept human contact with wolves to a minimum. Says Peterson, "For the Park Service, I think it has been an important place, because it is sort of the ideal of what parks aspired to be ecologically. For a lot of people, just knowing wolves are out here that aren't being heavily manipulated is an important thing. Isle Royale seemed to be the best possible example of letting natural management work things out."

But there is a second question operating here. One must ask not just what the wolves need, not just what the Endangered Species Act seems to demand, but what the land requires. Should we save the species for the sake of keeping types of animals? Or should we save the function they perform as part of the fabric of life? Peterson believes we ought to be looking at the ecosystem, not just one of the parts. "I think it would be irresponsible not to bring wolves back. We've got so many other parks with unrestrained ungulates. We know what's going to happen. Here, the moose will continue to eat the place to pieces. The balsam fir will disappear. This place is renowned for being a park with predators. Without predators it would be known as a place run down by moose."

Deciding on reintroduction would be very difficult for the Park Service. "If we had to put wolves back here, we'd have to decide what brand to use," says Peterson. Where would the new ones come from? Since genetic analysis has suggested that Isle Royale wolves are less closely related to those on the mainland shores than they are to wolves from much farther north, there is some question about whose genes these are.

One might simply reintroduce wolves from several different locations, but the individuals that prosper might do so because of momentary advantages, not because of long-term fitness. "I would tend to do it the cheap way," Peterson says. "Take some wolves from Minnesota that are due to be euthanized because of cattle depredation, examine them for disease, vaccinate them, and get them over here quick. They wouldn't necessarily be moose killers, but some would probably make it."

If the wolves vanish, the National Park Service may not have the luxury of depending on "natural management." Faced with the disappearance of balsam fir and perhaps moose, and who knows what else in the chain of consequence, they will have to make a decision. "We've got to decide what species we want here and what species we don't," says Peterson, "and that decision hasn't been made at Isle Royale before. There isn't a tradition of mulling these things over. Those easy days are gone."

The easy days are gone for wolves in general. As human activity continues to spill out into wolf habitat, wolves are increasingly going to be confined in biological islands. The reintroduced populations of red wolf in the American South are island populations that are likely to require a support program of captive breeding to avoid genetic depression. Reintroduction of Mexican wolf in the Southwest is likely to require similar support.

At Isle Royale, Peterson is anything but confident. "If I were a betting person, I wouldn't bet on this one," he says. "They are so flat; they're not doing what wolves do. One pair hasn't reproduced since 1988, but they've got eight hundred moose in their territory. All the experts are saying, 'We told you so.' It would be a big thing if they turned around." It would show that, at least here, a small population is viable.

At the Rock Harbor dock on Isle Royale, a ranger naturalist is giving a nature talk to visitors who are awaiting the arrival of the afternoon ferry to Copper Harbor. It has been raining, and the campers on the island are gloomy about prospects for summer ease on the long Fourth of July weekend. The talk, too, is gloomy: the ranger is explaining the decline of Isle Royale's wolves. Two young children suspend between them, like a clothesline, the sagging skin of an Isle Royale wolf. They are oddly unenthusiastic about the honor; perhaps it is a comment on the intensity of life in a living wolf that this relic seems not to elicit much excitement. An adult camper has asked why the Park Service doesn't deal with the population crash by transplanting more wolves to the island. The ranger tells them that wolves are territorial and that the new wolves might not get along with the existing wolves: "They might kill each other." Perhaps more important, she adds, "The Park Service has a hands-off policy," feeling that "natural management" or management without human intervention is the best course because it offers the best chances for authenticity and for health. Besides, the research question of what is happening to the wolves couldn't be answered if new wolves were brought onto the island.

Off in his cabin, Peterson ponders the likelihood of solving that question. Outside the cabin, the rain continues to fall. The ragged mist of clouds hugs the lakeshore. Raindrops slip quietly from the trees into the duff of the forest floor. It is achingly silent.

8

WILD ENOUGH FOR WOLVES

On a hot May afternoon, Michael Phillips of the U.S. Fish and Wildlife Service is checking a trap line near the newly designated Pocosin Lakes National Wildlife Refuge in North Carolina. The landscape is a tangle of maple, gum, bay, oak, and pond pines, all crowded so closely together that one must cut a way through with a machete or stay on the road. The road has been made by dredging up muck from the swampy ground and piling it into a causeway. Water lies so close to the surface that any digging makes a lake, and the trenches dug beside the roadways are blackwater canals that reflect the riotous greenery. Flies and mosquitoes buzz and whine in the air, and large black swallowtail butterflies dart between the thickets. Green herons squawk from the edge of a canal. A huge blacksnake glides across the road.

Phillips is looking for a female radio-collared red wolf that has drifted off the neighboring Alligator River National Wildlife Refuge, where since 1986 Phillips has been releasing captive-bred red wolves.

It is part of an attempt to restore a creature that has been absent from this swampy world for nearly two centuries.

The collar on the wolf Phillips is seeking needs a new battery. He checks the trap line daily, and this has already been a long day. After driving more than ninety miles from the refuge office in Manteo, he has checked one trap after another, and found nothing. In the heat, his near-shoulder-length blond hair sticks in ringlets to the side of his face.

As he approaches one of his sets at the edge of a canal, he sees something standing in the water. He cannot see clearly what the dark, spidery thing is. It is some kind of canid, but it is emaciated, black, and almost hairless. Caught in Phillips' trap, it has hobbled into the canal, where the drag hook has stuck. The animal is unable to lie down, and it stands helplessly in the shallows, staring vacantly, holding its head low. As Phillips gets closer, the animal offers no resistance. He easily restrains it and pries open the jaws of the trap. "This guy looks poor," he says, "but he's alive. Rascal is alive."

It is a coyote—its ears are too big and its snout is too narrow for a red wolf—but Phillips has to look closely to tell. Wet and muddy, it has lost 90 percent of its hair to mange, and probably suffers as well from heartworm, hookworm, and other intestinal parasites. Its sun-blackened skin hangs loose over its bones. When Phillips has eased the creature into a portable kennel, it sags listlessly. With its leathery skin and protruding bones, it looks more like a large fruit bat than a coyote. The eyes are dull and lifeless, and there seems to be a brown spot on the edge of one iris, a condition Phillips knows is common to such dog breeds as the German shepherd. He suspects it might be a coyote-dog hybrid.

For Phillips, questioning identities is reflexive. Biologists debate whether red wolves are a separate species of wolf, a subspecies of the gray wolf, or a hybrid resulting from the matings of wolves and coyotes.* Phillips leans to the view that it is a subspecies of gray wolf,

*A species is a genetically distinctive group of natural populations that breed with one another and are reproductively isolated from all other such groups. A subspecies is a group of natural populations within a species that differ genetically and are reproductively isolated from others of the species because of geographic barriers. Before genetic studies made it possible to compare genes between indi-

but if he were to say so at a scientific gathering he would be sure to start an argument.

The red wolf is the most puzzling of wolves. It was originally described in 1791 by John Bartram, who regarded it as a subspecies of gray wolf and gave it the name *Canis lupus niger.* In 1851, Audubon and John Bachman held that there was a gray wolf in the North, a black wolf in Florida and the Southeast, and a red wolf in Texas and Arkansas. In 1898, Outram Bangs designated the Florida wolf a separate species, *Canis ater.* In 1905, Vernon Bailey recognized the red wolf of Texas as a separate species and gave it the name *Canis rufus.* In 1937, Goldman combined the Florida and red wolves into a single species, *Canis niger.* But twenty years later, the International Commission on Zoological Nomenclature changed the name to *Canis rufus.* In 1967, Barbara Lawrence and William Bossert of Harvard University argued that the red wolf was a subspecies of the gray wolf. In 1970, David Mech held that the red wolf most likely originated from the crossing of gray wolves and coyotes. In 1972, Ronald Nowak, endangered-species coordinator for the U.S. Fish and Wildlife Service, examined historical and fossil specimens and concluded, "The red wolf is the surviving stock of the basic progenitor of all wolves. And it originated right where it survives today, in the southern United States." He believes the red wolf's descendants migrated north, crossed the ice-age land bridge into Asia, and there evolved into the gray wolf, which later migrated back into North America.

What makes the argument difficult is that by the early twentieth century the red wolf had vanished from the Southeast and the gray wolf from the Northeast. Nowak is a morphological taxonomist: he classifies animals on the basis of their skeletal dimensions, especially their skulls. Few skulls or other remains are available with which to appraise the identity of southern wolves. By 1970, the red wolf survived only in parts of Texas, Louisiana, and Mississippi. There are no

viduals, museum collectors judged the differences between species on the basis of such physical characteristics as size, color, and precise details of the teeth. Collectors noted that individuals of the same species might vary in size or color or other characteristics in different locations, and applied the term "subspecies" to such different forms. The term "race" is synonymous with "subspecies."

red-wolf remains from Virginia, North Carolina, South Carolina, Tennessee, or Kentucky. And there are few historical specimens of gray wolf from the eastern states. Nowak says he knows of only three complete skulls of gray wolf, and a half-dozen of red wolf, surviving from the United States south and east of Lake Michigan with proof of the time and place they were collected. "All the wolves were killed off before anybody gave a darn," he says. "They were just obliterated before anybody was interested in collecting."

Edward Goldman identified one historic Florida specimen as a gray wolf, but Nowak and Barbara Lawrence argue that it was a red wolf. The identification of this one specimen is an important question. Says Nowak, that is the only evidence in the literature of a gray wolf in Florida within historical times. If gray wolves didn't exist in the South, then red wolves couldn't be hybrids. And with perhaps only three specimens of red wolf from Alabama, one from Mississippi, and one from Florida to judge by, there is little baseline data upon which to argue the issue of hybridization.

Given so little historical evidence, the argument has turned on the relative sizes of coyotes, red wolves, and gray wolves. Do their average dimensions overlap? And if so, do they overlap because wolves and coyotes have hybridized, or because gray wolves, red wolves, and coyotes have been shaped by the various prey species they hunt?

The red wolf is almost invariably described with reference to the larger gray wolf or the smaller coyote. It has a thinner muzzle and a narrower head than the gray wolf, and its ears, though the same size as those of a gray wolf, appear larger by virtue of its smaller head. It has longer legs than a coyote. Observers sometimes think they are seeing coyotes because of the bigger ears and pointier nose. Even its behavior is described as intermediate between wolves and coyotes. For example, zoologist Vernon Bailey in 1907 held that the voice of the red wolf was "a compromise between that of the coyote and lobo, or rather a deep voiced yap-yap and howl of the coyote. It suggests the coyote much more than the lobo."

But there seem also to be clear distinctions between red wolves and coyotes. Color often distinguishes red wolves: they usually have a reddish tinge to their coats, and strong white highlights around their black lips; on many, the back is drably coyotelike, but some individuals are black and others have dark patches, like forest camouflage. At Alligator River, red wolves kill and eat deer, while coyotes tend only to feed on deer fawns and hunter-killed adults. The red wolves at Alligator Refuge also eat raccoons, which coyotes rarely attack. The red wolf forms packs, and is so intolerant of strangers that five of the animals released in reintroduction efforts have been killed by other red wolves. Coyotes, on the other hand, disperse before their second summer and seldom kill one another.

Before European colonization, coyotes were absent from the wooded areas east of the Mississippi. They inhabited only the fringes of wolf range, the open country of the American West and Mexico. Coyotes moved into wolf range after humans replaced the forest with fields and pastures and eliminated the large prey species that wolves depended upon. They arrived in Minnesota in 1875, southern Ontario in 1890, Pennsylvania in 1907, Isle Royale in 1912, Alaska in the 1920s, Tennessee in 1930, and Massachusetts in 1936.

Generally, wolves seem to view coyotes as competitors and kill them. The coyotes on Isle Royale disappeared three years after the wolves arrived. John Weaver found that one wolf pack in Jasper National Park killed four coyotes in a single month. But when dispersing wolves find no other wolves to mate with, they will interbreed with coyotes. In 1951, Stanley Young reported that two of the wolf specimens in the Royal Ontario Museum were coyote-wolf hybrids. In 1971, George Kolenosky, of the Ontario Fish and Wildlife Research Branch, reported that in the two preceding years a captive Algonquin Park wolf had mated with a captive York County, Ontario, coyote and successfully reared litters. Biologists presumed that if individuals mated across species lines the offspring would—like the mules that result from all but a few horse-donkey matings—be sterile. But Kolenosky's wolf-coyote hybrids proved to be fertile. The finding fueled suspicions that wolves were interbreeding with coyotes near Lakes Superior and Huron, where there were reports of a smaller race of wolves. This speculation was not confirmed until 1992, when Dr. Robert Wayne, of the University of California at Los Angeles, showed through molecular-biological tests that wolves around Lake Superior had coyote mitochondrial DNA. That meant that male wolves were mating with female coyotes.

There is no doubt that, by the middle of the twentieth century, red wolves and coyotes had hybridized in Texas. Howard McCarley, a biologist at Austin College in Sherman, Texas, began to wonder about red wolves in the 1950s. As he traveled around Texas, he would look at red-wolf and coyote carcasses ranchers hung from their fences, and he noted that most of the red wolves seemed to resemble coyotes. McCarley concluded that coyotes had replaced red wolves in Oklahoma, Arkansas, and eastern Texas. By 1964, he recognized that the red wolf was in a precarious position.

Roy McBride, who was trapping coyotes and red wolves in Texas ten years later, recalls, "When the coyotes came in, it was love at first sight. Hell, they were inseparable. You'd see tracks of a wolf and two coyotes traveling with it. You'd take pups out of a den, and one would grow up to be a sixty-eight-pound male and the other would grow up to be a thirty-five-pound male. They weren't coyotes. They weren't like wolves, either. They had strange habits. They killed small animals. They lived on nutrias. There were cows all over the place, but they didn't take them."

The question today is whether hybridization occurred all along or only after human persecution reduced red wolves to small numbers and coyotes moving into the emptying niche began to breed with them. Nowak found that specimens collected from west of the Mississippi River before 1930 were larger than those taken after 1930, and took this to indicate that the red wolf had hybridized only recently. He believes, with biologist Ernst Mayr, "By far the most frequent cause of hybridization in animals is the breakdown of habitat barriers, mostly as a result of human interference." He concluded, "Hybridization with the coyote did not begin until about a century ago, and the gray wolf was never involved."

When the wolves were first brought to Alligator River National Wildlife Refuge in North Carolina, there were no coyotes known to be in the area. Coyotes weren't present in the southern states when Europeans arrived. They began to turn up again in the 1920s and 1930s, after white settlement had eliminated wolves and pushed back the forest. The pattern of their reappearance does not suggest an enterprising canid gradually expanding into a niche vacated by the extinction of wolves: the coyotes appeared quite suddenly, here and there. For example, they appeared in Florida in 1925, but in Georgia not until 1929. They turned up in 1924 in South Carolina, but in North Carolina not until 1938. They were in Maryland in 1921, but in Virginia not until 1947. Almost certainly they were released in most of these places by humans. Possibly they were captives whose owners grew disenchanted with them, or they had been kept by hunters and were released to train chase dogs. Today, some hunters raise coyotes for just such purposes.

In 1986, a coyote turned up near Pungo Lake, west of the Alligator River Refuge. In that same year, two gray wolves and a cougar were released in the area by parties unknown. The cougar was shot and found to have an identifying tattoo in one ear. One wolf was shot by a Fish and Wildlife Service employee after it walked into his yard; on the basis of tooth wear, it was judged to have spent considerable time in captivity. Three coyotes were found in the area as well. The whole menagerie could have been a single release, perhaps the work of an owner of exotic pets who decided his animals would all be happier running around in the wild. Or it may have been an attempt to monkey-wrench the release of captive-bred red wolves. Says Michael Phillips, "I don't think it's a coincidence that they were released just when the [reintroduction] project was getting under way."

The Alligator River National Wildlife Refuge is the only refuge in the United States that permits hunting of deer with chase dogs. It does so because hunting with dogs is a tradition in these thickets, and because the reintroduction of red wolves depends on the support of local hunters, who would be a lot less cooperative if wolf reintroduction led the government to ban hunting. Hunting has been curtailed in parts of the refuge to protect black bears and migratory waterfowl, and while some hunters erroneously attribute the restriction to wolf reintroduction, so far as is known, none has gone after the wolves. One wolf drowned after being taken in a leghold trap, but it was not clear whether the trapper intended to take a wolf (illegally) or take a muskrat (legally). Another was shot by a man who claimed he mistook it for a wild dog. And whether the coyotes that have been showing up in greater numbers since 1986 are an act of retribution or just another dumping of exotic pets into the wild, no one is saying.

There is a risk that coyotes will interbreed with wolves, as they have in Texas and in the Great Lakes region. If they do, the mixture of genes could make red wolves ineligible for reintroduction under the Endangered Species Act, for a series of solicitors' opinions have held that hybrids are not protected by the act. Biologists say hybridization may slip the wolf in some way from the harness of nature and make it less fit. Others feel that, since the coyote arrived by human agency, hybridization would taint the wild wolves with human carelessness and purpose, would take away their wildness—would mean, in spiritual terms, that the wolves were no longer truthful.

The spectral coyote Phillips has caught worries him. In recent weeks, a male wolf, number 505, has been seen in the company of coyotes. If it mated with a coyote, that would mean the Alligator River red wolves had been tainted. The male coyote Phillips has just found is in an area being frequented by a female wolf. So far, all the known matings of wolves and coyotes in the wild have been between male wolves and female coyotes, and it is not clear whether male coyotes and female wolves can mate: it may be that female wolves are too tall or too aggressive for male coyotes. Phillips doubts the female wolf could be ignorant of the coyote's presence, and he wonders why she didn't kill it. He wonders what would happen if he took this coyote to a veterinarian and had it cleaned up and cured, sterilized, fitted with a radio collar, and then released it.

He lifts the portable kennel into the back of his truck and drives the ninety miles back to the Roanoke Island Animal Clinic in Manteo. The next morning, Phillips is on the phone with Gary Henry, red-wolf coordinator for the U.S. Fish and Wildlife Service in Atlanta, to ask for permission to rehabilitate and release the coyote. Says Phillips, "Part of putting a program together here that's doable means understanding how coyotes and wolves interact. I think there's going to be a problem with these guys maintaining themselves in the presence of coyotes. We can put out trap lines and give it a good go, but there's certain things that can't be done. The service is going to have to accept certain issues we can't control. I think they may have to accept a certain low level of natural hybridization, and that may not be bad."

Henry refuses. He doesn't want to risk the intrusion of coyote genes into the wild-wolf population. It could bring the whole reintroduction effort to a halt. In the end, the question will be moot: despite antibiotics and veterinary care, the coyote will die before the next morning.

The sick, spidery coyote Phillips caught poses an important question: is the world wild enough for wolves? A creature is not a wild creature unless it is being polished by the evolutionary forces that originally designed it. The wolf was shaped by unfenced landscapes, abundant prey, and the freedom to pursue it. We have added roads and radio collars and exotic animals to the forces of evolution. Have we so carved the world into the geometric shapes of possession, liability, privacy, and commercial haste that it cannot abide wolves?

A great deal of effort has gone into the reintroduction of red wolves. In 1967, the red wolf was designated an endangered species. When Curtis Carley became project leader for the red-wolf recovery program in 1974, he assumed that there was a pure red-wolf population in the wild and that recovery would mean protecting it there. But in 1972, Roy McBride and Glynn Riley, having been sent out by the Fish and Wildlife Service to search for red wolves, estimated that most of the survivors were in extreme southeastern Texas and southwestern Louisiana, and that even they were hybridizing. Once it appeared that there was no population of red wolves free from the threat of hybridization, recalls Carley, "we had to reorganize and rethink it. It was grab what information you can quickly. We didn't have time for extended study. In July 1975, we got authorization to capture the remaining genetically pure red wolves. Removal was problematic, because we, for all practical purposes, were making this animal extinct in the wild, and that can only be justified if you intend to put them back into the wild."

They set their traps on the raised-board roads leading to oil platforms and the dredge-spoil cow walks of marshy pastures. The last wild red wolves were taken from a marsh near an industrialized section of Galveston, Texas, in 1980. They were sent to the Point Defiance Zoo in Tacoma, Washington.

There was concern about the purity of the captives. Says Carley, "We caught four hundred canines down there from late 1974 to mid-1980. The majority of those were coyotes or obvious hybrids." He looked also at animals already in zoo collections which curators touted as additional candidates for the captive-breeding effort. Many of these animals, Carley saw, were hybrids. To protect the breeding stock, Carley eliminated all but forty-three of the purported red wolves. As he and his staff continued to breed the forty-three and look at the offspring of each pairing, they decided that most of them were coyote crosses, too. They eliminated all but seventeen individuals, which were deemed to be pure red wolves. Three of these never bred, so today the entire captive-breeding population of red wolves descends from fourteen individuals.

For years, all the known red wolves in the world were in captivity. And most of them were at the Point Defiance Zoo's Graham, Washington, red-wolf breeding facility, far from the zoo in Tacoma. The facility is on the grounds of a mink farm owned by Dale Peterson, a member of the Point Defiance Zoological Society's mammal committee, which has been interested in the plight of red wolves since the 1960s. Peterson's mink farm consists of several acres of sheds with rows of small elevated pens for mink, surrounded by a dense forest of fir and spruce. The breeding facility, behind a high chainlink fence and locked gates, consists of eighteen pens, some a hundred feet square, others fifty by a hundred feet, arranged in a five-acre rectangle. These pens have fir, spruce, oak, and alder trees in them, and most have a dense growth of grass and shrubs which provide cover for the animals. Here and there, the wolves have dug holes, and from time to time they unearth strange objects. Says Sue Behrns, caretaker of the facility, "Years ago, Purina owned the mink ranch. Somebody would take stuff out and dump it in the woods. When they put the pens in, some of it got buried. Sometimes, the wolves dig up old equipment, a wrench, part of a machine."

Such encounters are only part of the oddness of the scene. It's a little strange to think of this almost tropical wolf breeding in the somber, rainy gray of the Northwest. But, Behrns explains, the natural history of mammals these days often takes unnatural turns. Whether the turns have altered the wolves to the point where they can no longer be considered wild is the key question.

Behrns has bright-blue eyes, sun-reddened cheeks, and an easy, melodic laugh. As she drives up to the cyclone fence that surrounds the facility, the wolves howl spontaneously. But as soon as the visitors get out of their vehicles, the wolves dash to the backs of their pens or go into their dens, which are concrete septic tanks left standing above ground. They trot nervously along the back fences, and squat to urinate timidly. Some lower their bodies and walk in a sort of slow crouch, their heads and necks low, their ears flattened out. It's hard to say whether that is the nature of the animal or part of the conditioning of the program. From the start, the managers wanted to keep them shy of humans. Says Carley, "Every contact that a wolf has with a human should be a bad experience. You draw blood, give 'em shots, get in, get out. It's for the benefit of the wolf. They shouldn't ever want to get near a human being."

The aim of this breeding program is to share the genes of all fourteen founders as widely as possible by means of judicious pairings. The recovery-plan goal is to have 320 wolves in captivity. Currently there are about half that number, and they are now spread among two dozen facilities in various parts of the United States. The breeding is intensive and very selective. Every year, the captive-breeding team meets to decide which wolves should be tried for breeding, and every year young wolves are shipped from Graham to other breeding centers, or to Alligator River for release.

Most of these animals, Behrns says, are tame: some of the old wild-caught animals will chase the food truck along the fences or come to within a few feet of the fence and follow Behrns as she passes. When strangers come in, however, they get very shy. "If it's just one person, they may not be quite as shy," says Behrns. "If it's two, it may mean that somebody's going to get caught." They have all had the experience of having humans come in, put a noose over their necks or catch them in a big salmon net, and hold them down while blood is withdrawn.

With the emphasis on breeding, there is a fair amount of intervention into the lives of these creatures. A critical responsibility of Behrns' is figuring out exactly when a female is ready to breed. "We go in every other day to check them. As soon as we know they're close, we go in every day. We check for vulva swelling, and take smears and a blood test." The smear is examined under a microscope. When a female wolf comes into heat, the epithelial cells square off and their nuclei get very small. Blood samples are checked for the lowered estrogen and increased progesterone levels that indicate a wolf is ready for insemination. The tests are done at Puget Sound Hospital's gynecological clinic by a technician who does exactly the same tests on human blood samples. Behrns looks forward to the day when it will be possible to test hormonal levels with fecal or urine samples, because it will be less stressful to the wolves.

For all the handling, the wolves seem natural and alert. "At whelping season," says Behrns, "as soon as you drive in you know somebody's given birth." They all get excited, and walk around more warily. "They get very protective of the young. A female gave birth in a depression, fifteen feet from the fence. As soon as I walked in, the male was barking at me, even though he was in a different pen." She adds, "They all like puppies, which is something that works to our advantage. We can put a puppy a few days old into another pen if we need a foster mother," and the puppy will be accepted.

Parental devotion was not in itself proof that the red-wolf reintroduction would work. At the outset, reintroduction was still more a hope than a proven management technique. It had been done with peregrine falcons, black-footed ferrets, and a handful of fish species, but never had a large predator extinct in the wild been reintroduced. Moreover, because red wolves had vanished from the wild before any comprehensive study had been done of them, managers knew little about how they would act in the wild. When the project began, recalls Carley, "We thought we knew how it could be carried off. But we needed a laboratory to demonstrate that some of our theories were correct. The main thing was that we could control the wolf if it got loose—that we could recapture it." They decided Bulls Island, a five-thousand-acre island in Cape Romaine National Wildlife Refuge, South Carolina, was the laboratory they were looking for.

In 1976, a pair of red wolves was released on the island. Bulls Island is three miles from the mainland, but those three miles are marsh, rather than open water. At low tide, wolves can walk and swim across. Says Carley, "In our first attempt, our female got to the mainland in eleven days. When that happened, we thought it was the end of the world. We'd made all kinds of promises to kill the animal if it got to the mainland, so it was time to put up or shut up. The assistant manager and I chased after her for over forty hours." They finally recaptured the female by shooting her with a tranquilizing dart from a pursuing helicopter. The male also left the island and had to be recaptured. The project was a public-relations disaster.

"This called for a serious regrouping," Carley continues. "Over the next several months, we determined that a dog from an adjacent island had chased the wolf and caused her to flee. We also concluded that they had not been acclimated long enough.

"We had to fight with the administration, all the way to Washington, to get authorization for a second attempt. Because of our openness during the first introduction, we had a lot of following in the press. Due to our openness with the public, they became understanding and supportive and wanted us to try it again."

In 1977, another pair of red wolves was brought from Washington, and put into a pen on Bulls Island. They were soft-released—that is, they were kept penned for six months to acclimatize, and released early in 1978. For eight months, they stayed on the island without any problems. "But the agreement was that this was a short-term experimental introduction," says Carley. The wolves were recaptured and returned to captivity. "We would then start looking for a long-term release site."

The Fish and Wildlife Service looked for a release site that would support more than a pair of wolves. As Carley explains, "To get genetic viability and gene flow, you needed to have a population on the mainland." In 1979, they proposed to release red wolves on Tennessee Valley Authority properties at Land Between the Lakes in Kentucky. The proposal required detailed public-hearing and rule-making procedures, and the reintroduction proposal was not published in the Federal Register until 1983. In 1984, the Tennessee Wildlife Resources Agency, responding to hunters who feared that release of wolves would curtail their hunting in the area, rejected the proposed introduction. It looked as if the wolves had no place to go.

But in the same year, the Prudential Insurance Company donated 118,000 acres of freshwater swamp, pocosin, and brackish marsh habitat in northeastern North Carolina to the Fish and Wildlife Service. A subsidiary of Prudential had planned on mining peat from that land to manufacture methane, but commercial fishermen protested, fearing the effects of peat mining on water quality and nearby fisheries. The National Wildlife Federation, concerned about the loss of increasingly rare pocosin wetlands, sued to overturn the Corps of Engineers' issuance of a permit to drain these wetlands. Faced with growing opposition, Prudential gave up on the project and, with help from the Nature Conservancy, conveyed the land to the Fish and Wildlife Service for protection as a refuge. At the time, wolf reintroduction was not a primary aim of the transfer, but the site seemed perfect for wolves. It was surrounded by water on three sides. In the middle was a navy bombing-practice range, which kept nearly all human activity away. The neighboring farmers raised soybeans and tobacco, not livestock. Moreover, since wolves had been gone from this region for nearly two hundred years, there wasn't a local tradition of wolf hunting.

Warren Parker, who succeeded Carley as coordinator of the red-wolf reintroduction, saw that the only traditional use that had to be accommodated was hunting. He concluded that, with careful attention to the attitudes of hunters, hunting and wolf reintroduction could get along just fine.

The reintroduction could be tailored to the circumstance. Under the terms of the 1982 amendment to the Endangered Species Act, the Secretary of the Interior may designate an introduced population of a listed species as "experimental-nonessential" if loss of some of the population will not jeopardize the survival of the species. "Experimental-nonessential" regulations might, for example, allow ranchers to shoot wolves if they saw the animals attacking their sheep or goats. When the Fish and Wildlife Service designated the Alligator River Refuge wolves an "experimental-nonessential" population, it required farmers who saw wolves taking their livestock to contact Fish and Wildlife Service officers or state conservation agents rather than shoot the wolves. But if hunters or trappers unintentionally took a wolf and immediately reported it to the refuge manager, they would not be prosecuted. The regulation declared, too, that wolves could be taken in defense of human life. Nor would citizens who shooed wolves away from their houses and roads—an act usually regarded as hazing under the Endangered Species Act—be prosecuted. Fish and Wildlife Service officials hoped the less stringent protections would generate public support for a program that might otherwise have faced stiff opposition.

In November 1986, four pairs of red wolves were shipped from Graham, Washington, to the refuge. And in 1987, the first of the wolves were released. Two litters were born in the wild in April 1988, two more in 1990. Each year, the program released at least two pairs with litters of young too small to travel. Candidates for release arrived from Graham and were held in a camp on Sandy Ridge, a strip of higher ground that runs through the center of the refuge and provides a cool, pine-shaded setting for eleven fifty-by-fifty-foot holding pens. They bore litters there and acclimatized. By releasing adult wolves with sixteen-to-seventeen-week-old pups, the managers hoped to tie them to the release sites. It seemed to work. The number of wolves increased.

Things seemed to be going well. But in January 1989, at a meeting in Atlanta of the red-wolf recovery team, David Mech raised the old question of hybridization. Says Phillips, the people at the meeting were looking at the recovery plan's goal of having 320 red wolves in captivity when Mech said he thought that number was too large and the cost too great for an animal that might not really be unique. Feeling there were still questions about its legitimacy as a species, he urged Ron Nowak and Barbara Lawrence to review the literature. If there were still disparate opinions about the identity of the red wolf, he believed the recovery team should resort to the new genetic analyses being conducted at the University of California at Los Angeles by Dr. Robert Wayne.

Wayne had collected blood and tissues from 276 gray wolves and 240 coyotes, sampling animals over the entire range of both species. He extracted from these samples bits of mitochondrial DNA. Mitochondrial DNA is a circular loop of DNA found outside the nucleus of a cell. Since it is outside the nucleus, it is inherited only along maternal lines. Unlike the DNA inside the nucleus, mitochondrial DNA is not responsible for physical characteristics like size or blood type or resistance to disease that determine a wolf's fitness, so changes in it don't increase or reduce an individual's ability to reproduce. Because natural selection does not act upon them, mitochondrial DNA sequences evolve at rates five to ten times faster than the sequences in the DNA of a cell's nucleus. So there is more variety in the mitochondrial DNA than in the nuclear DNA of any given species. That variety makes it useful for estimating how closely related individuals may be.

Wayne cut the mitochondrial DNA into small strips with restriction enzymes, which snip DNA strands between precise sequences of the four kinds of nucleotide—adenine, guanine, cytosine, and thymine—that make up DNA. He then looked at the sequence of nucleotides on the fragments. Theoretically, gray wolves would have one set of patterns and coyotes another.

The panel identified thirteen gray-wolf and twenty-four coyote genotypes, or patterns of DNA fragments. Coyotes had three times the genetic variety in their mitochondrial DNA, a finding that was consistent with the archaeological evidence that, as a species, the coyote is roughly three times as old as the wolf. But four of the wolf genotypes were found to be identical with coyote genotypes, and three other wolf genotypes were remarkably similar to coyote genotypes. Wayne concluded there had been at least six instances of hybridization, and that the genes from these matings had been widely disseminated among wolves in the Great Lakes region. He found coyote-wolf hybrids only in northern Minnesota, southern Ontario, and Quebec and on Isle Royale, but no coyote genotypes in wolves north of the present limits of coyote range. He found that coyote genotypes among wolves grew more likely as one moved east, that half the Minnesota wolves had coyote-derived genotypes, and that all the Quebec wolves sampled had them.

That study cast doubt on the purity of Isle Royale's wolves, but it also seemed to Mech to offer a technique that could judge whether red wolves were hybrids. So Wayne and Susan M. Jenks, also of the University of California, undertook a study of mitochondrial DNA in red wolves and in coyotes. Taking samples of mitochondrial DNA from coyotes, gray wolves, and red wolves, they cut the DNA with restriction enzymes, as before, and compared the sequences of nucleotides in the fragments. The red wolves from four different lines of matrilineal descent had a single genotype, but that genotype was indistinguishable from a genotype found in two recent Louisiana coyotes. The finding strongly suggested hybridization.

In case the captives had only recently hybridized, Wayne looked at seventy-seven blood samples from animals taken during the 1970s. Morphological studies—studies based on the measurements of skulls—had classified 58 percent of these animals as coyotes, 31 percent as coyote–red-wolf hybrids, and only 11 percent as red wolves. But Wayne's analysis classified 84 percent as having coyote genotype, 7 percent having a northern-gray-wolf genotype, and 9 percent a genotype characteristic of the Mexican wolf. Wayne also looked at six museum specimens of red wolf collected before 1930 and found them all similar or identical to coyote or gray-wolf genotypes. He found no sequence that occurred only in red wolves. "It suggested there was no red-wolf genotype," says Wayne.

Wayne's results questioned the accuracy of the nineteenth- and early-twentieth-century collectors. He believes most of the features Nowak selected in his study of red-wolf morphology related to size, and he thinks it likely that the earlier collectors used the same criteria. Nowak, he suggests, may simply be confirming earlier misidentifications.

Wayne confesses that his study does not amount to conclusive evidence that the red wolf originated as a hybrid. For one thing, the six skins from animals taken before 1930 could have been coyotes misidentified as red wolves by the collectors. "Toward the end of red-wolf range," he says, "there aren't many red wolves; there are lots of coyotes; and there is a great deal of phenotypic [or physical] variation among coyotes in that area." Even if the blood samples and skins were indeed from red wolves, he could have missed distinctive sequences in the mitochondrial DNA in the course of slicing them up. Or the distinctive elements might have vanished from red-wolf mitochondrial DNA, possibly in recurrent periods of inbreeding in isolated packs long ago. But it is just as likely, he says, "that red-wolf phenotype [physical type, as opposed to genotype] could have derived entirely from hybridization between coyotes and red wolves."

Wayne hopes to confirm his hypothesis by looking at the DNA of the cell nucleus. Nuclear DNA is harder to compare, because there is so much more of it, and so much of it is shared, not just among subspecies, but among creatures as disparate as humans and ants. But if one found sequences of nuclear DNA in red wolves that are not present in gray wolves or coyotes, it would establish that the red wolf is a valid species. And if one found coyote DNA in the nuclear DNA of red wolves, it would argue that the red wolf is a hybrid.

For now, he points out, "the data we have presently can't distinguish between the three hypotheses: that red wolf is a distinctive species but we haven't found its distinctiveness; that it's a subspecies and we may not be able [through present genetic techniques] to observe a subspecies of wolves; or that the origin of the red wolf was as a hybrid. I still believe the morphological data can't be used to exclude that hypothesis." One morphological approach that might help answer the question would be a study in which gray wolves and coyotes are cross-bred for three generations in a laboratory, and the offspring examined to see whether they resemble red wolves. That, however, would be an expensive undertaking, and no one has offered to do it. Says Wayne, "We don't know really what coyote–gray-wolf hybrids should look like morphologically. We really need to know that."

Though Wayne did not declare the red wolf to be a hybrid, opponents of wolf recovery seized upon his studies and called for an end to federal protections for wolves. In 1990, after hearing early reports of Wayne's work, the Wyoming Farm Bureau Federation petitioned the federal government to delist the gray wolf on the grounds that it was "not a species" and that the Fish and Wildlife Service was "unable to distinguish pure wolves from hybrid wolves." In 1991, Wayne's findings prompted the American Sheep Industry Association to petition the Department of the Interior to delist the red wolf on the grounds that it was a hybrid.

The U.S. Department of the Interior's policy on hybrids is based on something called "the biological species concept," which holds that a species is a species because it will not breed with members of other species. Species and subspecies are precisely adapted to local conditions, and "the biological species concept" implies that, by introducing genes that evolved in other settings, we may reduce a creature's ability to survive, or cause it to have new and detrimental effects on its environment. The Endangered Species Act, which drives the conservation of plants and animals in the United States, was written with this species concept in mind, and it does not accept hybridization. In a series of rulings, Department of the Interior solicitors have refused to grant Endangered Species Act protections to hybrids. A proposal to add to the California population of southern sea otters by introducing sea otters from the Alaskan subspecies was rejected because of the hybridization issue. In 1981, the Fish and Wildlife Service refused to allow the last few dusky seaside sparrows—all of which were males—to be bred with females of another subspecies, Scott's seaside sparrow, even though it meant the certain extinction of the dusky seaside sparrow. The Fish and Wildlife Service refused to allow crossing of the Mississippi sandhill crane with the Florida sandhill crane, despite the presumption that the two subspecies had only been separated for perhaps two hundred years. The Mexican duck was removed from the endangered-species list because it hybridized with an expanding mallard-duck population. The Amistad gambusia and the Tecopa pupfish were removed from the endangered-species list because all the surviving individuals carried the genes of other fish species.

As a foundation for wildlife conservation, "the biological species concept" can be slippery, because it presumes species are fixed and unchanging. An evolutionary view holds that all species are in the process of change, and that hybridization is one of the possible ways species adapt to changing environments. In fact, the geographical barriers that give rise to subspecies can vanish through natural processes such as climate change, flood, or erosion, and subspecies may suddenly begin interbreeding. Intergrades between subspecies often appear where their ranges come together. Even different species can and will interbreed: species of mice, voles, and frogs have interbred in the wild; fish frequently breed across species lines and intergrade at the edges of their ranges, so that it is difficult to tell which species one has in hand; plant biologists long ago recognized that plant species interbreed, and that new species often form through hybridization in the wild.

Taxonomists, the biologists responsible for naming species and evaluating the evolutionary links between them, have tended both to lump and to split species over the facts of hybridization. In recent years, taxonomists have lumped together red- and yellow-shafted flickers, Bullock's and Baltimore orioles, and Mexican and mallard ducks because they so extensively interbreed. At the same time, they left separate black and mallard ducks, lazuli and blue buntings, and black-headed and rose-breasted grosbeaks, even though these frequently interbreed. Taxonomists are finding it more and more necessary to deal with the issue of hybridization and to bestow names on the basis of selectivity of mating, width of hybrid zones (the range over which two species mate and form hybrids), and the persistence of physical characteristics among the offspring of these matings.

The presumption against hybrids is especially strong when the hybridization is a result of human activity. We practice a kind of misanthropy, regarding coyote genes in wolf blood, or genetic depression due to inbreeding as stains of our own meddling and conceit.

At best, we fear we have altered the course of evolution, something we believe mere humans should never do—but something we do nonetheless on a massive scale. At worst, we fear we have made the wolves into human artifacts. It may be that we choose our hybrids with a view to how much other humans have been party to their creation, that the traits we fear in such creatures are not biological but moral.

To refuse protection to hybrids may in some cases amount to declaring that evolution must stop at the present. Says Wayne, "Species are not static. It's clear from the paleontological record that species are not set. They change." Increasingly, biologists argue that we must protect the biological processes, not just the species that result from them. What is important, Nowak believes, is to focus on whether the sharing of genes alters the organism's interactions with the environment. "One hundred percent of the gray wolves on Isle Royale have coyote mitochondrial DNA," he says. "And yet all of the people I've spoken with say they are 100 percent wolflike, morphologically and behaviorally." The genetic technique cannot tell whether the hybridization occurred once thousands of years ago or many times recently. Nor can it tell whether hybridization has affected an animal's fitness. Coyote genes may not have altered the red wolf in ways that matter to the wild landscape. What ought to count is not whether we have inadvertently lowered a reproductive barrier, but whether the wolves perform the same functions in the wild, and whether evolution still goes on shaping the animal as it did before.

There is increasing inclination at the U.S. Department of the Interior to adopt a more liberal hybrid policy. The federal government rejected the Wyoming Farm Bureau Federation petition to delist the wolf in part on the grounds that "mitochondrial DNA does not function in the production of observable traits" that are acted upon by evolution, and so it could not define the red wolf as a hybrid. It cautioned that Wayne's results constituted the first such look at wolves and were "subject to future reinterpretations." In 1992, the Department of the Interior denied the Sheep Industry Association's petition, saying, "Several different species concepts, including a revised biological species concept, are now dominating taxonomic thinking. These alternative concepts incorporate the idea of limited genetic interchange with other recognized species. The service is currently reviewing and evaluating possible alternative species concepts."

Until there is either a new hybrid policy or a consensus that the red wolf is a legitimate species, the fate of the red wolf will remain uncertain. Some biologists and wildlife officials will continue to believe the red wolf is an artifact and therefore not worthy of reintroduction. Phillips recalls a meeting to discuss canid conservation at Fossil Rim, Texas: "Every time the red wolf came up, there'd be a nervous laugh, and people would move on to other things." People who were avid about the reintroduction of gray wolves in Yellowstone, he says, would get up and leave the room when discussion turned to the red wolf.

The new genetic techniques were bound to create controversy. Molecular genetics is an upstart science weighted with technical terms that field researchers and ecologists find perplexing. Few of them know how the technology of reading gene sequences works, and when a geneticist speaks to them of "phylogenetically distinct mitochondrial DNA genotypes," a fog of unfamiliarity falls over the conversation.

It is not just a matter of unfamiliar terms, but of styles. The morphologist looks at color and texture, line and distance. He or she handles the specimen, runs a finger over the ridges and hollows of a skull, smooths out the fur of a museum specimen, imagines the eye that once looked out of the socket, perhaps feels the heart that once beat inside the skin. The geneticist, on the other hand, hardly ever sees the organism, or even a recognizable artifact of the living creature. What the geneticist sees are test tubes, beakers, and stacks of radioactive chemicals in acrylide gels. It is a long and blinding jump from the tactile and visual world of morphology to the abstraction and quantification of the molecular-genetics laboratory, a jump traditional scientists may be unprepared to make.

Clearly, the genetic techniques are useful. Says Wayne, "The techniques of molecular genetics promise new kinds of insights into the lives of wolves." One of the most exciting is the possibility of telling exactly how the wolves in an area are related to one another. Wayne has used a technique called "genetic fingerprinting" to look at the paternity and movement of individuals and packs in Alaska and Minnesota. Wayne analyzed mitochondrial DNA from three clusters of wolf packs and counted the DNA patterns shared between individuals. He judged from their similarities and differences which wolves were brothers and sisters, and which were unrelated. The technique can help tell which individuals have dispersed from other packs, and whether there is a sex bias in dispersal. Ultimately, the techniques may be used to show exact genealogical relationships among wolves in a broad area.

The new science produces results. One of these could be the revision of two centuries of taxonomic work and the renaming of plants and animals based upon comparisons of their genetic patterns. Already, genetics and morphology seem to be moving together toward a consensus that would reduce the number of subspecies of North American gray wolf from twenty-four to five. Early collectors gathered specimens on the basis of color or size, presuming that geographic differences inevitably led to subspecific differences. Our present classification may have been colored by subjective things that occurred to collectors years ago. Wayne says, "You can imagine the first explorers when they walked through Florida and saw a couple of black wolves, and of course that became a subspecies." So subspecies tended to multiply. In 1944, Edward Goldman listed twenty-three subspecies of gray wolf in North America. In 1981, Raymond Hall listed twenty-four (see Appendix 3 for the list). Subspecies are supposed to reflect geographic boundaries, but, given the demonstrated migration of radio-collared wolves, the geographic boundaries assumed by earlier taxonomists appear to be less and less meaningful. There has been growing agreement that not all these subspecies are valid, and this new view has tended to divide the North American subspecies north and south, roughly along the Canada–United States border.

Nowak compared the measurements of skulls of wolves north of the border with wolves south of the border. His analysis showed differences between the northern and southern subspecies, and led him to conclude that only five subspecies seemed legitimate. If other taxonomists agree with Nowak, a new list of wolf subspecies might look like this: *Canis lupus occidentalis* would be the northern wolf and would include the seven subspecies of Alaska, western Canada, and northern Montana. *Canis lupus nubilus* would be the southern subspecies, and it would encompass the twelve subspecies that range from southeastern Canada and southern British Columbia east to Texas and the Great Lakes, central Quebec, southern Greenland, and Baffin Island. *Canis lupus arctos* would lump together three subspecies of the Arctic Islands and northern Greenland. *Canis lupus baileyi* of Mexico and the American Southwest and *Canis lupus lycaon* of southeastern Canada and the northeastern United States would also be recognized as legitimate subspecies.

The most important characteristic Nowak found separating the subspecies was size, but there was no gradual change due to geographic range. The biggest wolves, for example, were from Alberta, not from Alaska. Nowak found that *hudsonicus,* the wolf living on the western side of Hudson Bay, *manningi,* the wolf of Baffin Island, and *beothucus,* the extinct wolf of Newfoundland, all fit nicely into the newly defined *nubilus.* Asks Nowak, "How can I refer animals way up on Newfoundland and Baffin Island to the southern group of species?" The answer seemed to be that successive waves of immigration of wolves from Asia pushed earlier arrivals east and south. "In Pleistocene, we had several reinvasions of North America via the land bridge. The subspecies furthest south—*baileyi* and *lycaon*—probably represent the earliest invasion. Then *nubilus.* Then, finally, I think the continent was invaded by fairly large wolves from ice-free refugia when the ice retreated."

The analysis seemed to Nowak to confirm the legitimacy of the red wolf, *Canis rufus,* as a separate species. He looked also at Eurasian subspecies of gray wolves and concluded that there was evidence to support five subspecies: *Canis lupus lupus* from Europe to Russia, *Canis lupus albus* in extreme northern Eurasia, *Canis lupus pallipes* from Israel to India, *Canis lupus cubanensis* in the Caucasus, Turkey, and Iran, and *Canis lupus communis* in the Ural Mountains region. He compared the three southernmost subspecies, *pallipes, baileyi,* and *rufus.* He found that *baileyi* stood apart, "but, despite one million years in time and ten thousand miles in space," he says, "we do have a tenuous overlap between *pallipes* and *rufus.* Here is evidence of this ancient migration of the ancestral wolf to Asia."

Wayne also surmised that there were fewer than twenty-four subspecies of gray wolf. "It was interesting to me," he says, "that subspecies had been defined that didn't correspond to any geographic boundaries. I suspected many of these subspecies didn't express real genetic boundaries." Genetic techniques had been used to inquire whether a subspecies is truly a separate group. DNA analyses have shown, for example, that deer mice that live near each other have similar gene sequences, but deer mice that live a long way from one another don't. That suggests that the distant populations are no longer interbreeding and may be regarded as distinct subspecies.

Wayne performed an analysis which showed a different pattern with canids. He says, "We found thirty-two different coyote genotypes in different localities. In each of the localities there were multiple genotypes, and we found the same genotypes in different sites." There seems to be mixing of genes of coyotes on a continental scale. With wolves, there are a few ubiquitous genotypes that are found more or less everywhere, reflecting gene flow over the continent. Wayne explains, "A wolf can disperse five hundred miles; one researcher estimates the zone of hybridization [the range in which dispersing individuals may form hybrids] is fifty times the dispersal distance, so that means wolves have a hybridization zone almost as big as the continent."

Wayne found evidence of five genotypes in the wolves of North America and seven in the wolves of the Old World. The clusters he came up with seem to converge with the clusters Nowak has found. "We did find evidence that the Mexican wolf had a unique genotype," says Wayne. "And we found some evidence that Alaskan wolves are different from wolves in the Northwest Territory. We didn't sample wolves in northeastern Canada. But we're coming close to agreement."

Such a convergence suggests that traditional morphology and modern genetics may yet make peace. "They've been seen to be at odds," Wayne says, "but the two complement each other." And as time goes on, he expects the two approaches to change the way we view species. "Our museum collections are based around the idea of a type," he says. "The truth is, there is immense variability, and somehow you've got to take into account the variability within a species." In some cases, he expects the genetic techniques to increase the recognized number of subspecies. "Smaller species with more local distribution, like pocket gophers and deer mice, will show more variation and more differentiation. On balance for the smaller species, molecular-genetics techniques probably will define more subspecies.

"But for highly mobile species, like wolves, the genetic information will show the high mobility has stifled any great degree of genetic variability," because wolves share their genes over such a broad range. So genetic techniques may end up reducing the number of subspecies in wolves.

Clearly, the genetic techniques are going to change the way we define species and subspecies. This redefinition poses real challenges to the way we look at conservation. We already argue over whether it is legitimate to try to save subspecies and local populations. When we start thinking about conserving the whole genetic range of a species, the task of conservation will grow much larger. "We're down to the point where every individual is genetically distinct," says Rolf Peterson. "So how are we going to save the whole gene pool?"

The new knowledge will urge us to integrate the saving of genes with the saving of ecological functions. Whereas today most people view conservation in terms of saving individual animals, future conservationists will have to think about saving ecosystems, and ecosystems will have to be defined in terms of the genes that constitute them. When we consider the genetic variety necessary to maintain an ecosystem, we will face new levels of complexity and conflict in management.

And how shall we accommodate distinctions we have long made—that the buffalo wolf of the Great Plains is distinct from the timber wolf of Minnesota, or the tundra wolf of Alaska is different from the wolf of the Alaskan interior? Zoos still register their wolves by subspecies, and breeders of wolves cling to and value these distinctions. Conservationists champion local varieties as unique and irreplaceable. Those who oppose reintroductions argue that the subspecies being reintroduced never inhabited the recovery area. Revision of the systematics of wolves is bound to confute many of these distinctions, and discomfort many of those who make them. It will add to the already contentious claims we make about the identities of wolves.

"The new knowledge we gain with all this new wolf genetics is not always easy to swallow," says Rolf Peterson. "It's such a new tool, and a marvelous tool. I think where it will take us is outside our frame of reference. Certainly it's going to take us to a new way of identifying species—that's a good thing, a new perspective—but it's going to be hard to accommodate."

In the future, wildness is going to be defined by the ways the environment modifies the expression of genes. And, increasingly, wolf genes—and wolf wildness—will be modified by the ways wolves interact with humans and the ways humans manage wolves. One of the first eight red wolves released came into the village of Manns Harbor and had to be recaptured in full view of the public. A woman in Hyde County complained that the wolves were killing cats, and someone saw a wolf with a cat in its mouth. Recently, some goats were killed on a farm outside Pocosin Lakes Refuge. Twenty-two of the released wolves have died—vehicles killed eight of them, one was shot and one killed in a trap—and seven were returned to captivity, some because they couldn't seem to stay away from the villages. Such interactions with humans and their possessions suggest to the villagers around Alligator River that the wolves are less than natural, not wild enough.

Says Orville Tillett, retired enforcement officer for the North Carolina Department of Fisheries, "I see 'em once in a while. To me, they look like a German shepherd." He thought the first wolf he saw was a dog and tried to call it to him; the wolf stood and looked at him, then drifted into the trees. "I believe they're German shepherds," he says, "or crossed with them."

Part of the suspicion that the wolves aren't wild seems to lie in the mere fact that the wolves aren't frightening. If they were wild, many locals suspect, they'd be bigger and more aggressive. "I've heard two or three [locals] say they're dangerous," says Tillett, "but I've never seen one offer to attack nobody. I saw three in one day this year. They're used to vehicles. They'll stand in the road until you get close to them. Then they run away from you." A white-haired lady with arthritic knuckles and a sweet smile stands in the Manns Harbor Post Office and says, "I had them in my yard. It got so I was afraid to go out at night." But she says that, after talking to a refuge employee who told her the wolves wouldn't hurt her, she no longer felt afraid.

Residents of the villages around the refuge believe the refuge managers are feeding the wolves and catching them regularly to deworm them. Says one hunter, "They trap 'em every month or so, detick 'em, give 'em shots, and feed 'em. They're not wild as far as I'm concerned."

At a meeting in Manns Harbor, Alligator River National Wildlife Refuge Manager Jim Johnson asked local hunters to offer suggestions for opening or closing areas of the refuge to hunting. The hunters were polite and attentive. And, although the closures have nothing to do with the reintroduction program, at the end of the meeting people started to ask about the wolves.

"Do they have to catch them up and worm them?" asked Tillett.

Johnson replied, "I've been hit a hundred times about 'the way you guys are feeding them critters.' That ain't true." He explained, "The critters you pick up in the wild very rarely will be carrying external parasites. That wolf may have been out there three years, and it has no heavy tick load." He reiterated that they weren't feeding the wild wolves.

"You ain't feeding 'em?" Tillett asked, still unable to believe it.

Michael Phillips believes the wolves are doing just fine. By 1992, there were forty-two red wolves in the wild. Says Phillips, "The story is, we got many, many animals that have four legs and are healthy. Sixty to 70 percent of our pups are wild-born animals. Of twenty-five or twenty-six puppies born in the wild, only four have died that we know of." Most of the animals on the Alligator River Refuge can simply be left alone. Phillips explains, "Animals that use Alligator River don't require much work, just a little monitoring. For these animals that are well established, we fly by and get a signal, and if it's not in mortality mode, we just keep flying by. We don't consider putting out parasiticides every thirty or forty days by dosing a piece of meat—we'd be forever driving around and dropping meat. Some people say heartworm will ensure that the red wolf doesn't make it. Heartworm is going to kill some wolves, but one of our most prolific pair has long been heartworm-positive and they've contributed three litters. The population is big enough that we're beginning to step back. We used to routinely replace radio collars. But take animal number 331: If his radio collar goes off the air, he isn't going anywhere. He's five years old. He's been here since 1989. He's going to die here."

There are also red wolves on Bulls Island in South Carolina, St. Vincent Island in Florida, Horn Island in Mississippi, and, since 1991, in Great Smoky Mountains National Park. The island facilities were created because it was thought that wolves born in the wild would be more likely to succeed at reintroduction. In practice, that hasn't always been so. Of four island-bred wolves released, two had to be recaptured after they raided turkey pens, and one was struck by a car. On the other hand, a male born at Graham, Washington, and released on Bulls Island cared for four pups two years in a row. Each of the females that bore these litters was killed by an alligator, and the male went on to raise the pups on his own. In 1989, Hurricane Hugo went over the island with a nineteen-foot surge of water. It tore out trees and destroyed the refuge headquarters. Refuge workers flying over the island after the storm spotted the widowed male and his four pups: they had survived.

In fact, living in the wild seems to awaken a liveliness and toughness in the wolves. Only a third to a half of the wolves manage to reproduce in captivity, but seventeen of the eighteen possible breeding opportunities in the wild produced litters. Of twenty-seven wolves born in the wild, all but two were surviving in 1993. "You gotta believe they like being out," says Phillips. "Free-ranging wolves come into camp and you hold them in the pens and they seem to get depressed." He tells of a female that was recaptured in 1989 after living wild for five months: "She never came outside of her box. In May, she had one puppy. The next day, there was no puppy. Her neck was rotting—we had to cut the collar from her neck, she had no hair on her neck. It wasn't the collar that was the problem; she had done well with it in the wild. She was depressed. Finally, we let her go."

Phillips believes the world is wild enough for wolves. "If you take a captive-reared red wolf in eastern North Carolina and you let it go in a good spot, when it is relatively young, it will do fine," he says. And he believes there will be more and more space for wolves. With new refuge lands, and with cooperation from private landowners around the refuges, says Phillips, "We should have access to one million acres in eastern North Carolina. We ought to get a hundred wolves out there."

. . .

The hard news about the program, however, is that, if it is successful, it will not end. Most people expect reintroduction to be a short-term effort, after which we will no longer need to manage the recovered creatures. But, says Phillips, "There is no end. We're irresponsible if we don't recognize there's no end.

"We're talking about fifty or a hundred years, hopefully forever. In fifty years, I'd guess you'll have two to three good trappers in northeastern North Carolina, dealing with wolves that get into chicken coops and goat paddocks. They deal with spot fires. There aren't that many ways the wolf is going to get in trouble with people. Depredating wolves won't be killed—they'll be put back into the captive population." Wolves consorting with coyotes may also be replaced, and there will be wolves captured and fitted with radio collars so that the population can be monitored. There will also have to be managers for the captive breeding and release. The program calls for 320 animals in captive-breeding programs in order to maintain another 220 animals in the wild. There will probably be four or five larger captive-breeding sites, with thirty to forty animals each. Humans will still decide which wolves breed, which go into the wild, and which are removed from the wild.

That suggests that the red wolf will never be free of human oversight and intervention. The degree of human manipulation is a persistent issue in reintroductions here and in the Southwest. Says trapper Roy McBride, "It would be a constant harassment to the wolves to have people monitoring them all the time, capturing those that get off the reservation, putting in new wolves to keep the gene pool stirred. I don't think that's right. I don't think it's right for the wolves. Sharks got to swim or they'll drown; wolves have to travel the country. I don't know how we're going to reestablish that. Are we going to get all the cities out and set aside three or four states?"

As long as wildness has a human dimension, we will argue over this. We may become astute technicians and learn to account for enough of the varied factors of landform and prey, migration barrier and weather to keep the red wolf out there. But behind it all there will always be computers ticking away, committees deliberating, biologists watching. For red wolves, at least, the wild of the future promises to be very different from the wild of the past.

9

THE RIGHT WOLF

The Sierra Vista Ranch house is a neat, white-stucco-sided building beside a tree-lined arroyo in the low, rugged hills of southern Arizona. The landscape is mesquite and cactus. A veranda looks north upon a sweeping view of the Altar Valley and the steep gray wall of the Baboquivaris. To the south are the low shapes of the Sierra Pozo Verde of Sonora, Mexico, to the east the Sierra San Luis. There is a huge vault of sky, with great white schooner clouds drifting over gray puddles of cloud shadow.

It is far from the city, from daily news and common convention. The space and silence set one apart, and the desert light seems to rob things of their substance. The hawk soaring overhead vanishes into the blinding light of the sun. The glimpse you get of the bobcat may only be its tail disappearing into the brush. The lizard you think you see scurrying over a rock may in fact be only the creature's shadow. It is a place where definitions shift and arguments grow. Some of the arguments are about wolves: whether there are any out there—and if there are, where they have come from.

Sitting on the veranda of the Sierra Vista Ranch house late on a summer afternoon in 1991, Joyce Vanelli heard a coyote howl. The sun had just gone below the hills behind the house. And when the coyote howled, Vanelli heard something else: "Something tried to join in," and howl with the coyote, she said. "But it couldn't. Something that could not have been another coyote tried to mimic the coyote. I couldn't say it was a dog or a wolf."

In October 1991, her hybrid wolf-dog, which was going into heat, ran off into the hills behind the ranch house. "She takes off all the time," says Vanelli. "I went after her, because it's a bad area in there for drug runners—they shoot dogs—so I went up the hill and down the other side." She caught up with her dog and was bringing her back when, she says, "I heard the wolf howl. I definitely, definitely heard the beautiful sound of a wolf. He was calling her back." She says she would have worked her way down through the brush to see the wolf, but didn't because "I didn't know what I would run into."

Next April, Vanelli was driving along the dusty ranch road toward the highway to Sasabe. When she drives, she says, she looks at the ground along the road for Indian relics. She had stopped because she thought she saw some bits of pottery, and as she looked down, she saw the footprints of a large canid along the side of the road.

"It was a wolf," she says with conviction. She had raised captive wolves in Utah. "I've been around these animals all my life. I lived with wolves for years and years and years. I know that had to be a big, big animal. It was running with a smaller animal, maybe a coyote. I got Dale, the caretaker at the ranch, and we went down together, and we saw where they had run a deer." They backtracked the prints to a clump of mesquite where the wolf and its companion had come out onto the road.

She called the Buenos Aires National Wildlife Refuge, which borders the ranch. The refuge superintendent, the regional director from Albuquerque, and the leader of the Mexican Wolf Recovery Team all happened to be at the refuge that day, and they sent out a technician to take plaster casts of the tracks. But the technician didn't take casts of the smaller print Vanelli said she had seen alongside the large one. "People are strange," says Vanelli. "These people at Buenos Aires are extremely uneducated about wildlife. They don't get out—they sit behind a desk. They don't see."

It is not for lack of effort that refuge officials haven't seen wolves. Steve Dobrott, biologist at the Buenos Aires Refuge, has photographs he took of wolf tracks on the Gray Ranch in New Mexico in 1984. And in 1990, a man cutting wood in the San Luis Mountains, just east of the refuge, reported that he had heard a wolf howl and that as it did so, his German shepherd crawled in fear under his trailer. He reported that he had seen the wolf, and its tracks. Says Dobrott, "He took me over to the wash and showed me the tracks. I couldn't tell—they were washed out. But they were big. We went back there in the evening with the fire truck. I cranked up the siren. We got three coyote groups to howl back. When I went back to the man's camp and told him I got coyotes to howl back, he said, 'No, didn't you hear that other thing?' So we don't know."

Refuge officials have heard tales of wolves in the area for years. There are plenty to hear. Emma Mae Townsend, widow of locally famous wolf trapper Hack Townsend, recalls that just two years before, when Hack was still alive, he was sitting on the porch of his home in Arivaca, a few miles from the refuge, when he heard a wolf howl. "He howled back and it sounded just like a wolf, and he got a howl back. And then he hollered, 'Emma, there's a wolf down there at the dump!' "

Danny Culling, Emma Mae's son-in-law, who works at the refuge, says he saw a wolf on the neighboring King Ranch with his father-in-law in 1985. He says, "There was a wolf spotted at Milepost Two, near Amado, a year ago," by a surveyor for the U.S. Forest Service. Feliciano Lopez, a ninety-year-old rancher with a deeply furrowed desert-dweller's face, says he heard wolves howl near Apache Wells the year before that. Carol Riggs, a Cochise County rancher, says she has watched wolves playing near Rucker Canyon.

And Dennis Parker, who studied wildlife biology in college and is now a wagon-maker in Patagonia just over the Santa Rita Mountains from Arivaca, carries around Ross Kane's three glossy color photos that are to all appearances of a wolf crossing a road in the Canelo Hills of southeastern Arizona. Parker says he himself photographed wolf tracks in 1984 in the Huachuca Mountains and saw scratches and tracks there again in April 1986. "I would say that, once we got out there in the wild and took a look, we'd find a lot more."

But neither the U.S. Fish and Wildlife Service, the agency charged under the Endangered Species Act with leading the effort to recover the Mexican wolf, nor the Arizona Department of Game and Fish is convinced there are wolves in the wilds of Arizona. Terry Johnson, nongame and endangered-species coordinator for the Arizona Department of Game and Fish and Arizona's representative on the recovery team, says he has heard secondhand reports of sightings and heard about Kane's photographs, but says neither the reports nor the photographs come to the department with enough supporting evidence to prove that there are wolves in the area. He declares, "There have been no documented occurrences of Mexican wolves in southern Arizona or in immediately adjacent northern Mexico during this last decade."

It is not as if the agencies are closed to the possibility that the Mexican wolf survives in the wild. Says Johnson, "We have sight-record cards we will make available to anybody who is willing to submit them. We are ready to accept any documentation that people are ready to submit, and then try to follow them up as best we can. But we haven't received any." And if there are only sporadic sightings of wolves in Arizona, says Johnson, "there are not enough to call it a viable population in the northern part of the range."

Canis lupus baileyi was the name given by taxonomist Edward Goldman to the Mexican wolf, the southernmost subspecies of gray wolf, which ranged from southern Arizona and New Mexico south into central Mexico. It was a slightly smaller wolf, as might be expected of a desert subspecies—desert races tend to be smaller than races from higher, colder altitudes. While it was not, strictly speaking, a desert wolf, but inhabited the wooded uplands above the deserts, the deer it fed on are among the smallest races of white-tailed deer in North America. Dennis Parker speculates that the northern margin of its distribution ran roughly along the present route of Interstate Highway 10 across Arizona and New Mexico, for north of that line the blue oak of the Sierra Madre habitat ends and gray oak, more typical of the Mogollon Rim country, takes over. North of that line, he speculates, the territory belonged to the subspecies *mogollonensis,* a larger wolf that fed on the larger mule deer of the uplands. And east of the Continental Divide, the neighboring subspecies was said by Goldman to be *monstrabilis.* Little record of that subspecies exists, for it was eradicated before much collecting was done. Both genetic screening and morphological studies suggest that, if the red wolf is either a hybrid or a separate species, *baileyi* may be the oldest North American subspecies of gray wolf, a subspecies that moved farther south as later subspecies crossed the land bridge from Asia.

For thousands of years, the Mexican wolf roamed the mountains of northern Mexico and the southernmost edges of Arizona, New Mexico, and perhaps Texas. But a century of shooting and poisoning did the Mexican wolf in. The last "documented" wolf in New Mexico was a carcass found in the Peloncillo Mountains by trapper Arnold Bayne in 1970. Two wolves of undetermined origin were taken in Brewster County, Texas, in December of the same year. The last documented Mexican wolf in Arizona was probably taken a few years later near Aravaipa Canyon by a private trapper for a $500 bounty put up by local stockmen. In 1976, the Mexican wolf was declared an endangered species.

If there is no viable natural population of wolves left in the wild, the U.S. Fish and Wildlife Service has, under the Endangered Species Act, a duty to recover them. Environmental groups have sued to press the service to reintroduce the Mexican wolf. Bobbie Holaday, who heads a 550-citizen group called Preserve Arizona's Wolves, explains: "The wolf is a part of our southwestern heritage. It belongs back in the wild as part of the fabric of life. We have taken away the predators, and the wilderness areas are now void of the real essence of the wilderness, the howl of the wolf. We have saved thousands and thousands of acres of wilderness in Arizona, but we don't have its essence. There is the wild spirit, and it's in us all. All of us have this longing for complete freedom, and what better symbol or embodiment of that than the wolf?"

There is, she believes, also a purely ecological reason to restore wolves. "The wolf has a very, very important role in the ecosystem. It's not just restoring one single species; we are trying to restore the whole fabric of life. We have taken it out of the ecosystem in areas where we now have ungulates seriously overgrazing. There are areas in Arizona with burgeoning elk populations. There are some areas, like the North Kaibab, where there is a similar situation with deer. The animals are starving in winter."

Holaday urges wolf reintroduction also because it may help to reverse a dangerous trend of extinctions. "We're losing species so rapidly," she says, "and we have to do something about stopping this. If we just keep letting species go right and left, sooner or later they're going to come for us." Eventually, she believes, the earth may cease to support human existence.

Says Holaday, "Biologically, there's no reason not to reintroduce the wolf. There's prey. There's space. There's water. People are the problem, and we have to do a lot of work to resolve that problem."

Ranchers are the heart of the opposition to reintroduction. Jim Chilton, who owns two ranches near the Arizona-Sonora border, says, "The hard evidence shows a high propensity in wolves to eat beef." If a wolf kills one of his cows, he wonders, will the government pay for the cow? What kind of proof will be required to get the government to pay? Will compensation be for the cow, or will it also pay for the time a rancher must spend riding the range looking for wolf tracks and missing cattle? If he sees a wolf killing his sheep or cows, must he call a federal animal damage-control agent to come deal with the problem, or can he shoot the wolf himself?

Ranchers fear also that reintroduction of wolves means restrictions on their use of private and public land, especially the federal lands they now lease to graze cattle. Already, 87 percent of Arizona is public land, and ranchers are seeing their grazing leases on Forest Service and Bureau of Land Management lands increasingly restricted. Ranchers fear that, if a wolf dens on a grazing lease, the area will be closed to grazing. Chilton purchased his two Arizona ranches in part because federal grazing leases came with them. There had been no wolves on the land for fifty years. If the government now introduces wolves, he says, it should be liable for any livestock losses and for any decreases in the value of the land he owns.

Fifteen sites have been studied in Arizona, and others have been proposed in New Mexico and Texas, for possible reintroduction of wolves. Ranchers in all of these areas fear that, once wolves are released, the Fish and Wildlife Service will ask that nearby lands be considered critical habitat, and that uses of both publicly owned grazing lands and privately owned ranches will be restricted. Even if they don't lose the use of the land, they fear the wolves will get into their herds, and they will be unable to prove their losses sufficiently to receive compensation. They fear having to spend hours or days or weeks in town satisfying federal regulatory requirements, chasing down federal officials, filing papers, waiting for bureaucratic decisions.

Some hunters also fear that reintroduction of wolves will mean fewer elk and deer to hunt. Trappers fear they will no longer be allowed to put out traps. Ranching families fear for their children. Says a rancher from Arizona's Chiricahua Mountains, "I'm afraid for my grandkids. We live in this little place out there. In the cities, they have these gangs and things that are very detrimental to society; you can't go outside at night. So now, with the reintroduction of this wolf, it's very highly possible that will mean I'm in that situation. What's going to happen in a drought, when all these wolves come down into my area?"

To make the controversy even more complex, some people oppose reintroduction because it promises to be cruel to wolves. Says an Arizona woman, "If wolves are reintroduced, they will be persecuted, slaughtered, and tortured to death exactly as they always were. Man's nature has not improved; he is still the same cruel killer that he has always been."

But if there are wolves in the wild, they must be considered in the reintroduction plan, and so far the plan assumes wolves have been absent in the Southwest for two decades. Reports of wolf sightings raise essential questions: Are the wolves that leave tracks and pose for photographs real wolves or hybrid wolf-dogs? If they are wolves, are they actually Mexican wolves that walked in from Mexico? Or are they captives brought from Canada and released for the purposes of taking photographs or leaving tracks? Who would release such an animal? Wolf proponents who want to see it back? Or ranchers who want to forestall a Fish and Wildlife Service plan to reintroduce Mexican wolves into the wild, and who believe that, if there are wolves out there in the wild and the wolves in captivity may be described as inbred or tainted with foreign genes, then the reintroduction will not take place? So far, most of the reports of sightings come from opponents of reintroduction.

Reports of sightings are apt to raise more questions than they answer. The plaster casts taken from Vanelli's reported wolf were not conclusive. Because wolves walk in such a way that their hind legs swing in line with their forelegs, the back paws step into the prints of the forepaws. Dog hind legs swing inside the forelegs, and the hind paw print appears next to and inside the forepaw. But only individual paw prints were cast at Vanelli's find, so the pattern of the tracks in the dirt wasn't recorded. Some of the plaster casts were sent to Roy McBride, who for many years trapped wolves on Mexican ranches. McBride looked at the casts and replied, "I didn't think they were wild-wolf tracks. Wolves and coyotes get their toenails real worn down, and this one had real long toenails. So, if it was a wolf, it was someone's pet. The heel was more like a dog's. The toes weren't parallel. The outside toes pointed out, like a dog. Probably it was her own dog got out."

McBride's is only one opinion among many contending voices in the land of the Mexican wolf. Deciding what is out there in the hills of southern Arizona is just part of the problem. The nature of the wolves being bred for possible reintroduction into the wild is also an issue.

In 1971, Norma Ames was assistant chief of game management for the New Mexico Game and Fish Department, drafting laws and regulations that would go to the legislature or the Game and Fish Commission. It was largely a desk job. She had a degree in biology from Smith College and had done graduate work. Energetic, talented, and intelligent, she wanted to be a field biologist, but going out to count deer or quail was a job reserved for men in the department. "I came into the kind of work I was doing too early for opportunities to be much for females," she says. "I was confined to being a paper biologist and indoors." One of Ames' duties was to issue permits to scientific collectors. As she talked with biologists collecting specimens along the border, she became interested in wolves.

In the early 1970s, the realization was just dawning that the Mexican wolf was on the verge of extinction. There were very few in zoos. The Arizona-Sonora Desert Museum had acquired a male taken from Tumacacori, Arizona, in 1959 and a female from Yécora,

Sonora, in 1961. It had successfully bred them, and as the fate of the Mexican wolf in the wild grew grimmer, the museum made the young from its successful breeding program available to other zoos. A number went to the Ghost Ranch, a Department of Game and Fish facility in Abiquiu, New Mexico, which bred the offspring and, as was common practice at the time, sold or gave surplus animals away.

In 1971, Ames mentioned her interest in wolves to a colleague in the department, who knew that the Ghost Ranch wolves were about to have a litter. Seven pups were born, but they were confined to a cage with a concrete floor, in full view of visitors, and the mother, in panic while trying to move them out of view, killed four of them. Ames' colleague called her one day and told her that, if she wanted some wolf pups, she had better get out to Ghost Ranch fast. Ames was then a fifty-year-old divorcée living alone, her daughter off in college. She had 240 acres of land in a mountain valley ten miles outside of Santa Fe, behind locked gates at the end of a road and surrounded by national forest—"a lovely place for raising wolves," she felt. She had often kept wild animals that other people had found injured or orphaned. It was still five years before Mexican wolves would be listed by the federal or state governments as endangered, and there was no legal bar to possessing them. Ames went out to Ghost Ranch and returned with two pups; she later acquired a second pair from Ghost Ranch.

She built an eight-foot-high chainlink enclosure adjoining her house, and set about raising wolves. One door of her house opened directly into the wolf enclosure, and the living-room window looked out into it. She socialized the wolves so that they would regard her as an intimate friend.

She found the wolves lively and full of surprises. "One winter day, I heard this banging against the wall. The eldest female was running the length of the enclosure, leaping and bounding off the wall. She was hitting the wall seven feet off the ground. She was having a ball."

Before 1971, Ames had been interested in wolves because she felt they represented wildness and freedom and beauty, and seemed to link the worlds of biology and spirit. But once she had them, other motives came into play. She had read Joy Adamson's books about reintroducing zoo-bred lions into the wilds of Kenya and saw that, as Mexican wolves vanished from the wild, her wolves had increasing value as stock from which the wild population might be replenished. She wondered whether her wolves might be returned to the wild. The main question, she thought, was, would they hunt?

One moonlit night, she thought she might have the answer. She had left the door open to the wolf enclosure outside. "I woke up and thought, 'What is going on?' Hovering over me, standing on the bed, was one of the wolves. It was offering me a much-chewed-up ground squirrel it had caught out there." She found that anything that wandered into the wolf cages—squirrels, ravens, or mice—would be caught. "People said wolves raised in captivity can't cope in the wild," she says. "I think they can."

Until then, she had treated her wolves as pets. But "after that, I was into serious wolf breeding, to save the Mexican wolf, and from that point I didn't socialize with them."

She found that raising wolves was demanding work. Each wolf had to be understood as an individual. "There were a couple I wouldn't turn my back on. The first female had a litter in a concrete structure in the enclosure; she dug a den and moved the pups into it. One day, I had to shoot through the enclosure back into the house and without thinking I turned my back. In an instant, she knocked me down with a bite to my leg. If she'd wanted to attack, she could have, because she had me down. But she didn't."

Only that one time was she bitten. Sometimes males would challenge her, staring hard with hackles up, and she would back down. The wolves would fight with one another. "It's the sort of thing, if I had had lots of money, I would never have let happen: you detect the first signs of long-lasting enmity and take steps. In the wild, I'm sure it wouldn't have continued, because the wolf that was getting the worst of it would leave. But here they couldn't leave. If there were fights, I did not want to treat injured animals, so I'd separate them. I'd have to go out to prevent further damage, and, as angry as they were with each other, they'd look up and smile, as if to say, 'You here, too?' " She would distract the wolves, open a gate, haze the wolf through it, and slam the gate when the wolf had gone through.

The wolves quickly took up all her time and money. "It was all personally supported. Each time I allowed the female to reproduce, that meant building more enclosures. My pay was awful low, but it was something I had to do." She couldn't leave the wolves in the hands of a caretaker, because no one else knew them well enough to read their needs, and no one else could safely enter the pens to doctor or clean up after them. Even her veterinarian was afraid of the animals. "To gather up the wolves and take them in to him was a nightmare. It reached the point where he said, 'You come in and I'll give you the medications you need.' " She had to do everything herself.

So she fed and doctored, mended fences, and buried the dead. The wolves filled her life. She had to give up her position on the Western Region Environmental Education Council; she had previously published two romance novels, but when her publishers called to ask her for another, she told them she just didn't have any time. When the department finally offered her fieldwork, she said, "I wouldn't give up what I'm doing to go count deer for the Game and Fish Department."

"It changed my life in many ways," says Ames. She felt as if she was slipping away from humankind and deeper into wolfdom. When her veterinarian asked her how the wolves were, "I'd say, 'They're getting easier. They're getting humanized.' " But he would reply, "You're getting wolfized." And she thinks today he was right. "I tend to be, as a whole, a loner. And living with those wolves made me more so." She found that when people drove up her driveway she would think, "Here comes another human being you're going to have to cope with."

Her work with wolves remained separate from her work in the New Mexico Game and Fish Department, but because she knew so much about wolves, in 1979, when the U.S. Fish and Wildlife Service formed a team to write a recovery plan for the Mexican wolf, she was selected to represent New Mexico's department.

The first problem faced by the team was deciding whether to try to save the wolf where it still seemed to exist, in Mexico. Says Ames, "We used to say, The first priority was to save the wolf in Mexico and all the recovery will be done there." But, she says, "it became clear they weren't going to be protected in the wild. The land was being broken up into farms. Cattlemen were putting out poison. The man who was the Mexican game warden in the area where the wolf existed covered three states. The last Mexican wolf we caught in the wild in Mexico was caught because he came in to breed with a ranch dog. There was little hope of protecting the wolves or setting aside habitat for them. We came to the conclusion that protecting them where they still existed wasn't feasible."

The recovery team concluded that wolves would have to be reintroduced somewhere in the United States. When the recovery plan was approved in 1982, it called for maintenance of a captive-breeding population and reestablishment of a self-sustaining population of a hundred wolves in the wild. But where would the wolves come from? Zoos had not been active in breeding Mexican wolves. By the time the recovery team approached the American Association of Zoological Parks to ask that a species survival plan be written for the Mexican wolf, a step aimed at getting zoos interested in their propagation, they were faced with a serious problem: there were few captives, and all posed questions of genetic health or integrity.

There were questions particularly about the purity of the founding Ghost Ranch wolves, which had originally been bred at the Arizona-Sonora Desert Museum. The female had been purchased from a family in Magdalena, Sonora, by a Canadian tourist who brought her into the United States on the back of a motorcycle. No one could say where that wolf had really come from. Had she been bred of wolves and dogs on a Mexican ranch? Had she been taken from a den in the wild? The male had been trapped by a cowboy on an Arizona ranch. Some observers said these founder wolves looked to them like dogs. Dan Gish, the former wolf trapper, said the male caught in Arizona had been hanging around ranch buildings and howling during the daytime—both, in Gish's mind, indications that it wasn't a wild wolf. Surviving photographs of the male show a stressed or drugged animal on a veterinarian's examining table. In 1964, the male escaped from the Arizona-Sonora Desert Museum, and it is presumed that someone shot him. No tissue sample remains from him, so there is no way ever to tell whether he was a hybrid.

Gish never actually saw the male, but he saw several of its offspring and concluded, "They'd been mixed with dogs. I don't remember any of them that looked like a Mexican wolf." He thought they had "smallish front paws," wide chests, and "less ranginess to the overall profile." The ears of the single female born into a 1978 litter became semi-erect at about eighteen months of age, and both hybridization and inbreeding were suggested as reasons.

A study of the skulls of Ghost Ranch wolves by Michael Bogen and Patricia Melhop in 1980 concluded that they "show tendencies toward dogs, but whether these result from dog genes or the effects of captivity is unknown." The single doglike characteristic the researchers found was one wolf's shorter muzzle. But shortened muzzles may merely be features of domestication, not of hybridization. Bone is trained by muscle, and muscle by the exigencies of life. Just as an old man's jaw may recede if he loses his teeth, or an athlete adds bone mass by lifting weights, a pup raised on soft foods in captivity might have a shorter muzzle because it develops less muscle—and therefore less bone to anchor the muscle—than its wild relatives. Even wild-born lion cubs taken into captivity show shorter, broader skulls than wild lions. Similar changes have been observed in purebred wolves in zoos. A 1941 study concluded that "skulls of wolves reared in captivity are shorter, broader and higher than those of wild wolves" and "nearer to those of domestic dogs."

Even if there were dog genes in the Ghost Ranch wolves, Ames argued that historical evidence suggested there had been dog genes in the wild population for a long time. Gish had reported that federal trappers frequently trapped hybrids or saw wild wolves running with feral dogs. "If dog genes did indeed exist in the *baileyi* populations in the past," Ames asked, "what objections can there be to releasing *baileyi* individuals today that might be similarly tainted? How can we prove, for that matter, that dog genes *didn't* exist in the wild population before?"

Despite Ames' defense, the Ghost Ranch wolves were increasingly viewed as tainted. In 1977, the U.S. Fish and Wildlife Service sent trapper Roy McBride to capture wolves in Mexico. He came back with three males, and a female that proved to be pregnant. They were placed in the Arizona-Sonora Desert Museum, where the female gave birth. Pups from that and successive litters have been shipped to other breeding facilities, such as the Wild Canid Survival and Research Center in Eureka, Missouri, and the Rio Grande Zoo in Albuquerque. That provided a rival line of breeding stock to the Ghost Ranch wolves, with which McBride wolves have never been crossed. Since then, the McBride line has produced more than 170 pups, and there are descendants in fifteen facilities, including the Phoenix Zoo and the Chapultepec Zoo in Mexico City. By 1993, there were more than ninety wolves in the "certified" McBride or Arizona-Sonora Desert Museum line.

Once the McBride wolves were breeding in captivity, those who had them were even more critical of the Ghost Ranch line, and the Arizona-Sonora Desert Museum stopped breeding Ghost Ranch–line wolves. To avoid charges that all captive-bred wolves were tainted, recalls Ames, "the U.S. Fish and Wildlife Service wanted to kill all the other breeding stocks." The service recommended at least neutering all the Ghost Ranch wolves. Everywhere, the line was devalued, disparaged, or destroyed. The Living Desert State Park, in Carlsbad, New Mexico, euthanized five young males in 1979. When the Arizona-Sonora Desert Museum got down to the last of the Ghost Ranch–derived wolves, a female that had failed to breed, they kept her in an indoor facility with a sign on the door saying "The Thing." The Ghost Ranch asked to have the wolf, but the Arizona-Sonora Desert Museum staff decided to euthanize her.

This new line of certified wolves was not without its own problems. One of the males never bred in captivity, so the founding stock of the McBride line was four individuals. With only four founders, the McBride wolves risked inbreeding depression, or loss of genetic resources to the point where they were less likely to survive. There was an enormous desire to add to the gene pool by adding wolves from other lines, but the additional possible sources all posed problems. The Ghost Ranch line was regarded as tainted. There was another lineage of Mexican wolves, at the San Juan de Aragón Zoo in Mexico, the founders of which came from the Chapultepec Zoo, but the lineage cannot be traced back to a wild ancestor. And it was rumored that some of the Aragón wolves had been bred to a dog by a keeper who wasn't thinking about returning them to the wild. There was much discussion of whether to try to capture more wolves for the captive-breeding program. Says Ames, "Some people said, 'No, no, no, don't take any more out of the wild. Save them in the wild.' "

For Ames, it was a difficult situation. She couldn't become an advocate of including the Ghost Ranch wolves in the breeding stock, "because I saw the argument that Mama wants her child on the team. I had to play my personal involvement with them down quite a bit. I'd get criticism from the Fish and Wildlife Service. They wouldn't say anything directly about me, but they'd say, 'There are all those people raising wolves in their backyards, thinking they're doing a favor for the species, and they're not doing anything for the species.'"

Several studies suggested that the Ghost Ranch line ought to be brought into the breeding stock. Work by John Patton, of LGL Ecological Genetics of Byron, Texas, found the same distinctive Mexican wolf–DNA markers in Ghost Ranch wolves and in the certified line. Ulysses Seal, of the Captive Breeding Specialist Group at the Minnesota Zoological Gardens, recommended that the Ghost Ranch population be included in the breeding population. "Most of us were convinced that these were pure Mexican wolves," says Ames, "but because there was that question of possible dog blood in it, it became politically unwise to add it.

"I hoped at first my wolves would go back into the wild. I thought I was helping to save the Mexican wolf. I kept raising them in such a condition that, if they were ever certified, they wouldn't have been conditioned to people." But it became increasingly clear that these wolves were not going to be approved for release.

"About 1980, when I could see the way things were likely to go politically, and that there was not going to be a place for wolves of this lineage to go, I decided I would have to stop breeding them." She began to separate the females in breeding season. No pups gladdened her summer days. One by one, the wolves aged and died, and she would carry them out of the pens in deep sadness, and bury them. By 1987, she was down to one last wolf. She had retired from the department, and she was about to remarry, sell her place in New Mexico, and move to another part of the country. There was no breeding program to accept the animal. She could not bring herself to exile it to solitary life in a cage in a zoo. She decided she would have to euthanize the last wolf.

It took her a month actually to do it. She went out one morning to "get it over with. I had the one wolf alone in one of the large enclosures. All through the years I kept them, I found it difficult to provide the wolves with medication, because they're suspicious of things. Vets said, 'You just sprinkle this medication on their food,' but it doesn't work that way. All through the years I had them, I bought immense quantities of frankfurters. Every day, I went out and broke frankfurters into pieces and tossed them to the wolves." The wolves would catch them in midair and bolt them down to keep other wolves from getting them. When she needed to immobilize a wolf, she would put tranquilizer pills in the frankfurters. On this day, she tossed the last wolf a bit of frankfurter and tranquilized it. "I went out and injected a killing drug. It took seconds.

"It was hard to do," she says. She had loved the wolves, and there were few people to share her grief. "You had to be careful about who you told about it, because they'd say, 'She doesn't really care about wolves, she killed her wolf.'"

It wasn't the end of the Ghost Ranch line, however. In 1992, there were sixteen survivors of the lineage, some at Ghost Ranch, others at the Navaho Nation Zoo on the Navaho Indian Reservation, the Hillcrest Zoo in Clovis, New Mexico, and two private facilities in Colorado. But they were still shunned by the Mexican Wolf Recovery Team.

Ames moved with her new husband to Colville, Washington, but she still has cause to be intensely interested in wolves. Wolves are beginning to drift down from Canada into the Selkirk and Cascade mountains. "People come and talk to the wolf lady and tell her about the wolf they just saw," she says. Many of the sightings she thinks are coyotes or dogs, many she supposes are hybrids. Meanwhile, local ranchers circulate rumors that the U.S. Fish and Wildlife Service has secretly released wolves. "They say, 'We heard some howling the other night.'"

Mistrust now hovers over the McBride line of wolves. With only four founders, its detractors say the wolves are so inbred that they ought not to be released. In Dennis Parker's view, "One female and two males is insufficient to sustain a viable population." If such criticisms are valid, we ought to be able to see evidence in the McBride line of wolves that has been bred in the Rio Grande Zoo in Albuquerque, New Mexico.

At the zoo, the Mexican wolves are enclosed in a lodgepole-pine stockade that circles a half-acre habitat. The exhibit is shaded by large cottonwoods and small junipers. It is leafy and green with grass and low shrubs. In the center, under the shade of the cottonwoods, is a mound of worn red earth where two wolves sprawl. They are unconcerned at the cries of peacocks, the roar of lions, or the squawk of parrots. When the first human visitor of the day arrives, however, the female is up immediately, glowering, her tail pressed low between her legs. Then she turns and takes off at a hurried, loose-limbed trot. Moving around the perimeter of the enclosure, head low, lips parted, glancing up apprehensively at the visitor, she slinks, trying to outrun her apprehension. And one can see in the cold yellow of that over-the-shoulder glance how fear must have crossed and recrossed this exchange of gazes over the centuries. With each circuit, she uses a well-worn trail in the underbrush. She pads on paths abraded by this ragged desire to be somewhere else. It takes her half an hour to settle down. Always when she stops at the mound, she turns and stares at the cage door, where a keeper might appear. If a human enters this enclosure, the wolves move as far away from the person as possible. "If we don't let 'em get away from us, they'll jump on the walls," says Kent Newton, Rio Grande Zoo's curator of mammals. He sees the wolf's nervousness as evidence of its wildness.

Newton stands outside the enclosure and looks at the wolves. His face is young, but his hair shows hints of gray. His neatly trimmed mustache and polished cowboy boots suggest that he is divided between city and country. As guardian of these wolves, he must see himself as both herdsman and liberator. At times, he must wonder what these walls do to the wolves and to his own wild nature.

For their part, the zoo visitors spend little time in front of the wolves: the maximum stay is about thirty seconds. People look into the trees and bushes to find the shape and register in their brains the word "wolf," then move on to the polar bears and lions. To see a wolf in a zoo is to see a picture of a wolf. It's not doing anything. People howl at it or shout "Lobo!," but it doesn't lift its head, doesn't even crack open an eye. The wolf has heard all this a hundred times a day for years, and is bored by it. And people aren't entranced by their own reflected boresomeness. The wolves are too available, but not sufficiently dramatic. "I'd rather see them in the wild," one woman says to another.

Of course, there are none to be seen in the wild. The pair in front of them is the most accessible pair of Mexican wolves in the world.

They have borne two litters in the past two years. Some of last year's pups are in a separate facility here, and others have been shipped to other breeding centers. The six pups currently in the den come up twice a day to be fed.

The Rio Grande Zoo has produced forty-four McBride-line pups, and all of them so far have survived. It was not a seamless tale of success, however. At first, the wolves were all kept in adjoining pens, and dominance struggles through the fences seemed to keep them from breeding. Only when they were put into secluded pens did they begin to breed. Since then, much care has had to go into the pairings. Wolves thrown together for the purpose of breeding may not bond, may not like each other, may not know the right gestures to unlock the mysteries of sexual union. They may be neurotic from living inside fences and being stared at all their lives, or they may be distracted or silly, aloof or bored.

Normally, zoos like to speed up the breeding of such creatures by tinkering with the technology of reproduction. There is extra interest in speeding up the reproduction of Mexican wolves because the recovery team has decided that until there are one hundred animals in the breeding population, none will be released into the wild. Also, the larger the breeding population, the smaller the likelihood of inbreeding depression. But there are limits to what can be done. Capturing females to inseminate them may cause such stress that they cannot conceive. Accustoming them to handling to counteract the stress may eliminate traits that are necessary for survival in the wild. Says Newton, "Since these animals are potentially release animals, our philosophy is hands off."

Parker believes captivity has already altered these animals. "Look at the old-type animals," he says. "They're higher-shouldered in the front, they're longer-legged, they have a broader head and broader snout." The captives, he says, "have bat ears that resemble a German shepherd. They're actually creating an animal in captivity. When you have captured animals, you are fostering a whole new set of parameters." He suggests that raising these animals in cold-climate breeding centers like the Detroit Zoo or the Point Defiance Zoo in Tacoma, Washington, may alter them in unknown ways.

Newton thinks little of Parker's charges. To his eye, these wolves look authentic. He points to a study of the wolves' DNA conducted by Dr. Steven Fain of the U.S. Fish and Wildlife Service's National Fish and Wildlife Forensics Laboratory, which indicates that the present McBride-line wolves have 90 percent of the genes of their founders. He says, "We've had Dr. Robert Wayne and Dr. John Patton study the genetics of the captive wolves. They reached the conclusion that the Mexican-wolf population is as heterozygous [genetically diverse] as the northern gray-wolf population."

Newton adds that the wolves in front of us are behaviorally every bit Mexican wolves. "The animals pair-bond, they copulate, they dig natural dens, they birth their offspring in the dens, they nurse, they regurgitate food for the pups, with aunting behavior by the other adults. There is a peck order that is established as the pups are growing up. At one time we had eleven animals in here, adults and two litters, with little aggression at all. So, in general, we haven't seen any loss of genetic ability to maintain a cohesive unit. When the female goes in the den to whelp, the male stays out in a high place as a sentinel. The adult male will even carry food to her when she is first having the pups. Security guards at night hear them howl. The female vocalizes to the pups. She calls them out of the den. She will make a whining noise to call them back into the den when she feels security is compromised. They have the so-called edginess or perceptive ability which has been lost in domestic animals. We see all of the classic behavior text mechanisms that you would read about in a book. It's all there. We haven't seen any indication of inbreeding depression through testing or behavior."

Dr. Steven Fain is perhaps a little less sanguine than Newton. "We have genetic variation," he says. "We don't have much to work with. This is an example of a very, very genetically depauperate species." He describes what is happening. Of twelve specific DNA sequences identified in two of the founder wolves, only six were passed to the next generation. In the next pairing, he found seven variations, but one offspring only inherited one of them, and another inherited five. "The variation is still there," he says, "it's just averaged over a larger number of individuals." Careful breeding may conserve enough of the genes for the wolves to survive.

When Fain looked at the same DNA segments in the Ghost Ranch and Aragón Zoo wolves, he found that the Ghost Ranch line had more variability than the McBride wolves. He concluded that bringing the Ghost Ranch and Aragón lines into the breeding program could double the genetic variability of the Mexican wolf.

Whether or not the various gene lines are brought together, the very existence of these wolves seems to demand a future release into the wild. For, in the end, a wolf in a zoo is not a wolf. It is the interaction of genes and environment that makes a species, and the whole complex interplay of thousands of such processes that makes an ecosystem. If we lock wolves up in zoos, we stop the interactions. We change their evolution.

So Newton proceeds all along as if these wolves are aimed at eventual release. He wants them to dig dens, to stay shy of humans, to keep their social habits. The wolves aren't named, because, "when you give them a name, you want to talk to them. The first thing humans want to do is relate to an animal. I don't want any relating to these animals at all—no talking, no play. In the long run, that will benefit reintroduction if that occurs. My philosophy is to give them numbers. That keeps the human psyche away from them."

But where are we going to put the wolf? Newton can't do much about this question here at the zoo, and it is a question that much troubles him. He looks back at the stockade. The female wolf is up again, running the barrier path, looking apprehensively over her shoulder.

The Mexican wolf's prospects south of the international border are at best dim. "There is very little hope for the Mexican wolf," sighs Julio Carrera of Antonio Narro University in Saltillo, Mexico. Carrera is a round-faced man with dark, inquisitive eyes and a neatly trimmed white beard. He has a look of pained resignation about him. "My whole life I am interested in wolves," he says. He has tried to be the Mexican government's point man on the subject. But the government hasn't embraced Carrera or his efforts, so he has had to seek private backing for his project.

He has been looking for wolves in Mexico. *Canis lupus baileyi* once ranged south through Durango and Zacatecas, almost as far as Mexico City. In 1977, the U.S. Fish and Wildlife Service arranged for McBride to go into Mexico to ascertain the status of the Mexican wolf in the wild. He talked to cattle buyers, bankers, and other people who would be likely to hear of wolf predation; he looked at the habitat and talked to ranchers. In his estimate, there were only fifty to a hundred wolves left. Today, Mexican wolves may survive in Chihuahua, Sonora, Durango, and Zacatecas. There were reports in 1991 of three wolves seen traveling along the Chihuahua-Sonora border; American biologist Charles Jonkel says he saw wolf tracks in that area at that time. In 1988, there was a report of a wolf killing cattle on a ranch in Sierra de las Tunas. A forester in the Sierra del Promontorio said he heard howls in 1991. And Carrera spoke with a Mexican trapper who claimed to have seen a wolf drinking from a watering trough near El Salado in Zacatecas.

For six weeks in 1992, Carrera searched remote ranching country on the Sonora-Chihuahua border. "I found only one heifer that was killed by a wolf near Río Negro," he says. He heard about a wolf killed over the carcass of a cow on an *ejido* near Tres Ríos in Sonora. Though he questioned one of the suspected wolf killers at Tres Ríos, "he didn't want to talk and he sent me off." He heard that the man might have the wolf's skin, and offered a reward for that, but as yet he has not found physical evidence to prove that the wolf, rather than an occasional hybrid, is surviving in the wilds of Mexico. In 1993, he went back, this time with funding from the U.S. Fish and Wildlife Service, Wolf Haven, and private Mexican supporters, but still found no hard evidence.

If Carrera finds wild wolves, what then? How would they be protected? The terrain is remote, and Mexican wardens do not even have vehicles with which to patrol. Ranchers still put out poison. There is a law calling for fines and imprisonment for anyone killing a wolf in Mexico. Says McBride, "It's on the books, but I don't know where you'd go to find the book." Surely not in the remote ranches of northern Mexico. Surely not in the hearts of Mexican cattlemen.

There are captive wolves in Mexico. The Chapultepec Zoo, the San Cayetano government facility near Mexico City, La Michilia Biosphere in Durango, and a private ranch in Chihuahua each have a pair. The Aragón Zoo has six uncertified wolves, but recent genetic studies have revealed in some of them mitochondrial DNA characteristic of wolves from northern Canada, and that raises the possibility that zookeepers in the past bred other subspecies into the line. In all, there are only fourteen certified pure Mexican wolves in captivity in Mexico. Even if they can be bred, Carrera worries about what might be done with them after that. Both uninhabited land and native prey are disappearing. Says Carrera, "I don't think many places will support the wolf very much longer."

Carrera fears that the enormousness of the task of doing anything for wolves in the wild makes it seem far easier simply to catch the wolves and bring them into zoos for captive breeding. But without wolves in the wild, Carrera fears, it would do little more than allow us the illusion that something is being done while we watch their extinction.

If prospects for recovery are dim in Mexico, hopes for reintroduction must focus on the United States. And with the fear that captivity will shape the Mexican wolf into an animal that can't be reintroduced, there is a sense of urgency about finding a place in the United States in which to release the wolves. When the U.S. Fish and Wildlife Service asked Texas, Arizona, and New Mexico to suggest places where wolves might be reintroduced, Arizona identified fifteen sites that had the potential to support wolves. Some, like the Catalina Mountains just behind Tucson and the Santa Rita Mountains nearby, were rejected because urban growth clouded the future for wolves there. The Department of Game and Fish looked closely at four of the areas and in 1992 began working on a reintroduction plan for the Blue Range Primitive Area in east-central Arizona. Information being developed in that plan will go into a federal Environmental Impact Statement being developed for the Mexican-wolf reintroduction.

Texas ducked the issue by declaring that it had no areas of public land of suitable size. A number of people since then have suggested that Big Bend National Park and adjoining state wildlife areas would be a good reintroduction site, but McBride disagrees. "The only deer in the park are in the Chisos Mountains," he says, "and I just don't see the wolves going up and down trying to get them. You'll hardly ever see wolves killing stuff on the side of a hill. They get dragged and kicked when they're hunting. Furthermore, those deer are really utilized by mountain lions." He doesn't think there ever were many wolves in the area, or in the Sierra del Carmen, across the border. "I never did see any sign of them, or hear any talk of them there."

New Mexico suggested only the White Sands Missile Proving Grounds, thirty-four hundred square miles of land administered by the Army. Not far from Las Cruces, New Mexico, the San Andres Mountains stretch north a hundred miles, like a tilted bench, rising to the eastern sky. They are sparsely wooded with piñon and juniper, mountain mahogany and yucca. The eastern escarpment looks down into the white gypsum sands of the Tularosa Basin. East of Las Cruces, Highway 70 climbs toward San Augustin Pass, passing the Star Wars Deli, the Moon Gate Cafe, and a billboard advertising the Space Museum in Alamagordo, which invites the traveler to "Dare to Dream." A military jet screams over the pass, rolls right and left, and then sweeps east over the chalky haze of White Sands. An F-14 shrieks overhead and does a barrel roll. Two buzzards circle, too old to be concerned with all this martial pride and haste. Signs by the road say "Warning, U.S. Government Property—No Trespassing."

It is because it was federal property that the New Mexican government suggested White Sands: the state wouldn't have to spend money on studies or on public hearings in which ranchers and environmentalists belabored one another, and it could let the two agencies of the federal government fight out the wolf issue among themselves. The Army, not wanting to subject itself to the demands of other federal agencies, withdrew the site from consideration, but reversed itself after the Mexican Wolf Coalition, the Wolf Action Group, the Sierra Club, the National Audubon Society, the Wilderness Society, and the Environmental Defense Fund all sued. A decision on reintroduction has not yet been made.

The Fish and Wildlife Service concluded that White Sands would support wolves comfortably. Says Peter Siminski, leader of the captive management breeding team for Mexican wolves, "The habitat's there. Wolves are pretty adaptable. If you get them through the first breeding and pups, they'll survive." Others disagree, however: Parker argues that there is no documented evidence that wolves ever lived in White Sands. And Roy McBride says, "I wouldn't try to stick wolves in White Sands. You could raise camels in a used-car lot, but what's it going to cost you? What are the wolves going to eat there?"

Though he hasn't walked the San Andres or Oscura mountains, he has flown over them many times, and he says, "They don't look like any places I've ever caught wolves."

All around White Sands is cattle country, and ranchers are passing rumors that wolves have already been reintroduced. Pete Gnatkowski, a rancher in Carrizozo, saw a wolflike animal after a friend had shot it; the hair around its neck was matted, as if it had worn a collar. Another friend shot another wolflike animal after it killed some sheep. Gnatkowski worries that, if wolves are introduced into White Sands, they will leave the reintroduction site and raid neighboring livestock herds. Ranchers have turned out in numbers to oppose reintroduction.

But ranchers may not have the last word. The bulk of the United States population live in cities, and they hear in the howl of the wolf the call of wildness, the sound of what was once right in nature. It is yet another identity that the Mexican wolf will bear.

West of Carrizozo, Highway 380 snakes up over the northern end of the Sierra Oscura, which rises like a board lifted up to the setting sun. Sometime long ago, the Sierra Oscura probably met the San Andres Mountains, which rise out of the west, and today they look like pillars that once supported a huge, high, mountainous dome. As the highway climbs over these hills, it runs through grassland with a lot of piñon pine. The land looks as if it would support a deer herd and afford wolves the cover of low-growing piñon and a maze of arroyos.

Between the Oscuras and the San Andres Mountains is a low, sandy desert valley that was the Trinity site, the place the first atomic bomb was tested. The mountain walls east and west lean toward each other, dark with piñon pine. The valley in between is not the flat, white creosote desert you might imagine the Trinity site to be, but greener, more mountainous, a bit closer to heaven. To think that someday wolves might pad across the valley, to imagine the immense silence of this place stirred by the howls of wolves, is to think about the righting of wrongs and the redemption of the human heart.

It is also to think about shifting definitions. What are all those wolf watchers seeing out there? Are they wolves or dogs, coyotes or hybrids? Are they wild creatures or the product of squint-eyed meddling? Are they beasts of ravening desolation, the shadows of government tyranny, or the shapes of hope and freedom? Are they atonements for the angry, mushroom-shaped cloud that has for half a century darkened our lives? Things are not always easy to see in this landscape.

10

THE PERSISTENCE OF WOLVES

Bob Ream flies his Cessna 633 south and west from the Missoula Airport, over the Idaho-Montana Divide. He flies over the White Mountain fire lookout and the edge of Big Burn, an area that has remained almost treeless since a 1910 forest fire. To the east is the snow-covered mass of the Continental Divide. Below is a wide, glacially cut canyon, the walls of which are cloaked in Douglas fir and lodgepole pine. This is good elk-and-moose country—good wolf country, too, because there are no cattle here. As Ream flies over Kelly Creek, the ragged spur of rock known as Kelly Thumb points up at the sky. Ream adjusts his radio and fishes for the signal of a radio-collared wolf.

The last government-trapped wolf in Montana was taken in 1936. But for the packs in Glacier National Park, wolves probably haven't denned in Montana since the 1940s. By midcentury, most people thought wolves were extinct in Montana and Idaho. Twelve years ago, Mike Schlegel of the Idaho Department of Fish and Game took a picture of a wolf in the valley below. Seven years ago, a researcher found wolf tracks on the hillside. There have been reports of wolves here since then, and not long ago Ream, a University of Montana biologist and member of the state legislature, flew over Kelly Thumb, looking for Wolf 9013, a gray radio-collared male, and found him, exactly where, nearly twelve years before, Schlegel had taken a photograph of a wolf. "When I found him," says Ream, "that whole slope was covered with elk trails, and I saw two moose standing right on Kelly Thumb." A few weeks ago, Wolf 9013 was at Lolo Pass, and then he was sighted ten miles south of the pass, on the edge of the Selway-Bitterroot Wilderness. Ream located him in a small meadow on Fish Creek in Montana, but when he flew a few days later, the wolf was thirty miles away, on Burdette Creek in Idaho.

As Ream finishes recounting the past wolf sightings at Kelly Thumb, the radio begins to ping like a dripping faucet. Wolf 9013 is indeed down below. We are at ninety-one hundred feet. We turn south, banking gently. The signal grows faint. Ream switches from right antenna to left antenna and back again, back and forth, listening for the signal. He peers out his window, looking for a gray wolf below. We circle, corkscrewing down, closer and closer to the ground, dropping to sixty-eight hundred feet. Now we are looking up at Kelly Thumb as we go around and around it. There is a latticework of long shadows, the trunks of burned lodgepole in the morning sun below. We circle, straining to find the wolf in the webwork of trees.

We don't see him, only his raven flying off to the south. One might imagine the wolf has changed shape, turned into a raven, and flown off. More likely, he is growing accustomed to airplanes and is simply lying very still in the shadows below, pressing against the ground, waiting for the gnatlike whine of the airplane to go away.

Wolf 9013 was caught and collared inside Glacier National Park two years ago. In the past four months, he has wandered down here to the Idaho side of the Continental Divide. He is not old enough to be either the wolf Schlegel saw twelve years before or the wolf that left tracks seven years ago, and yet he is in exactly the same location. How did he end up here? Ream is much interested in this question. Thinking about the travels of the wolf below ripples the watchful quiet of his boyish face. The persistence of wolves stirs his sense of wonder.

We head north, toward Missoula. The molar white of the Mission Range looms in the distance, and beyond it, in a blue-green haze, the mountain ridges of Glacier National Park, where Wolf 9013 was collared. The park is 150 miles away and across two main highways. "I think this pretty well proves these interstate highways have no impact on wolf movements," says Ream. Pointing off to the southeast, he adds, "There's as much wild land between here and Yellowstone. So I just don't think there are any barriers to wolves repopulating Yellowstone."

Yellowstone is the heart of the matter. Wolves have been gone from Yellowstone for decades. Of all the places in the world to which Americans would like to see wolves returned, Yellowstone is their heart's desire. It is the largest national park outside Alaska, and the emblem of American wildness. To millions of Americans, a return of wolves to Yellowstone would be a sign that nature is still alive, persistent, mysterious, and beautiful. Reintroduction of wolves there has been proposed repeatedly since the 1930s, but real momentum has gathered behind the idea since the early 1980s. There are a variety of ways it might happen. Wildlife officials might capture wolves in Canada and release them in the back country of the park, or wolves might simply walk from Canada to Yellowstone on their own. Ream sees the wolf below as one of the pioneers in that recolonization. "I've really been intrigued by this whole dispersal thing. I've said all along it's going to be the mechanism by which recovery will be accomplished. And that seems to be what's happening."

With Missoula below, Ream turns northwest and follows Interstate 90 a few miles. The highway veers off to the left, and we are flying over the Ninemile Valley. A dirt road accompanies a shallow creek up the valley, crossing a succession of fenced pastures. The slopes rising from each side of the creek are forested, but patched with clear-cuts. There is a wide green meadow, and at its edge an old barn. Last year, a litter of wolves was raised in that meadow. This year, there are more wolves.

Ream flicks the radio switch with his thumb. The radio pings, announcing that there is another radio-collared wolf below. We turn right, and follow a side canyon; and the signal grows louder. The source of the signal is a narrow ridge between two creeks draining down into the Ninemile. As we circle and corkscrew down again, it becomes clear that the signal is coming from a location perhaps less than two hundred yards across. But we see no wolves. They hunker down, press against the ground, and don't move.

The wolf below has also appeared where previous wolves materialized. In 1989, four wolves—two adults and two young ones—that had migrated out of Glacier National Park were trapped near Marion, Montana, because they had begun to attack livestock. They were returned to the park, but the two adults left the release site, and the two young ones starved. The male adult stepped into a trap outside the park. Even though he was treated by a veterinarian, he was later found starving and suffering from gangrene. A Fish and Wildlife Service officer shot him to put him out of his misery. The black female adult headed south across the Swan Range, down the Swan Valley, and into the Ninemile drainage. She found a gray male wolf there, they mated and in the spring of 1990 had a litter of six pups. They raised the litter in the pasture below, paying no attention to the cows grazing there. In July 1990, however, the female was shot to death. In September, the male was struck and killed by a vehicle on the interstate. The six pups were left on their own in the meadow below.

They survived until spring, when several of them went over Squaw Peak to the north and killed two steers near Dixon. Three were trapped and removed; the other three disappeared. A few months later, a female radio-collared wolf that had vanished months before from Glacier National Park appeared in exactly the same pasture in the Ninemile Valley. In 1992, Ream was flying up the Ninemile and located her. He decided to fly over the Idaho Divide, trying other radio frequencies, and there he found the male, Wolf 9013. Both wolves had traveled more than 150 miles from where they had been collared, and both were in locations that had been frequented by other wolves.

Ream has been looking for wolves in Montana for nineteen years. In the 1960s, he had heard stories of wolves being shot in Montana, and he wanted to find out whether there were wolves still in the wild. In 1973, the year the Rocky Mountain gray wolf (*Canis lupus irremotus*) was listed as an endangered species, he began to collect reports of wolf sightings in Montana and Idaho.

For decades, people had been saying the wolf was gone. But in fact, all along, ranchers and back-country rangers as far south as Wyoming had repeatedly seen glimpses. In 1976, George Gruell, then a biologist on the Bridger-Teton National Forest, in northwestern Wyoming, gave the Rocky Mountain Wolf Recovery Team a list of fifty purported sightings. A range rider for the Hansen Ranch, near Jackson, Wyoming, observed wolves in the woods. Gruell talked to a man who had seen a wolf run across the ice on Jackson Lake in the 1940s. Says Gruell, "He was flabbergasted at its size and thought it was a moose." He said the man did not tell any authorities because he didn't want to be accused of making up stories.

Through the 1970s, the reports increased in Montana. Shortly after Ream started collecting sightings, Jerry Desanto, a Glacier National Park ranger, walked into a meadow near Polebridge, on the North Fork of the Flathead River, and saw a wolf. By 1977, Ream had collected 315 credible sightings. Three skulls of recently killed animals were confirmed by taxonomists as belonging to wolves. In 1979, Joe Smith caught Kishinena, the wolf Diane Boyd followed for two years, just north of the Canadian border. With all this evidence that wolves were poised to make a comeback in the Northern Rockies, Ream's Wolf Ecology Project got funding enough to hire Diane Boyd and Mike Fairchild to track wolves on the North Fork.

When the wolves began denning in Glacier National Park, Ream's effort to find wolves turned into an effort to study them. The Wolf Ecology Project got funding from the National Park Service, the U.S. Fish and Wildlife Service, the U.S. Forest Service, and the University of Montana. When funding ran short, Ream got private donations to keep volunteers in the field doing winter tracking. His studies on the North Fork of the Flathead River have given a detailed picture of the recolonization. A genealogical chart hangs on the wall of his office. The Wigwam Pack, which formed just north of the Canadian border in the early 1980s, was poisoned out. The Magic Pack, which denned in Glacier in 1986, split in two and became the Camas and the Sage Creek packs. The Sage Creek Pack consisted of two adults and five pups until British Columbia opened a hunting season and all but a mother and one pup were shot. Before the shootings, two males had left the pack, and they joined with a dispersing female to form the Headwaters Pack. In 1990, two females in the Camas Pack bred and the pack divided into the North Camas and South Camas packs. A female from the Camas Pack dispersed and joined with a surviving male from the Wigwam Pack to form the Spruce Creek Pack.

With radio collars on pack members, Ream's researchers have been able to follow the wolves over vast distances. They have watched them disperse south to the Ninemile Valley and the Idaho-Montana Divide. One wolf from the Magic Pack went 550 miles north to the Peace River Country in Alberta. Clearly the wolves of Glacier could reach Yellowstone.

Ream knew how intolerant people might be of the returning wolves, and that biologists still argued about the effects of wolves on prey. Once it was clear that wolves were coming back, he began to address questions that would be posed concerning recovery. One of the first was, what effect might wolves have on elk, moose, and deer? To find out, Dan Pletscher of the University of Montana School of Forestry radio-collared thirty moose, thirty elk, and thirty deer. All the collars had mortality transmitters: if the collar lay still for four hours, the radio pulse rate doubled to a hundred beeps per hour, and Boyd and Fairchild would go in quickly to determine what had killed the animal. The study showed that, among white-tailed deer, two were killed by mountain lions, two by humans, and two by wolves. Among moose, two were killed by bears and one by wolves. Among elk, nine were killed by mountain lions, two by humans, and three by wolves. Says Ream, "This shows very nicely that the wolf is just one of a number of predators. It certainly is no worse than any other predator."

The research in Glacier National Park kept Ream at the center of wolf issues. He was a member of the recovery team that wrote the original recovery plan in 1977 and then revised it in 1987. When the 1987 revision called for the reintroduction of wolves into the wild, the state of Montana, angry that the federal agencies made all the rules about grizzly recovery but did not pay the costs of state responsibilities for the bears, declared it would not participate in endangered-species recovery unless the species was delisted. Wolf recovery was growing more and more contentious.

There are two different philosophies urging the return of the wolf in the Northern Rockies.

One view, which Ream holds, says that the wolves are already back, and that they are headed straight for Yellowstone. "I think dispersal is going to be the mechanism by which wolf recovery is going to occur," says Ream. "We've had nine long-range dispersals. Most of them have been north into Canada, but the last three have been south into Montana and Idaho. We still have two dispersers that haven't been found. We still have a lot to learn about dispersal."

The other view says wolves are going to be reestablished in Yellowstone only by human agency. That view is given greater authority by the politics of the situation. Ranchers fear that wolves that get to the park on their own may be beyond control, because the Endangered Species Act makes it illegal to kill them. If wolves are introduced into Yellowstone by human agency, they can be deemed an "experimental-nonessential" population and subject to greater restrictions if they leave the park. Ranchers catching such animals in the act of killing livestock, for example, might be permitted to shoot them. Therefore, many who see the return of the wolf as inevitable are fighting hard for reintroduction. Says Ream, "In the last year, I've seen a flipflop. Livestock people were totally against reintroduction; now they're saying that doesn't look too bad."

At the same time, there are people in the agency who look upon the wolves arriving by dispersal as passive and uncertain. They see that as something less than management, whereas they see reintroduction as something clear, predictable, and tangible. Ream explains, "The U.S. Fish and Wildlife Service is more management-oriented and more control-oriented." And they want to spend money on management, not on research.

In 1988, the recovery team was disbanded, and its members were told that the U.S. Fish and Wildlife Service would carry out the plan. In place of the recovery team, there would be a Wolf Working Group, consisting chiefly of U.S. Fish and Wildlife Service, U.S. Forest Service, National Park Service, and Indian-reservation biologists. The Wolf Working Group was more interested in reintroduction than in natural recovery, partly because its members were managers who felt that reintroduction was a way to control where the wolves go and where they don't, partly because of a lawsuit filed by Defenders of Wildlife, charging that the Secretary of the Interior would be remiss if he did not proceed with the introduction. In 1993, a draft environmental-impact statement proposed reintroduction into Yellowstone National Park.

With reintroduction in the forefront, Ream's federal research money dried up. He would like to keep the studies going. "I don't think the study area is full yet," he says. "I think we'll see some change in home range as the population increases. Since we've been in since the first wolf arrived, I feel we should follow it until it reaches saturation. But they think we've gotten enough data."

Ream feels he has been pushed into the periphery. "When I was on the recovery team, I felt I was very much in the loop," he says. "Now . . ." he shrugs his shoulders. But he is still out there pulling for the wolves—flying his own aircraft on his own time, at his own expense, following the radio signals of collars now maintained by the U.S. Fish and Wildlife Service and the National Park Service, monitoring the progress of the wolves.

It is clear that wolves are returning to the Rockies. Ream looks down at the landscape below. The peaks of Glacier rise dark and green, ridge after fading ridge, to the north. To the south and east are the Bitterroots, the Sapphires, the Beaverheads, and, beyond the Continental Divide, the Tobacco Roots and the Madisons. There are no broad croplands or extensive urban areas to cross, but plenty of wild country. Ream traces the route with his eyes, off into the haze-shrouded east. Wolves go where wolves have been before. And if there is a wolf highway down below, invisible to human senses but broad and beckoning to wolves, it is aimed at the nation's jewel of wild places, Yellowstone National Park.

It is clear to Ream that down below, in the Ninemile Valley, something momentous is taking place.

There is a full moon over the Ninemile Valley, but it peeks through tatters of rainclouds that have given the sky a ragged blackness. When the clouds part, the silvery moonlight illuminates the pines and spruces in grays and greens. It casts dark shadows behind small pebbles on a dirt road that winds up the north side of the valley. Mike Jimenez gets down from his battered blue government pickup truck and stands in the night, listening to the rush of a creek down the hill and the sigh of wind in the pines. At forty-four, he has wrinkles at the corners of his bright-brown eyes. With a generous smile and an expression of pleasant surprise, he looks as though he was placed on earth just to put people at ease.

Jimenez tests the silence, and then he howls—a short, deep note that rises, hovers, and then falls again. He waits a few seconds and howls once more, starting two notes higher, and rising higher before the tremolo and the falling note. He runs through the sequence more rapidly than most howlers; it is more like conversation than like an invocation of spirit.

Almost immediately there is an answer from down in the valley, a repeated bark, at once peevish and questioning, as if to say, "Who's out there? Come on, there's someone out there, and I don't like it." We listen. Jimenez says it might be the wolf, no matter how doglike it sounds. Its pitch is as deep as a wolf's, and wolves sometimes do respond by barking back, sounding annoyed at Jimenez's intrusion in their night quiet. Pups may squeal and yammer, and then he'll hear an adult make short summary barks that seem to silence the young ones. But this barking, though deep, seems to go on too long for a wolf's, into the undignified and ineloquent repetition of dogdom.

He switches on the radio receiver and holds up the directional antenna. The signals ping loudest from just the direction of the barking. Maybe it is a wolf; maybe it's a dog barking at the wolf. We drive down to the valley floor and howl again. Now there are two animals barking from a pasture a quarter-mile down the valley. One of them is clearly a dog, its voice without music, its yammer expressing, if anything, discomfort with the night. Jimenez tries the radio again, and the signals come from up the slope, away from the dog.

We go up the valley a couple of miles to the suspected rendezvous site. The Ninemile is a cafeteria for wolves. As we drive up the dirt road, there are deer in every pasture; from the clustering of eyes shining in the night, one can judge that there are hundreds of them. "Deer and cows both use the fields," says Jimenez, "and the wolves hunt the deer among the cattle." Even though cattle would be easier prey, no wolf has yet been shown to have killed a cow here.

We stand in the dark awhile, listening. The night air is cold. The wind is soft, but the trees are full of murmurs, and the creek is full of drowning voices. We howl again. A hundred yards away, a coyote looses a cackling bark, a kind of loony mirth on the heels of Jimenez's more dignified howl. Another coyote chimes in with a high, hysterical squeal, repeated over and over again without cadence or dignity. Still, no wolf howl lights up the night.

Driving back down the valley, we see wolf scat on the road—almost certainly fresh, since we didn't see it when we came down the road a few minutes before. It's hard to avoid the inference that this is a message intended for us. Perhaps a critical view of Jimenez's howl? He collects it in a Ziploc bag. We look on the road for fresh tracks. In the headlights of the truck, our shadows on the trees become lumbering sasquatches.

We take the dirt road back up the hillside and radio-locate the female. Her signal is faint, and then strong, and then it vanishes, suggesting that the female is moving behind rocks and trees, and perhaps behind the crest of the ridge above us. Jimenez howls. There is silence. Then, in the valley, like a far-off wind, the faint but unmistakable howl of a wolf, low and rising and mournful, rides the air.

Jimenez is out here checking because, twice this weekend, people have seen the wolves and the wolves have not, as usual, quickly melted away into the forest. Two days ago, Jim and Chris Farrington, who have a weekend cabin here, were out cutting wood. While he was operating the chain saw, she walked into some dense brush, and there encountered two pups and an adult wolf, seventy-five feet away. The adult wolf stood its ground and barked, then howled; that spooked Chris, so she left. Later the same day, she and Jim both saw the pups, but the pups didn't move off, either.

That behavior bodes ill for the wolves, especially with hunting season only a few weeks away. In October 1991, a hunter shot one of the wolves. Jimenez discovered it while he was tracking three of the wolves through the snow. "The back end of a stump was all blown out, and there was blood in the snow. The wounded wolf's tracks went off, and there were human tracks following it. Whether it lived or died, I don't know."

Jimenez has been tied to the wolves of the Ninemile Valley for two years. Born in California, he took a degree in biology from San Francisco State University, and had been visiting the North Fork area of Montana since his college days. He met Ream, Boyd, and

Fairchild and became interested in the wolf study. He enrolled at the University of Montana, wrote a master's thesis on the Glacier wolves, and went to work for the Fish and Wildlife Service in Helena. When the first wolves turned up in the Ninemile Valley, he was detailed to monitor them. He was a highly fortuitous choice for the job.

The wolves had moved from a den somewhere in the hills to a rendezvous site in the pasture down in the valley, the pups staying in the woods at the edge of the pasture while the adults went off to hunt. When the adults brought food back, the whole family would be out loafing or cavorting in the meadow. At night, Jimenez would hide in the old barn at the edge of the meadow and watch the wolves through an infrared scope, which illuminated the animals in long-wave light invisible to their eyes.

Part of Jimenez's job was to protect the wolves, part to be on the spot to look after the ranchers' rights; most of it was public relations. "There's a really common myth that we're dumping wolves everywhere," he says, "so a lot of it was to establish rapport, telling people this is what's going on, and telling them to come to us if you have problems." It was not a simple assignment. Jimenez's baptism in the Ninemile was a complaint by a resident that the male wolf in the pack had killed her dog. "There were cattle all over," says Jimenez. "The wolves would go in and out of cattle. The male wolf would cross the pasture, taking the pups to and from kills. The male wolf was taking the pups over to a moose kill and he ran into this dog. It was not a threat. It was an old dog. It didn't bark." Just as the wolf attacked the dog, its owners came up the road and turned into their driveway. When the owner turned, her car headlights spotlighted the wolf on top of her dog, killing it. She called Ed Bangs of the U.S. Fish and Wildlife Service in Helena at 11:30 p.m. and held the phone out the window to let Bangs hear the wolf howling. Jimenez was sent out to the Ninemile Valley the next morning.

"She wanted to vent at someone," says Jimenez. He talked to her for a couple of hours, buried the dog for her, and held a little graveside ceremony for which he even made a cross. "We sat and talked on the injustices of life. She was a whole lot more pro-wolf before she lost that dog," but she still invites Jimenez back to barbecues in summer.

The encounter made it clear that the job entailed more than

Jimenez or the Fish and Wildlife Service had supposed. "All of a sudden," says Jimenez, "the Fish and Wildlife Service needed to do something more than just monitor." He saw that he would have to "try to come up with solutions to help ranchers live with wolves," and figured he could best protect the wolves by trying to see to it that the ranchers' experience with them was benign.

His approach was to try to meet as many of the residents of the valley as he could, and to approach them with a kind of "I got some good news, I got some bad news" routine. "Basically, I'd listen and not talk down to them," he says. "Some of them were, 'Thank you very much, we don't want 'em.' It's hard to change people's attitudes. Those ranchers see wolf tracks on the road, and twenty feet away they have calves—it's a legitimate concern. It's a tough place to have predators."

He sees that they have problems of their own. Once he was called up to check on a report that the wolves had killed some lambs in the valley. "These two elderly people had five lambs and an ewe killed by predators. He's got a heart condition, and each lamb had a grandchild's name. The lambs were their winter money. I expected to get torn apart. It ended up being a coyote that killed them."

The ranchers actually saw the wolves rarely—they saw tracks, or caught quick glimpses of wolves darting across the road. Jimenez found he could help both wolves and ranchers by letting the ranchers know where the wolves were and what they were doing: "A lot of people's tolerance is improved if you just make sure they know what's going on." He would take ranchers out tracking, and he would urge them to let him know where the wolves were. "If you're a rancher, you feel better if you have some input. I tell them, 'Give me a call.' "

Jimenez saw that the essential issue was that the ranchers wanted to have control over their own lives. "A lot of the resentment for predators is focused about that. They can't do a lot about markets or about disease, but a predator shows up and that's something they can do something about." He recognized that living with wolves would require of the ranchers a very untraditional sense of forbearance. He hoped somehow to convince them that they didn't *have* to do anything about these wolves. He hoped, too, that the wolves would cooperate by staying away from livestock.

Jimenez would go out and check up on the wolves and on the ranchers every couple of days. When the male was killed on the highway in September, Jimenez had to make a decision about the orphaned pups. He decided he would have to bring them food, and proposed to feed the pups road-killed deer for the next six weeks, until hunting season. After that, he thought, successful hunters would leave enough gut piles in the woods, and unsuccessful hunters would leave enough wounded deer behind, to let the wolves fend for themselves. He had to find the roadkills. In the beginning, he spent a lot of time cruising back roads at night, looking for dead deer. Then he discovered that bicyclists often knew, from having cycled through the pungent smell, where animals had dragged themselves off the road and died in the bushes; he got cyclists to tell him where to find roadkills. At first, the state of Montana agreed to let him have the necessary permit to take roadkills, but when stockmen outside the Ninemile area put pressure on the Montana Department of Wildlife and Parks, the state revoked the permit. Ironically, local people were beginning to support Jimenez: "People in Ninemile would call me up and say, 'I know where a roadkill is.' "

Part of Jimenez's job was to keep the wolves from becoming habituated to people. "I've always tried to keep them from associating with me. Once in a while, they did see me, and I would be really obnoxious with them, shoot over their heads, make a lot of noise. I used rubber gloves to keep human scent off carcasses." He put the roadkills out only at night, so that other humans wouldn't follow him and linger to watch the wolves. One of his worries was that the wolves would use the roads too often during hunting season. "The first week of hunting season brings out these road hunters, people who drive around in trucks and shoot out of the windows. There'd be gut piles on the road, and wolf tracks coming down to the gut piles that night. I'd go around at night with a shovel and pick up the gut piles and put ammonia down on the dirt to hide the smell."

To try to discourage hunters from taking shots at the wolves, he made his presence plain, dressing in his Fish and Wildlife Service uniform and driving up and down the road. "And again," he says, "most of the hunters were really interested and asked me to come see the tracks they had found."

By Thanksgiving, the wolves were moving around more. By Christmas, they were killing deer on their own. It looked as if they would not begin killing cows. Says Jimenez, "We had 'em around cattle all the time they were growing up. Shirley Hager had this old cow with a prolapsed uterus, and it was nerve-damaged and crippled." Tracks in the snow indicated the wolves passed within six feet of the cow but left it alone. "They could have had it for breakfast and they just weren't interested." In the end, Hager did what the wolves would not and put the cow out of its misery.

But in the spring, the wolves went over the mountain range to the north and killed two cows near Dixon. "As long as wolves aren't killing things," says Jimenez, "ranchers are incredibly open-minded. When something gets killed by a wolf, it's an absolutely irrational response. When those two heifers were killed over near Dixon, there was daily communication with people over there calling people in Ninemile saying, 'We knew this was going to happen, and you let us down.'"

If the Fish and Wildlife Service didn't act, it would jeopardize the whole reintroduction effort. Jimenez says, "It seems like it's a fair assumption, once they kill cows, they will kill cows again. Usually when they start it, it's like a dog killing chickens: they get better at it. And then you get into the situation of wondering what kind of a gene pool do you want out there?" There seemed to be no alternative: the decision was made to remove the wolves. Three were trapped and taken to Glacier National Park, but there all three dispersed in different directions. One was shot in the Swan Valley by a rancher who mistook it for a dog. Another was shot by persons unknown and found afterward floating in Mud Lake, near Columbia Falls. The third was recaptured and sent to Wolf Haven in Tenino, Washington. The three remaining Ninemile wolves disappeared from sight. For a time, no wolf howls hovered over the moonlight of the Ninemile Valley.

And then, months later, another radio-collared female turned up in the same Ninemile Valley pasture. Diane Boyd came out and, trying the radio frequencies of wolves that had gone off the air, identified her as a Glacier National Park wolf that had vanished in January. By 1992, people in the Ninemile Valley were seeing the tracks of three wolves. Jimenez surmised that the Glacier female had joined two members of the previous year's Ninemile pack. That spring, the newly formed pack produced a litter of pups.

The new wolves were not as shy as the previous pack: there were reports of wolves standing and barking when humans approached, instead of running off. One of the males in the pack, a black wolf, would go off by himself for a few days at a time and then return. A farmer along Interstate 90, near Frenchtown, saw the black male walking among his cows without attacking them. Jimenez went out with the farmer one night. "We were out there by the interstate, howling, with trucks roaring by," he says. "It was a new dimension in wolf research."

On May 20, 1992, the wolves were found to have eaten off a dead steer in the Ninemile Valley. It was not clear whether they had killed the steer or just scavenged an animal that had died of disease. Following the Fish and Wildlife Service's announced depredation procedures, Jimenez captured a male and the female within a few days. He put a capture collar on the male and released him. "If they kill another cow, they will probably have to be removed," he said, "but it isn't clear they killed the cow, only that they ate off it."

Months passed, and no other cows were lost. Jimenez hoped that this meant either that the wolves had not killed the cow or that, if they had, it hadn't made them habitual cattle killers. The latter is a novel idea, but it hasn't been tested, because, once depredation occurs, the Fish and Wildlife Service is generally obliged to remove the predator. "In the past," says Jimenez, "when it occurred, we went in and destroyed our data base."

He feels parental about these wolves. "You get pissed off if they come near cattle. You want to say, 'I told you once, I told you a thousand times!' " He feels responsible both for the wolves and for the people who live out there.

But the dead steer makes Jimenez's job harder. "I keep everybody informed where they are and tell them, 'If they're around cattle, let me know.' The ranchers' attitude is, 'We appreciate knowing where they are, but we slept better not knowing.' "

The tenacity of these wolves engages Jimenez on a level beyond biology. "There's the attraction that they've been wiped out and you have this perspective of reestablishing things, and of righting the wrongs. They keep hanging on, they keep knocking." And they keep knocking in the same places. The mystery of that entrances Jimenez: "There's a side of me that just kind of smiles. We know some things, but we don't know everything. The Ninemile is not that unique. Given they could go in any direction out of Glacier, why would a wolf end up there? I don't understand it."

In a sense, Jimenez practices the art of consolation. Consolation has a way of turning up, much the way wolves have a way of turning up, again and again in the same setting. It is tempting to see Jimenez as a noble experiment in coexistence. But if his work here is an experiment, who designed it? Jimenez? Circumstance? The wolves?

Bob Demin drives his truck very slowly down the Ninemile road. On a horse behind the truck rides Arch, a grizzled cowboy, whose face is set in that half-wince, half-sneer that men affect when it is important not to show pain. After Arch come the cows, about sixty of them, and calves and three bulls. They amble slowly, their sides heaving right and left. Behind them, a motley group of riders—schoolgirls, housewives, and retirees—drive the cattle. An old man in a bright-orange hunting cap, riding along in the cab of a truck, brings up the rear. It is the fall drive. Demin is bringing in the cows and calves he has summered on Forest Service grazing allotments above the Ninemile Valley. The horses in Shirley Hager's pasture race over to the fence to celebrate the passage.

There is a fresh dusting of snow on the Ninemile Divide, to the southwest. Clouds hug the peaks. It is wintry cold, and the riders are bundled in sheepskin jackets, hunting coats, and ski parkas. It has rained off and on, and is snowing even now a few miles up the valley. This drive has gone well: Demin is short only one cow. He thinks that, when he brought it off the allotment in the woods, it got mixed in with a neighbor's stock, and will turn up.

This is an important day. In a business that is 70 percent worry, Demin is getting this summer's calves into his own pasture, right outside his kitchen door. He has already sold the calves via a videotape auction, and they will be trucked off to market in a few weeks. The cows will remain under his watchful eye over the winter.

The cowpokes herd the last of the cows into the pasture and latch the gate, and then head on over to the house for a barbecue. Demin has cooked beef in a pit in the ground and there are casseroles, elk meat, and fruit pies. He introduces me to Arch, the sun- and wind-sculpted old cowboy, and tells him I am writing about wolves. Arch turns and walks off on bowed legs, nodding his head to acknowledge the introduction, and saying, in a voice as creaky as saddle leather, "Balance of nature, I guess." It is both summary and farewell: it is all the hospitality he can muster. His response expresses the conservatism of ranching, and the suspicion with which outsiders and newcomers and wolves are all met in the rural West.

Demin is sixty-one, no-nonsense straight, purposeful, and focused. In a way, he bridges the gap between the Old West and the urban world that applauds the return of wolves. He grew up on this ranch, then went off to join the infantry and was away for twenty-eight years. He was stationed for years at San Francisco's Presidio, and saw that city through the tumultuous years of the hippies and the rise of gay rights. In 1979, he retired and came home to take over management of the ranch from his aging mother.

The ranch requires watchfulness. Cows die and disappear. "One year," he says, "my mother thought we lost one to a human predator. One year, a neighbor down the valley, it seemed like all his cows had twins and everybody else's cows didn't have any calves." Hunters are also a problem. In two weeks, the opening of deer season brings an assortment of hunters, some of whom shoot before they know what they're shooting at. A rancher must also live with the flat-out stupidity of cows. "One year," says Demin, "I had a cow that got into a ditch and rolled in it and got on her back and couldn't get up and died." And predators are a concern. There are bears, mountain lions, and now wolves. Nobody can say for sure whether the wolves killed the cow they were feeding on in May, but Demin thinks they did.

Demin is especially aware of the wolves. His cows were grazing in the pasture the wolves used as a rendezvous site, and it was his Forest Service grazing allotment that the wolves denned on this year. "The ranger down there didn't even want to tell me where the thing was," says Demin. "His comment was, 'It's up there in the middle of your allotment. You probably don't want to put your cows up there.' I said, 'Where am I going to put my cows?' "

When he found out he had wolves near his cattle, he went to see Montana Congressman Ron Marlenee, a fiercely outspoken opponent of wolf reintroduction. "I talked to Marlenee. He said, 'Now, we're not going to have any wolves.' I said, 'Wait a minute: they're here!' and I showed him pictures. He said, 'Get rid of 'em.' I said, 'That's easier said than done.' The wolves are here and they're going to stay here. There's nothing I can do about it, or anyone else can do about it.

"Before they were here, I never thought about wolves. I wasn't too happy when that female showed up, the one that was involved in the predation over at Marion. She didn't last too long—somebody shot her. And the male, I guess he was a model citizen. In the next meadow, he killed an elk, and he was going back and forth to the pups, and I had some cows in that pasture—he didn't bother the cows."

If the wolves would stay up in the hills, he says, he would have little problem with them. "Some of the people here are much more shoot-'em-on-sight people. As long as they don't bother me, I've got no problem with 'em. I could live with 'em. It's just another hazard. But if they're out here killing my cows, and I've got a rifle, if I can kill them, I would."

To make clear his resolve, he talks about coyotes. For ten years, he had no trouble with coyotes. Four years ago, just when the cows were calving, coyotes took to hanging around his pasture. "I'd go out in the morning and see them wandering around where the cows were. I'd take a rifle and shoot around them to scare them off." But the third day, he saw a coyote walking with what seemed to him to be purposeful looks behind a freshly born calf. "I said, 'Buddy, you don't learn,' and I shot him."

Ranching is a form of pessimism. "We've been lucky so far," says Demin. "Before, we had a pair of wolves and their pups, and they didn't kill anything over here. This year, we had the same thing, and they took that one yearling. What's going to happen next year, when you're facing up to seven wolves? Things are going to be a little different. When you talk about a pack of ten to twelve wolves, it's going to be a different situation. Right now, there's so few of them."

There are two issues here. One is the likelihood of wolf predation on cattle—if the wolves take his cows, who, if anybody, will compensate him? The other issue is what controls federal agencies will put on land uses once wolves get established. The potential for restrictions on Demin's Forest Service allotment is relatively small. "I don't turn out until the first of July," says Demin, "and by then, the wolves have probably moved to a rendezvous site. I don't think the Forest Service would say not to use the allotment because of the wolves."

What he worries about is the Fish and Wildlife Service coming up with restrictions. Conversant enough with the world to know that a large number of wolf advocates see the animal as a symbol of wilderness, he worries about losing his grazing allotments to a demand for wilderness for the sake of the wolf. "Our experience in the Ninemile shows they can be just like coyotes, as far as living with man. It's not a wilderness. They seem to like these roads—it's easier to get around on them. They say they need these corridors of wildness to move around on. That's bullshit. That first wolf den was right on the edge of a clear-cut and within a half-mile of two houses." Now, he says, "the Fish and Wildlife Service makes a point of saying you don't require any special management. But you never know.

"Wolves are a part of the planet . . ." he sighs. Though he leaves the rest of the sentence under his breath, maybe it is this: ". . . but I wish we didn't have to have them right here."

It must say something about the wolves that, in settling into the Ninemile Valley, they didn't choose a pasture owned by a Ron Marlenee but one watched over by Ralph and Bruce Thisted. The Thisteds live on the edge of Ninemile Valley, overlooking the pasture of the farm they operated for fifty years. They raised cattle and today the comfort and lively curiosity in their faces testify to their apparent success. A few years ago, actress Andie MacDowell bought the old farmhouse and pasture on the valley floor, and the pasture is now leased to other ranchers. The Thisteds built their new home on a hill overlooking the pasture, and now live there, two bachelor brothers, each amiably contesting what the other says about just about everything. There is a broad flagstone fireplace, wood-paneled walls, and picture windows giving expansive views of the spruce and fir trees and the pasture below. Bruce—white-haired, straight-spined, trim, with intense blue eyes—sits in a chair in the middle of the room with a pair of binoculars in his hands. He says Ralph saw a bunch of magpies take wing down in the meadow and has gone out to see what alarmed them. "Being we heard the wolves howl in the meadow last night," he says, "it may be the wolves got some deer."

Says Bruce, "I believe there's wolves been coming through here forever. We heard one howl ten or fifteen years ago. If you'd have said it was a wolf then, people would have said, 'Oh, yeah, sure.' It was like a bigfoot or a UFO."

Ralph walks in. He is lanky, fit, and bubbling with enthusiastic curiosity. "Magpies, ravens, and a coyote," he says, reporting on what's going on in the meadow. "That coyote won't leave—I got within twenty feet of him. I can't imagine what those birds were up to."

In July 1989, the Thisteds saw the first wolf. They were fixing a fence by the road when their pointer dog, Heidi, began to bark. "Her hair was standing on end," says Bruce. "We saw something in the trees, just a shadow moving. I had the .22 in the truck. I got it, and I looked through the scope sight. I said, 'It's not a coyote.' It was something we didn't know what it was. You see things like that in the woods all the time.

"Thanksgiving, we started seeing the wolf tracks. Then a lion hunter out here asked me, 'You see that wolf track around here?' That's when Ralph became pretty sure it was a wolf."

And when the litter appeared in their old pasture below, the brothers were delighted. "We could sit here with the scope and watch the wolves play by the hour," says Bruce. The grass in the meadow was a couple of feet high. People coming along the road couldn't see the wolves through the grass, but the house was high enough that the Thisteds could see down into it. Ralph would go over before daylight to the barn and hide with a video camera. As it grew light, he would videotape the wolves. "All the time I was there," he says, "they never realized I was there. Even the magpies didn't know I was there. If you can fool a magpie, you can fool anything. After the sun was up, they'd go back into the timber."

"The cattle were here," says Ralph. "There'd be cows out in the field, and wolves at the same time. The wolves were playing, and the cows didn't even look at 'em. They never bothered anything. I have videos of them a hundred feet from the cows. You can hear the cows, and the wolves are barking a little bit, and the cow doesn't pay them any attention."

They watched the first litter grow and shared their videos with Jimenez. They were philosophical when the wolves killed the cows near Dixon, and Ralph helped Jimenez to trap the wolves in the pasture and transport them to Glacier.

In April 1991, the female with the collar appeared. They saw her first through the trees. She was hunting mice in the meadow. "It was in the exact spot in the pasture as the other pack," says Bruce. "A week before that, there was another one, over here." He points down valley a little. "It had a collar." Diane Boyd came out with her radio receiver and went through some frequencies, and up came the collar. "We'd get the signal from this living room."

Bruce tells me, "Ralph is more in favor of wolves than I am. I think they ought to have a chance, but if they got to slaughtering cows, well . . ."

Ralph jumps in: "Boy, some people have no concern about anything out in the woods," he says. "Logging doesn't bother 'em, mining doesn't bother 'em. But mention the wolf and they go through the roof."

The Thisteds know that some of their neighbors would like to see the wolves removed. "There's probably not a place for wolves in the Ninemile," Ralph says. "It's a small area, and they're bound to get in trouble sooner or later. Anything that's found dead, the wolves are just going to be blamed for it."

Says Bruce, "Years ago, a farmer would lose a cow and he'd drag it out on the railroad track. It's going to be the same with the wolves: if they lose a cow, they'll blame it on wolves.

"I think it's ridiculous," Bruce continues. "We lost calves every summer from logging trucks, and the sheriff wouldn't even come out and investigate. Logging trucks would come running down the road sixty miles an hour. One summer, we lost a half-dozen calves to the logger camp up there. We know they butchered 'em up there—we had someone tell us. When we lost a half-dozen calves to coyotes, we didn't have reporters here. Why are you reporters so interested? Let me turn this around and ask you that."

Ralph, who has fallen silent, figures out his magpies now. He thinks it was so cold the grasshoppers in the field couldn't fly, and the magpies descended on them, and the coyote perhaps came for the same feast.

Returning to the wolves, he says, "It's an adventure, because there's so many people interested in it—because so many people are excited about it. We had a rancher come down to talk to us from the Paradise Valley. They've got wolves there, too. He wasn't upset about it."

That the wolves should have picked the Thisteds' pasture as a rendezvous site is at least a fortuitous coincidence. "It's an ideal spot for them in the valley, because they had a place where they weren't harassed," says Bruce. Almost anywhere else, he says, "I just don't think wolves and people mix. This is the only full-time residence at the end of the valley, so they haven't been bothered.

"I like wildlife," says Bruce. "Always have. Heck, when I see a bull elk out there, I watch him till he's out of sight."

"In high school, all you thought about was going hunting," says Ralph.

"You get older, you mellow," says Bruce. "It's a thrill for me to see a wolf."

Montana is getting older and mellower, too. Wolves are coming back, and they are coming back in part because people aren't shooting them on sight. They go where the country beckons, and perhaps part of the welcome sign is a sense, not just of the willingness of the landscape to take them in, but of the willingness of people. There are places like the Ninemile Valley here and there around the Northern Rockies. The Fish and Wildlife Service has invited people to call in their wolf sightings. The Forest Service provides mailers so that hikers and joggers can write down sighting reports and send them in. From the fall of 1991 to the fall of 1992, there were more than three hundred reported sightings. A pack of five, with two collars, was sighted near Murphy Lake. An elk calf was killed by a wolf on the Idaho border, near St. Regis. Another Glacier Park disperser was found down in the Bitterroots. There was evidence of a wolf near Darby, and howling near Dillon, and there were sightings even in Yellowstone National Park.

Wolves go where wolves have gone before. They seem to be

headed to a rendezvous with wolf destiny in the place most Americans think of as the symbolic heart of wilderness, to Yellowstone. And as they go, Bob Ream is up there in his airplane, soaring over the mountain fastness of the great divide, the sun smiling down on his small aircraft, the wolves pressing close to the ground below. And Ream is listening.

11

YELLOWSTONE

In August 1992, a crew of filmmakers set up their equipment several hundred yards from a bison carcass in the Hayden Valley in Yellowstone National Park. The bison had died in a rutting battle, and already grizzly bears and coyotes had found it and were scavenging a meal. Something else was there, too, standing off to the side, patiently watching the bears—a large dark animal, bigger than the coyotes, black with a white blaze on the chest. The animal sauntered over to the kill just as a sow grizzly and her two cubs were leaving. They passed close by each other, the strange animal not seeming to give the bears much notice. But the bear cubs stood on their hind feet and swiveled their heads around in an ursine double take, as if to say, "Was that a wolf?"

The strange animal slinked over to the bison, where another bear was still feeding. It stole up to the carcass, took a piece of meat, and trotted away with the loose-limbed gait of a wolf. A coyote waiting nearby stood almost mouth to mouth with the black animal as it bolted the meat, but its proximity didn't seem to bother the black animal.

The National Park Service, the media, and the wider community of wolf fanciers, however, were bothered. Wolves had not been trapped or shot or found dead in the park for sixty years. The last photograph of a wolf in Yellowstone, a jittery home movie that itself remained a piece of controversy, had been taken nearly twenty-five years before. Now the U.S. Fish and Wildlife Service was in the midst of preparing an environmental-impact statement which would recommend reintroduction. If wolves follow the same routes and cover the same ground year after year, as is suggested by Montana's Ninemile wolves, this might be the start of a naturally occurring breeding population. But if this were a wolf and it produced young in Yellowstone, said Ed Bangs, the U.S. Fish and Wildlife Service's coordinator for the environmental-impact statement, that would preclude reintroduction of an "experimental-nonessential" population. It would also mean that wolves were back, and because they weren't experimental-nonessential wolves, ranchers would be forbidden even to frighten them away if they caught them in the act of attacking cows or sheep.

A lot rode on the identity of the mysterious dark animal. Was it a wolf or a hybrid wolf-dog? Park rangers might try to trap the animal and examine it for obvious dog characteristics, but, explained Wayne Brewster of the National Park Service, even this scrutiny might not prove conclusive. Since no pattern of DNA unique to dogs has yet been found, not even DNA testing could tell definitively whether it was a hybrid. Even if it were judged to be a wolf, the question would arise: how had it gotten there? Had it walked? Or had it arrived in the truck of a clandestine wolf-restorer? "And if you do capture it," said Brewster, "there's a risk of injury or mortality. Right now, we don't think it's worth the risk." So the Park Service ruled out capture.

Two weeks later, at the International Wolf Symposium in Edmonton, a group of wolf biologists assembled to view the videotape. John Weaver, who was completing a study of wolves in Jasper National Park, said, "If it does have dog blood, it's not much." But, clearly of two minds, he added, "The impression I have of wolves in Jasper National Park in June and July is they almost look like greyhounds. This one looks too well fed." David Mech said, "I don't like that steep pitch in the head. I don't think we can say what it is." He also expressed concern about the black-and-white coloring, but Diane Boyd said she had seen similarly colored wolves in Glacier. Robert Ream smiled quietly at the thought that this animal might prove that wolves can indeed make it all the way from Canada to Yellowstone without human assistance. Roy McBride said, "That's a dog. It's got no muscle on it. He's kind of narrow between the ears. He acts like he knows what a bear is. But I think somebody turned him loose." Ron Nowak, who has done more morphological study of wolves than anyone else, said the contrasting black-and-white color and the steepness of the head gave him doubts. "I think people who know behavior would be in a better position to say," he sighed.

Months passed without a satisfactory answer, though the animal was seen again several times in the park. A retired Forest Service supervisor watched it for fifteen minutes through binoculars near Blacktail Ponds in December.

Meanwhile, another wolf was shot, outside the park, near Worland, Wyoming. A hunter said he had seen it running with a pack of similar animals, and thought at first it was a coyote. Tests of the animal's DNA, however, showed that it was related to the Ninemile wolves, and could have been one of the young from the earlier pack or a wolf from the same ancestry that had come all the way from Glacier. "This demonstrates that they have the capacity to get to Yellowstone on their own," said Steve Fritts, U.S. Fish and Wildlife Service's coordinator of wolf recovery for the Northern Rockies.

And that meant, to many ranchers and hunters around Yellowstone, that the return of the wolf was all but inevitable. Either wolves would walk into the park fully protected by the Endangered Species Act, or officials would release them and perhaps dole out some conditions under which a rancher might harass, remove, or even shoot them to save his livestock.

It looked as if the long debate over returning wolves to Yellowstone was over.

Wolves have had a troubled life in Yellowstone from the time the first whites arrived. Reservation of Yellowstone as a national park in 1872 did not protect wolves. The cavalrymen who patrolled the park from

1886 to 1916 killed several dozen of them. In 1914, passage of the law establishing the Predatory Animal and Rodent Control Service made it national policy to eliminate predators from all federal lands. In 1915, a year before the birth of the National Park Service, Vernon Bailey of the U.S. Bureau of the Biological Survey was sent to Yellowstone to develop a program to eradicate predators. He found the number of wolves in Yellowstone "alarming" and warned that unless park officials started "getting wolves and coyotes out of that region" they would consume all the elk. A year later, when the National Park Service assumed management of the park, it invited predator-control agents of the Bureau of Biological Survey into the park to take wolves, coyotes, and mountain lions. In ten years, the predator-control agents killed at least 134 park wolves. Joseph Grinnell, a professor of biology at the University of California, and his former student George Wright objected. But Park Service Director Stephen Mather feared that, if control of coyotes and wolves didn't continue, the predators breeding in the park would spill out into neighboring cattle and sheep ranches, and the ranchers would run to their congressional representatives crying that the national park was bad for their businesses. The national-park idea was then young and not widely accepted. Mather saw his first responsibility to be the building of a constituency to support the parks.

In 1926, what appeared to be the last wolf in the park was taken. Though reports of individual wolves, and even of occasional packs, trickled in over the next five decades, there was no evidence that wolves were breeding in Yellowstone. In 1933, the Park Service announced the policy that no native predator should be harmed because of its utilization of other animals in a park, except when the animal threatened the existence of another species. But the new policy came too late. In 1938, when Adolph Murie completed a study of the role of coyotes in controlling elk numbers in the park, he concluded that wolves and mountain lions had been eradicated. Predator control stopped inside the park, but the Bureau of the Biological Survey set up poison-bait stations along the park boundaries, in effect erecting a wall of cyanide to keep park coyotes from wandering onto the ranchlands beyond the park.

Today, it is not clear whether wolves survived in the park or in the back country of the neighboring Bridger-Teton and Shoshone national forests. In the 1940s, 1950s, and 1960s, forest rangers and range riders now and then heard howls or saw animals in the neighboring national forests. Yellowstone Park ranger Ben Arnold said he saw four wolves feeding on an elk carcass in the park in 1934. Glen F. Cole, supervisory research biologist in Yellowstone in the 1960s, combed park records and came up with 104 observations of 156 wolves in or adjacent to the park between 1930 and 1969. He judged sixty-nine of those observations to be reliable sightings of wolves.

Others doubt that wolves survived, however. John Craighead, the distinguished biologist who studied grizzlies in the back country of Yellowstone from 1959 to 1973, says that, with one notable exception, "there weren't any wolves in there, absolutely. Wolves, when they are present, cover a lot of ground, and they make a lot of tracks. In winter, wolves can't really use the whole park, because the snow is six to ten feet deep," depriving elk and bison of forage: all the wintering elk and bison concentrate around the thermal areas, and if wolves survived in the park, they, too, would have concentrated around the thermal areas. In years of traveling those areas in winter, Craighead says he never saw wolf tracks or heard howls.

Without a wolf population in view, scientists were never able to appraise the role of wolves in the Yellowstone ecosystem. Today, that role is hotly debated. There is no compelling evidence that predators ever regulated ungulates in Yellowstone. It has been suggested that there are so many different species of predator and prey that no one species of predator can control prey numbers. Yet many people have believed that the eradication of wolves knocked Yellowstone out of ecological balance.

Much of what people think about wolves in Yellowstone is inferred from what they conclude about elk. Elk are especially controversial in Yellowstone. People debate how many there used to be and how many the park can support today. Park scientists don't know what the prehistoric elk population may have been, and so cannot tell what a normal density might be. Current views of the elk are colored by faulty estimates of past numbers. Milton Skinner, who was park naturalist in the 1920s, declared, "Back in 1900 when I first knew the region, there were over 75,000 elk in it. . . . But in 1925, there were only 30,000 left." No one today can say how Skinner arrived at his early estimate. Recently, Yellowstone National Park research biologist Douglas Houston looked at the park's past elk-census efforts and concluded that early managers tended to overestimate the number of elk. Between 1911 and 1920, for example, they allowed counting routes to overlap, so that the same elk could be counted twice. In addition, they estimated the elk they didn't see on the basis of the way they felt the population might be going. In 1915, when there was no census, it was estimated that there were 35,000 elk in the park. The next year, it was assumed, because the winter had been mild, that the number was even higher, even though only 12,455 elk were actually counted. The fall of 1919 started with cold temperatures and heavy snows and the ensuing winter was remembered as severe, even though records show that December, January, and February were unseasonably mild. And since people expected elk to succumb to hard winters, it was concluded that as many as 14,000 had fallen to starvation and hunters. After reviewing the historical evidence, Houston declared, "Nothing resembling a population eruption and crash occurred."

But the Park Service continued to believe that elk numbers were originally high and then fell precipitously. Skinner thought he saw a clear decline, and believed its cause was that wintering grounds to the north of the park had filled in with farms and fences, livestock which stripped the ground of winter forage, and hunters who mowed down the weakened elk. "The only hope for saving the elk," he wrote, "lay in keeping them in the mountains on protected lands rather than letting them migrate down into the lower valleys fairly swarming with rifles." The Park Service put out hay to try to keep elk in the park in winter.

However, with the drought of 1931–34, the Park Service's view of elk changed: now the elk appeared to be too numerous. Early photographs of the park showed thick groves of aspen and willow. In drought years of the 1930s, the park's plant cover suffered. Soil was trampled bare by elk herds, and shrubs and grasses were nibbled to ground level. Park biologist Rudolf Grimm believed that human settlement north of the park was holding elk on winter range inside the park, and that "the continued heavy use of this range by excessively large numbers of animals had detrimental effects upon the growth and succession of plant cover." He held that the process had been going on for decades but had only become apparent during the drought. Park scientists put up "exclosures," fencing around groves of aspen or plots of sage or grass, and watched to see how plants responded when protected from grazing. They saw native vegetation return inside the exclosures while plants outside continued to be nibbled to the ground. In aspen exclosures trees grew taller, while outside they lost height because of browsing. The number of trees inside the fences increased as suckers growing from older trees put forth new stalks that weren't eaten by herbivores, while outside the fences the number of trees declined. One could see a braid of ecological consequence unraveling as the vegetation changed. As aspen disappeared, so would the beavers. Without beavers building dams, the streams would cut deeper and water tables would drop. Streamside willows and cottonwoods would disappear. Even the trout would vanish. Grimm believed elk were the heart of the problem. He estimated the northern-Yellowstone herd numbered about 12,000 in 1939 and urged that it be reduced to no more than 7,334, "to prevent serious injury to the range plant cover and needless suffering among the animals."

The hunting had already begun. From 1935 to 1968, the Park Service shot elk inside the park in an attempt to regulate numbers. Habitat quality did not improve, and by 1962 the Park Service sought to reduce the elk to about 5,000. The elk hunts were intensified. Elk were hazed with helicopters toward waiting teams of hunters, who blazed away at them from point-blank range. Before the shooting stopped, the northern-elk herd was reduced to about 5,725 animals. When the scenes of slaughter were shown on television, the reaction was predictable: the American public was outraged, most especially because the killing was taking place inside a national park.

Biologists generally accepted the view that the elk were overpopulated. Wrote Durward Allen, "This situation has been brought about by the elimination of important natural enemies and the failure, in establishing the park, to include adequate year around range for big-game animals. At the time of Lewis and Clark it was to some extent true that the wolf and cougar allowed the grass to grow." Murie had suggested, in 1938, that the best solution would be to restore predators. In 1963, an Advisory Board on Wildlife Management assembled by Secretary of the Interior Stewart Udall and headed by A. Starker Leopold, professor of zoology at the University of California, issued a report focused in part on the Yellowstone elk hunts. It supported the hunt, but argued that, if humans had not altered ecological balances in the first place, the hunt would probably not have been necessary. The Leopold Report urged that national parks should, as a matter of principle, re-create and maintain the conditions that existed when whites first arrived. Udall directed the parks to try as much as possible to implement the principles of the report.

But was the changing face of the park really the result of predator control? In the 1980s, Douglas Houston undertook a more exhaustive study of elk-population estimates, climate, and vegetation and concluded that there had been no elk eruption before 1980 and that the earlier studies revealed effects of drought rather than overgrazing. He set up exclosures and found that, even where grazing was excluded, the aspen stands did not resemble the groves in photographs taken during the 1890s. He was convinced that the decline of aspen was due to the suppression of fires, which formerly stimulated low, dense growth of trees from suckers. The decline of willow he believed may have been due to reduced soil moisture because of climate change and fire suppression. More recent studies of pollen in lake sediments suggest that there has been little change in aspen cover in the park over the years. The view of the Park Service today is that climate, rather than predators, regulates elk numbers.

But in the 1960s, the Park Service believed elk densities were unnaturally high. Some voices clamored for more human predation on elk, others for restoration of wolves. In the 1960s, Yellowstone National Park Superintendent Jack Anderson may well have thought wolves were the preferable of the two.

In 1967, wolves suddenly started appearing again in Yellowstone. At first, there were sightings by park visitors and back-country rangers. Then, in December 1967, Marshall Gates, a seasonal ranger back to visit in winter, filmed a wolf running away from him on an eight-millimeter movie camera. Glen Cole set up a monitoring system whereby visitors who saw what they thought might be wolves could report the incidents to rangers. In two years, he claimed to have 126 observations of 214 wolves. "The greater number of observations since 1968," he held, "is partly due to an established system for reporting sightings of wolves and intensified efforts to see the animals," implying that the wolves had been in the park all along. He suggested that the increased sightings after 1969 were due to "one and possibly two pairs of wolves producing young." By 1971, he believed there were "a minimum of ten and possibly 15 different animals" in the park.

But, gradually, those sightings faded. John Varley of the National Park Service mapped the sightings year by year, and showed that the animals moved west and out of the park, where, one by one, they seem to have been shot by ranchers. Between 1973 and 1975, John Weaver searched the park for wolves but found none. Varley guesses the failure of the wolves to stay in the park may have had to do with the fact that the elk hunts had reduced the elk population to a quarter of their previous levels, and that prey populations may have been too low to support the wolves. By 1975, Yellowstone's nights were again empty of wolf song.

What accounted for this cluster of sightings? The evidence is strong that someone introduced wolves into Yellowstone National Park in 1967. A Park Service employee reported seeing wolves in cages in the back of a truck in a Park Service garage at the time. Craighead, who worked in the park during those years, says emphatically, "There *was* an introduction. The behavior of Superintendent Anderson at the time that that first wolf was sighted and some statements he made to me at the time" led Craighead to conclude that the wolves had been released by Park Service personnel. He suspects Park Service managers may have manufactured earlier sightings of wolves in park records to give the appearance of a natural return. And Stanley Hathaway, former Wyoming governor and later Secretary of the United States Department of the Interior, recalls seeing one of the animals while on a winter snowmobiling tour of the park with Superintendent Anderson. They had stopped for lunch at West Thumb, and Hathaway looked out onto the frozen surface of Yellowstone Lake and saw a large, dark animal chasing a wounded goose. "I said to Anderson, 'When did we get wolves in Yellowstone?' He said, 'That's not a wolf. It's a coyote.' I said, 'You can't fool me. That's a wolf.' I can definitely tell you it wasn't a coyote. It was too big and too dark in color.

"He said, 'Yeah, we put a few in here a few years ago. We brought them from Alaska to see what would happen.' He said they brought five or six, as I remember."

Other Park Service personnel adamantly deny that they introduced wolves. Glen Cole, now retired from the service, says, "It wasn't done. A group of professionals in a federal agency just don't do that. Your professional reputation is at stake. You don't pull that kind of crap."

Says Mary Meagher, who has been a member of the park's scientific staff since 1959: "No, we didn't dump off any wolves. If we thought it would have worked, we would have. I do not believe wolf reintroduction could be pulled off without my knowledge. And I have no knowledge of one."

Cole suggests that there were always wolves in the area, and that they staged their own small renaissance. Or, he says, "It's conceivable someone would have dumped a pet wolf. We did see some animals there, and they surely didn't behave like wolves. It's conceivable someone could have dropped one in. But I wouldn't have given a plugged nickel for its chances."

Whatever entered the park in 1967 didn't last. Said Cole, "There's intense competition for food there, both from avians and mammals. A carcass doesn't last very long. It was a difficult place for wolves. Coyotes were in there very strong and would pack up and kill moose." Perhaps it was because they were unable to form packs and thus unable to compete with the coyotes that the wolves drifted west to wither under the gunfire of ranchers outside the park.

Cole says he long ago decided that coyote densities of two per square mile in the winter elk range north of the park would prevent Canadian wolves from recolonizing Yellowstone: "I think a wolf would have to fight his way through coyotes to get there." He believed that if wolves survived in the national forests around the park they would not serve as a source of future populations. Having worked with the Fish and Wildlife Service to get its predator-control hunters to pull their Compound 1080 bait stations back from the park boundaries, he looked upon the removal of the bait stations as a test of the idea that wolves could naturally recolonize. But wolves didn't reappear. He concluded, "It kind of looked after a while like reintroduction was the way to go."

Cole was convinced wolves should be returned to Yellowstone. "It was part of your objective if you were going to follow the congressional mandate to portray a representative fauna," he says. "It was not so that you could control the elk—the prey controls the predator, rather than the predator controls the prey. The wolf was missing from the system, and it was your mandate to restore it."

So Cole worked to bring about a public and legal reintroduction. In 1975, he wrote an environmental assessment for a project to "restore a viable wolf population in Yellowstone National Park by introductions." He proposed: "Between 15 and 20 wolves of the proper subspecies from viable populations in Canada would be introduced in a manner designed to reestablish two or three viable packs." Whole packs would be released so that they could compete with the coyotes—"I think you could release single wolves in there and you'd never hear from them again," said Cole. The wolves would be soft-released—kept and fed in enclosures in the Lamar Valley, where they would form social bonds that would keep them together once they were freed. Such wolves would not scatter, as the wolves of 1967 seem to have done. Anticipating that the wolves would probably slip outside the park and kill livestock, Cole suggested that owners of livestock be compensated for losses. He urged "research to determine how both livestock and wolves can occur on these public lands with minimal conflicts."

Even if the National Park Service had agreed to Cole's proposal, success would have been far from guaranteed. Wolf releases were still experimental: there had been only four publicly acknowledged attempts, and all had failed. The first failure occurred on Isle Royale in 1952. In 1960, four nineteen-month-old pen-reared wolves were released on Coronation Island in Alaska and they survived to produce at least one litter; but by 1968, they had exterminated the deer on the thirty-square-mile island and died out. In 1972, five pen-reared wolves from Alaska's Arctic Research Laboratory were released two hundred miles away, near the Colville River. Before the release, experimenters hung dead caribou bulls by wires in a standing position inside their cages to see whether the wolves showed any inclination to hunt. One of the males seized a dead caribou by the rump; that seemed to suggest they had the instincts to survive. But once they were released, all the wolves returned to civilization. One showed up in the company of a wild wolf at the garbage dump of an oil-drilling camp, twelve days after being released. Two others hung around the village of Umiat's dump. One was shot near Umiat. Another was shot when it approached a native hunter's camp near Teshekpuk Lake. None of them showed any ability to catch wild caribou. The fourth release took place in 1974, when the Michigan Department of Natural Resources captured two male and two female wolves near International Falls, Minnesota, and released them near Huron City, Michigan—an experiment to see whether wild-caught wolves would stay and breed where they were released. Three of the wolves departed at once and took up residence fifty miles from the release site. Deer hunters opposed to the release offered a $100 reward to anyone who shot a wolf. Two of the released wolves were shot. One died in a trap. The last was struck and killed by a car on a road.

Reintroduction was at best experimental. But in 1975, there were other reasons the National Park Service would not even come close to approving a reintroduction. Yellowstone was laboring under a heavier burden: recovery of grizzly-bear populations. Sheep ranchers had lost grazing allotments where grizzlies were known to summer, and one rancher near Yellowstone had actually been prosecuted for shooting a bear. Logging and mining operations were modified for the sake of bears, and parts of Yellowstone closed to hikers to keep them from encounters with bears. The park was already embroiled in conflict, and it was no time to invite new controversy.

Cole's environmental assessment for a wolf release remained unsigned by the regional director of the Park Service. In 1975, Anderson retired and Cole transferred to Voyageurs National Park in Minnesota. John Townsley became the superintendent of Yellowstone National Park. Says Norman Bishop of the National Park Service, "John Townsley said to me in about 1980, 'We aren't going to mention wolf recovery until we get the grizzly on a good footing.' He was very sensitive to the fact that if he even mentioned wolf recovery it could shoot grizzly recovery."

. . .

In 1977, the first recovery plan for Northern Rocky Mountain wolf populations was written, but it made no specific recommendations for reintroduction. It was argued that a healthy population of wolves existed just across the border, in Canada, and because of that, species that seemed down to their last gasps, such as California condors or whooping cranes, were more deserving of attention. The argument that wolves should be returned to Yellowstone simply to restore the historic fauna was not in itself enough, but the argument would change.

Renée Askins would help to change it. When Askins was a student at Kalamazoo College in Michigan in the 1970s, she wanted to work with wolves. With Erich Klinghammer's Indiana Wolf Park only two hours away, she arranged to do a behavioral study there. Watching wolves for as many as eighteen hours a day, she saw in them something powerful and moving, something reflective of the human need to balance impulse and order. She wrote a paper about the ways different religious traditions viewed wolves. One day, John Weaver was visiting Wolf Park to talk with Klinghammer about the behavioral implications of wolf reintroduction in Yellowstone National Park, and Askins met and talked with him. She visited Yellowstone and realized, "The longer I was with the captive wolves, the more painful and difficult the compromises of seeing them in captivity grew. Wildness joined what I experienced in Yellowstone and what I experienced with wolves in captivity. I made the commitment that I would spend time working to see them in the wild." In 1981, she moved out west to work toward the return of wolves to Yellowstone.

Askins arrived in Yellowstone at a time when plans for reintroduction were becalmed. The Reagan administration had taken office with a vendetta against anyone who argued for programs which might restrict a private landowner's property rights or use of public lands. The ranching community of the West was tied both ideologically and politically to the new administration, and ardent preservationists were ferreted out of the Department of the Interior. In 1982, Russell Dickenson, the director of the National Park Service, told Congress that the Park Service had no intention of introducing wolves into Yellowstone.

Askins wanted to get the issue rolling again. She heard about an exhibit prepared by the Science Museum of Minnesota entitled "Wolves and Humans," a rich celebration of wolf complexity and personality. She decided that bringing it to Yellowstone would help create a broader constituency for wolf reintroduction, and that the exhibit might serve as a platform from which those already supporting reintroduction within the National Park Service could launch a renewed effort. Working with the Yellowstone Library and Museum Association, the National Park Service, and Defenders of Wildlife, she got a grant to bring the exhibit to Yellowstone, and organized a symposium on wolves to coincide with the opening. The Park Service was so impressed with her dedication and effectiveness that it hired her to work for the summer as coordinator of the exhibit. The exhibit was a huge success. Whereas the Park Service had doubted that thirty thousand people would visit the exhibit in the course of its display at Yellowstone, in fact 215,000 came to see it. "We just couldn't handle them all," recalls Askins. "We had to let people in fifty at a time."

Blue-eyed, with long brown hair and a ready smile, Askins conveys at once innocent optimism and youthful enthusiasm. She has an affability and a likeness to people that draws her into conversation and inspires others to listen to her. On Askins' second day on the job, William Penn Mott came to Yellowstone to be inaugurated as the new director of the National Park Service. After the ceremony, Yellowstone Park Superintendent Robert Barbee introduced Askins to Mott. Askins, unabashedly a persuader, set about convincing Mott not only that wolf reintroduction was ecologically desirable, but that it was an act with far-reaching implications for the human spirit. Mott was especially interested in the way parks might meet spiritual needs. He and Askins talked for an hour. And Mott listened.

Askins perceived that the debate about wolves is not just about historic faunas or ecosystem functions or loss of livestock. She realized that the debate is so laden with hidden meanings that it is almost wholly symbolic. "You can never predict the way people are going to connect to the animal," she says. "They're so wholly a metaphorical animal. They are a creature of dawn and dusk. In Minnesota three years ago, I was flying with Dave Mech, and there were thirteen wolves out on the ice on a little peninsula. We came around to see them again and—*bump!*—they were gone. The closest forest was a mile away—they couldn't have gotten there. They just disappeared. I think they offer a vehicle for us to talk metaphorically about the things in our lives that are not here or we wish were here.

"Wolves represent something far greater than the consummate predator in an ecosystem," she says. "When I talk about the wolf issue, I talk about the importance of wildness in our lives. It's wildness that heals us. We need contact with it, regardless of whether we live in the city or in the Alaskan wilderness.

"Wolves offer that sense of wildness—the way wolves move, the way they play, their unpredictability, their living on the edge of their endurance, savage and surviving out there." To see such things, she says, helps us to find ourselves.

Askins went on to Yale to earn a graduate degree in wildlife ecology, and twice during that year, Mott spoke at Yale. Both times, Askins lobbied him to work for wolf reintroduction. By the time of their second meeting at Yale, Mott was listing wolf reintroduction as one of his priorities. He had wolf buttons made, and he had gotten the Zion Natural History Association to produce educational materials about wolves. Says Askins, "He made an extraordinary issue out of wolf reintroduction. He started right from the beginning to mention it in his talks and put it on the agenda." But once Mott put it on his agenda, he was undermined by Congress. When the Park Service sent Yellowstone Park biologist Norman Bishop out to do educational programs about wolves, the Idaho and Wyoming congressional delegations complained to the Reagan appointees at the top of the department, and the Park Service had to reel Bishop back in. When Mott indicated the National Park Service might move forward and write an environmental-impact statement on wolf reintroduction into Yellowstone, Senators Simpson and Wallop of Wyoming and Symms of Idaho prevented it. But Mott kept talking about wolf reintroduction. "He was gutsy," says Askins.

After Yale, Askins returned to Wyoming. In 1986, she started a project aimed at the restoration of wolves in Yellowstone under the auspices of the Craighead Research Institute. And in 1990, she formed a separate group, the Wolf Fund, which would work toward the goal of reintroducing wolves into Yellowstone by seeking out key individuals and convincing them. Askins would meet, for example, with Interior Secretary Donald Hodel or Assistant Secretary William

Horn, to try to get them to think about the issue. "We find people who comprise pivotal points in the way things work," she said. "I think that's how it's done. This has to do with people and shifting bedrock. We try to move people."

People *were* moving. Three hundred miles to the northwest of Yellowstone, in Missoula, Montana, Hank Fischer, of Defenders of Wildlife, was also working toward the reintroduction of wolves. Fischer had come to Montana from Ohio in the early 1970s, worked for a time as a free-lance writer. He took wildlife-biology courses at the University of Montana and then went to work for Defenders. The National Park Service's 1980 management policies called for the restoration of species lost to parks through human agencies; that suggested that Yellowstone might be considered as a site for reintroduction. When a series of warm winters in the 1980s dramatically increased the elk herd from twelve to nineteen thousand, that seemed to Fischer to lend weight to the argument that predators were needed to help restore ecological balance. He began attending the meetings of the Northern Rocky Mountain Wolf Recovery Team to urge them to call for specific reintroductions. "Just getting to the point where we had a recovery plan that suggested reintroduction took four or five years. Getting the U.S. Fish and Wildlife Service to sign the recovery plan took another one and a half years." The revised plan, approved in 1987, called for reestablishment of the wolf in three areas: in Glacier National Park and the Bob Marshall Wilderness, in central Idaho, and in Yellowstone National Park. Lacking the wholehearted support of the U.S. Fish and Wildlife Service, it was signed not by the director of the U.S. Fish and Wildlife Service, Frank Dunkel, but by John Spinks, deputy regional director.

Once the plan was approved, Fischer sought to build enough public support to convince the government to implement it. Defenders of Wildlife sponsored the "Wolves and Humans" exhibit in Yellowstone, and also manned a "Vote Wolf" booth in Yellowstone, at which volunteers informed visitors about wolf issues and urged them to write letters and sign petitions in support of reintroduction. Fischer wrote articles, talked with ranchers, and lobbied.

In 1987, he helped persuade Utah Representative Wayne Owens to introduce a bill calling for reintroduction, but the Wyoming and

Idaho congressional delegations blocked it. In 1988, Congress called upon the National Park Service to determine whether wolves would affect prey and grizzly bears in the park, and whether or not reintroduced wolves would be controlled in and out of the park, and to reports its findings. Completed in 1990, the study found that wolves would pose no threat to either prey species or grizzlies in Yellowstone. A day after the report was published, Senator James McClure of Idaho introduced a bill calling for the removal of wolves from the endangered-species list, and subsequent reintroduction into Yellowstone. In effect, he was conceding that wolves were inevitable, and that the ranchers would be wise to compromise in order to protect their ability to deal with depredations. Congress, however, rejected McClure's bill as an attempt to circumvent the Endangered Species Act, and instead called for the preparation of an environmental-impact statement for wolf reintroduction into Yellowstone National Park. That amounted to tentative approval of reintroduction.

Many of Yellowstone's neighbors were outraged.

The Yellowstone River flows out of Yellowstone National Park into Montana's Paradise Valley. From Gardiner to Livingston, the valley is a land of great swooning hills of grass and forest. Overhead is an enormous overarching sky. The snow-covered serrations of the Absarokas line the east. There are dark pyramidal peaks with spidery webs of snow dusting their summits, and deeply gouged avalanche chutes diving down into dark forests of spruce and pine. Below them are long sensuous rolling meadows where elk bugle in the fall. It is a vast, lush, heart-throbbing country, the kind of landscape that's apt to invite immodesty and passionate devotion to causes. It was once dedicated entirely to ranching and hunting. Today, suburban ranch houses sprout on the hilltops. Close up against Yellowstone National Park, the Church Universal and Triumphant sprawls across a vast bulldozed construction pad, a jumbled city of trailers and buildings and school buses apart in style and spirit from the lonely ranches up and down the valley. It is a parable of the modern West: migrants from the urban world have come to find peace or deliverance or beauty, but they bring with them urban styles that challenge the setting and discomfort the older population. In such places, the Old West finds itself fighting for its life. The wolf has become a symbol in the conflict.

Frank Rigler has a ranch just north of Corwin Springs, overlooking the Yellowstone River. His family has lived on this land since 1908, when his grandfather, a former miner, settled here. He recalls an uncle who was working for Canyon Inn in the park who saw wolves there in 1908. At forty-nine, Rigler is muscular and compact, quick-moving and intense, almost a blur of motion. Beneath his close-trimmed beard, his face lights up now and then with a very youthful whimsy. He is not a keeper of secrets or a quiet plotter; his feelings are right near the surface and he is critical unto sarcasm, emotional unto anger. He tells of coming out of his ranch house one day to see a stranger taking pictures of the apple trees growing next to his sheep pasture. When he asked what the fellow was doing, the young man identified himself as representing something called "The Grizzly Project," and told Rigler that apple trees attract grizzlies and that grizzlies that visit ranchyards end up getting shot. Rigler didn't appreciate the lecture. "Sometimes you just get mad," says Rigler. His eyes narrow, and he coils as he finishes the story, then unfurls a punch just two inches from the nose of the listener. "I hit him as hard as I could. Then I told him, 'You see that "No Trespassing" sign out there? You get off my property and stay off!'"

Both a rancher and an outfitter, Rigler keeps sheep, cows, and horses on eight hundred acres of his own land. "There's families been ranching here for 100 years, ever since the 1880s," he says. "Something's gotta eat this grass." In the fall, he leads hunters out into the hills behind his house to hunt elk "from here to Yellowstone." Kevin Gallagher, who is helping Rigler build a new house, says he has seen elk so numerous that they looked like waving grass moving along the hillside behind Rigler's barn. Rigler tells of a cousin who lived in California and had been a hunter, but decided that hunting was beyond his ethical limits, gave away his guns, and became a vegetarian. When the relative came to visit and they got into a disagreement over the virtue of hunting, Rigler told him to hit the road and never come back. "You can choose your friends, but you can't choose your relatives," he sighs.

Rigler once had nearly four thousand acres, all on lands lying north of Yellowstone National Park, in corridors used by elk, bison, deer, and antelope as they migrate out of the heavy winter of the higher ground in the park. In the warm winters of the 1980s, the number of wintering elk north of the park grew and grew. To provide more wintering ground, the state of Montana started buying up winter range for elk outside of the park. Montana approached Rigler for three thousand of his acres. "After a while," he shrugs, "you get so many elk, you throw up your hands and say, 'Hey, let 'em have it.' It wasn't pseudo–elk range like they have in Jackson. I thought these elk could migrate out of the park and they wouldn't overgraze the park."

Clearly, Rigler doesn't regard Yellowstone National Park with any sense of neighborliness. "They don't take care of it," he says. Among other things, he believes it is overgrazed. "If I had a piece of ground above Yellowstone National Park that looks like the park does now, they'd have taken my land—they'd have condemned it. The last beaver I saw in the Hayden Valley was dragging sagebrush. There's no trees left out there." In winter, he says, the elk all move out of the park onto private lands, and he believes that is because the park managers are inept. In 1988, fire burned one-fifth of the park; park managers look upon the conflagration as a natural event, but Rigler sees it as a waste of resources. He thinks park managers were foolish not to extinguish the fires promptly. "It's these same people who are promoting the recovery of the wolf," he says.

In 1987, he got involved with wolves by going to a meeting inside the park. Like many people who live near mountain parks, Rigler wants the roads to open as early in the year as possible. He attended the meeting, he says, because it was on his mind that budget cuts were causing the park to open the roads later and later each year. "The park didn't have money to plow the roads. But at the same time, they'd hired the University of Wyoming to do a study that said you gotta educate people that wolves are good. It cost the taxpayer a lot of money." On the visit, he came upon the Defenders of Wildlife wolf-information booth, in which volunteers talked about wolves, passed out fliers, and put on puppet shows for children. Rigler looked upon the booth as a Park Service entity, and he got angry. He was particularly outraged by a puppet show he watched that day. "They would tell visitors that the Little Red Riding Hood view of wolves was all wrong. They had a puppet Little Red Riding Hood and a puppet wolf. The Little Red Riding Hood puppet said, 'I was wrong. Wolves have changed.'

"Wolves haven't changed!"

He fears what wolves will do on his ranch as they follow the herds of elk and bison and deer out of Yellowstone in winter. "They'd run the horses through the fence, into the highway. I had some dogs do that." One horse ran through three fences. When Rigler caught up with him, an artery on the horse's leg was spurting blood. "You think wolves are going to do anything different? Hell, wolves are just the same as dogs. It isn't what they'd eat, it's what they'd chew up.

"That's the pantry up there," he says, gesturing toward the park. "Well, a lot of 'em leave the pantry. The wolves aren't going to stay on that Forest Service ground, either. They travel sixty miles in a day."

Even more, he fears that he will lose the ability to do what he wants with his land. "The biggest thing is property rights. My place right now is in prime grizzly-bear habitat: the grizzly has more rights than I do." He fears that, once wolves become established in the Paradise Valley, he may be prevented from hunting on his own land, or even stopped from developing his land to compensate for the loss of ranching and hunting. Talking about the land he just sold for elk habitat, he explains, "If you get a bunch of wolves denning down there, they'll shut it down to fishing." He notes that parts of Denali National Park have been closed to keep people from disturbing active wolf dens, and says, "One-third of Yellowstone is now shut down due to grizzly bears.* How much will be shut down for wolves? The hikers won't even be in there!"

Rigler has become a spokesman for the outfitters and ranchers that oppose wolf reintroduction. At hearings, he offers grim aphorisms and slogans. For example: "What do wolves and AIDS have in common? Once you got 'em, you'll never get rid of 'em." Speaking about damage to livestock and closure of public lands, he asks,

*Norman Bishop of Yellowstone National Park says that only 18.5 percent of the park has ever been closed at any one time to protect grizzlies, and that most of that area was restricted only to certain kinds of use during certain seasons or at certain hours of the day. Much, for example, was restricted in March, April, and May, when few visitors were in the park.

"Who's gonna pay for it? We're short of money in education. We're short of money in health. We're short of money in so many things. Let's get back to basics before we start to think about the wolf."

The media, he says, have sold out to the other side. "They don't tell the whole story. We never get our side in the press. These press people are a bunch of bunny huggers any more. You can quote me." He says a crew came out from the television program *48 Hours* to film a segment on wolves. "These people are very sympathetic to wolves. I'm an outfitter, and I sent 'em down to my cousin. He had pictures of cows lying there with their guts out, still alive. The wolves are sitting there, all blood, watching these cows. One got its bag tore off and its guts hanging out and its eye tore out." *48 Hours* wouldn't use the pictures, but they wanted Rigler to lead one of his hunters out with their film crew in attendance so that they could film the hunter bringing down an elk.

"Hitler educated people that Jews were bad. I've got a *Weekly Reader* here with a story about a government scientist. That's how they've educated my children. They've changed the image of the wolf—all at government expense." He rummages around his kitchen until he comes up with a copy of *Kids: The Weekly Reader Magazine,* and then he reads from it, "'But scientists are helping the wolves. They're finding safe places in national parks for wolves to live. They're teaching kids the truth about wolves.'"

Rigler bangs his fist down on the kitchen counter. "Bullshit! These scientists—paid for by my tax dollar!—they're just reading each other's work. They're not doing science, it isn't science they're dealing with. They're simply promoting an image of the wolf.

"Guys like Mech, guys like Bangs—guys like that, they'll build an empire around wolves."

"It's going to change my way of life. It'll flat break me. My land would be totally worthless. Where am I going to put my horses? Where am I going to put my cows?" He bangs the countertop again in frustration. "People are after your ass! After your way of life!"

The heart of the opposition to wolves is the ranching community. At a public hearing in Helena, one rancher compared releasing wolves in the park to dumping hazardous waste in a suburban neighborhood. The president of the Montana Woolgrowers Association declared, "Anything that has blood in its veins will be a target for wolves!" A rancher from Jordan, Montana, said, "On our ranch, we lose from ten to one hundred lambs every year to coyotes. Nobody in their right mind would introduce the wolf, which is a far worse predator than a coyote. . . . We in agriculture will protect our livestock and our private property from all predators, by any means possible. No wolves, nowhere!"

Some ranchers were children when the last wolves were trapped and poisoned. They recall the sense of triumph in the killing of wolves, the ardent belief that eradicating wolves made the land more productive. The killing of wolves is still a symbol of the iron-willed, independent capitalism that defined the Old West, and there are men and women today who feel it is a duty to shoot a coyote on sight, and would just as quickly shoot a wolf.

That older view of the West is deeply embattled, because ranch life is changing and rural life is in trouble. Sheep ranchers are not doing well in Montana. There were one million sheep in the state in 1972; there are only about half that many today. Montana sheepgrowers lost about 20 percent of their sheep to weather, disease, predators, and other causes in 1992; coyotes accounted for three-quarters of all the predator losses. Some individual operators have lost as much as 30 percent of their sheep to coyotes alone.

Predators, however, are only part of the sheepgrowers' problem. Grazing permits cost more. Federal wool subsidies have been discontinued. Foreign competition is tough: Australia has a nearly twenty-year supply of wool in storage, waiting for a market to materialize. Rural sons and daughters are drawn to the cities by more remunerative careers. Even if they want to ranch, when their fathers or mothers die, children cannot afford to pay inheritance taxes on the land, so the land is sold. Much of eastern Montana is, as a result, losing population.

Increasingly, the land is bought by outsiders. In western Montana, the Bitterroot, Flathead, and Paradise valleys sprout suburban ranchettes where cattle once grazed. The new owners don't herd cows. And they don't feel threatened by wolves.

For example, Kevin Gallagher inherited a ranch in the Paradise Valley. He had grown up in Iowa, the son of a small-town weekly newspaper editor. When the ranch changed owners, the Forest Service canceled its grazing lease, and effectively put it out of business as a cattle operation. Gallagher supports the ranch by working as a carpenter. "I'm lucky I have a trade that allows me to keep my ground," he confesses. "My only other viable thing would be to put in campsites. The only thing we got here is tourism. All you gotta do to see what it's going to look like in twenty years is to go to Jackson Hole, Wyoming."

He loves his land and what he sees on it. Last winter, there were 350 elk, 150 deer, sixty antelopes, and a dozen bison in his pasture. One winter morning in 1990, he looked out over a light cover of snow on the ground and saw a wolf. He didn't think it was bothering anything. "I'm live-and-let-live until I'm crossed," he says. "I don't think wolves are going to be that big a problem."

The newcomers don't buy seed or farm equipment, so the businesses that once supported ranchers languish. The old ranching community shrinks. Because the newcomers have come for the scenery and the wildlife, they support new restrictions on logging, mining, and grazing. That makes life harder on the ranchers. Says Bangs, "These ranchers like being ranchers. They don't like going to meetings. But now they have to. This week they've got this wolf meeting. Next week, the water rights meeting."

If wolves are reestablished, some ranchers will lose stock to them. In May 1987, two Glacier National Park wolves killed a steer near Babb, Montana. Two ranchers witnessed the attack. It took federal Animal Damage Control trappers two months to catch the wolves. One was shipped to Minnesota, the other radio-collared and released. The trappers followed it back to a rendezvous site where they found two other adults and three pups. The trappers waited to see whether the wolves would kill livestock again, and in August they found the radio-collared male feeding on a dead sheep, and shot it. Animal Damage Control trapped two of the wolves and shot two more, but four months after the first killing, there were still stock-killing wolves wandering around the area, and that made ranchers angry.

Such incidents are not expected to become widespread if wolves are reestablished, however. Not all wolves develop a taste for livestock. In Alberta, between 1974 and 1990, ranchers lost 76 cows, 159 calves, and 33 sheep a year. In northern Minnesota, where there are more than 1,500 wolves, annual depredation on cattle averaged 4 cows, 23 calves, and 50 sheep a year between 1979 and 1991. Wolves in northern Minnesota attacked on the average 1 cow in 2,000 and 1 sheep in 1,000. Only 28 to 30 of the 7,200 farms in Minnesota wolf range suffer losses in a year. The state of Minnesota has established a compensation fund to pay such victims. To qualify for payments, Minnesota farmers must get federal officials to identify the lost cow or sheep as a wolf kill, but all ranchers with verified claims have been paid. The cost to the state has been as little as $8,000 and as much as $43,000 a year.

Fischer recalls that, after the Babb incident, "We wondered, 'How can we make this better?'" Fischer decided that the most useful thing Defenders could do was to compensate the ranchers for their loss. It would show that those who championed endangered species were willing to pay their costs. "Mott had always told us the most important thing we could do was set up a compensation fund," says Fischer. "It would take the economic argument out of the issue." So Fischer went out to raise money. Within a few days, he had several thousand dollars. The ranchers in Babb were paid, and the situation cooled down.

The idea of compensation grew. Fischer set a goal of raising a $100,000 fund with which to pay ranchers who lost stock to wolves. Defenders sold wolf posters and solicited private donations: singer James Taylor put on benefit concerts to help raise the money. In eighteen months, they had $100,000. Between 1987 and 1994, Defenders had only had to pay out about $15,000 of the fund. It proves one thing, says Fischer: "This isn't about money."

Still, losses to an individual rancher can be considerable. And even if they are compensated, and if Animal Damage Control hunters catch and remove the offending wolves, a rancher will have to depend on government services, deal with the paperwork, and spend more time talking with bureaucrats. Ranchers suffering losses will bear the burden of proving that the loss was due to wolves, a difficult thing if the rancher discovers the carcass weeks after the killing. And most ranchers don't like the idea of resigning themselves to losing stock to predators, for they have a long history of asserting themselves to protect their stock. Accepting compensation from a government agency or a conservation group for livestock killed by predators seems to violate the tradition of independence and self-defense that is so much a part of the West's view of itself, and it deprives ranchers of the feeling that they are valuable because they feed people. "We're not marketing our livestock through predators," says Joe Helle, a Dillon, Montana, woolgrower. "It's just not the way people do it."

To reduce the likelihood of losses, ranchers can avoid leaving dead stock in fields, where scavenging wolves may acquire a taste for beef or mutton; they can herd their sheep onto bedding grounds at night and keep a shepherd with them; or they can put out guard dogs. European shepherds have effectively used guard dogs for centuries. Ninety percent of the sheep ranchers in Idaho already use guard dogs to try to keep coyotes away. When they first started using the dogs, coyote predation almost ceased. But the coyotes adapted: they now try to sneak around the guard dogs, or to test them. Some ranchers are shifting to bigger, more aggressive guard dogs, but they aren't certain that these bigger, more aggressive dogs will prove effective against wild-wolf packs. And U.S. ranchers' range practices are quite different from European practices: their flocks are larger than those traditionally guarded by dogs in Europe, and they are accustomed to leaving their cattle unattended on summer ranges. Increased watchfulness and the training and feeding of large guard dogs will add to their costs.

The ranchers observe that most of those promoting wolf reintroduction are, like Askins and Fischer, relatively recent migrants to the West. They see that the real force behind the effort comes from people who live in cities and suburbs far from the Northern Rocky Mountains. The ranchers point out that such outsiders don't have to bear the cost in lost cattle and sheep, reduced hunting opportunities, and increasing headaches dealing with bureaucrats. But if it comes down to that, taxpayers in the distant cities are increasingly convinced that they have been paying subsidies for logging roads and grazing leases on public lands, and that the management of the public lands they pay taxes to maintain ought to reflect their views as well as those of ranchers.

It all comes down to the issue of control. Says Fischer, "The debate is over who has primacy over public land. Basically, what it shows is, if these people can bring wolves back here, it shows that the grazers and the forest products industry aren't calling the shots any more."

Says Askins, "Wildness isn't just waving grain. Wildness is about giving up control. It's about making room." In the end, she says, "This is not about controlling wolves, this is about controlling the West, and the wolf has simply become the metaphor for control. Having something that is beyond our control and perhaps beyond our understanding is healing. It creates an understanding and a humility in ourselves."

But deference and humility simply run against the grain of the Old West, and many opponents can't quite believe Askins. If they dismiss the idea of wildness as bogus, they are apt to look for more sinister motives. They frequently charge that conservation organizations are using the wolves as a marketing tool. Troy Mader, who runs the Abundant Wildlife Society, a Gillette, Wyoming, organization devoted to stopping wolf reintroduction, says, "Wolves are the most lucrative asset for raising money in the animal kingdom. I'm talking about animal-preservationist groups. The main thing on their agenda is create a crisis so they can make some money. They're not doing it for wolves; they're doing it because they want to make money and control land." Montana Congressman Ron Marlenee held that wolf reintroduction was "a tool for overzealous environmentalists and bureaucrats to implement federal land-use control on private property and restrict access to public lands."

But the fact is, a vast number of Americans see wildness slipping away from them, and from the West as well. To them, wolves embody a wildness they want, and a wildness modern civilization can sustain. An overwhelming majority of people commenting on reintroduction, both inside and outside the Yellowstone area, favored it. Support for reintroduction was so strong that in the summer of 1993, when the environmental-impact statement was published, it called for reintroduction of wolves into a 20,700-square-mile area of central Idaho and a 25,000-square-mile area centered upon Yellowstone National Park. The two reintroductions would be conducted differently. In Idaho, there would be a "hard release." Wolves would be captured in the wilds of Canada and released directly into the national forest that makes up most of the recovery area. In Yellowstone, there would be a "soft release." Packs or pairs with young captured in Canada would be brought to Yellowstone, put into holding facilities in the back country, fed deer and elk carcasses, and as far as possible kept from human contact. After a few months, once the young are used to the area, the gate to the pen would be opened and the wolves allowed to leave. Supplemental feeding would continue as long as needed. The only land-use changes contemplated by the environmental-impact statement were possible closures of one-square-mile areas around active dens, and the delay of timber harvests until after June 15. Even then, closures seemed to Bangs unlikely, because closures announce where a den is and increase the likelihood of harassment by curious wolf-fanciers or destruction by someone with a grudge and a gun on his hip.

Says Bangs, "We tell everybody, 'There's no big government going to come down on you unless you kill one illegally.' If we ever go into a more restrictive management mode, it will be because of some jerk who got loose with a gun."

In both areas, the wolves would be designated "experimental-nonessential" populations, an administrative measure which would allow the Fish and Wildlife Service to forgive inadvertent killing, such as by highway accident, and to allow ranchers to harass or kill wolves caught in the act of attacking livestock on private land. Though the federal government would not provide compensation to ranchers who lose cattle or sheep to wolves, it would rely on private parties, such as Defenders of Wildlife, to do so.

Loss of livestock was, at any rate, no longer the issue. Said Kim Enkerud, of the Montana Stockgrowers Association, "I remember a one-sentence resolution: No wolves, no way, no how. Now our resolution is: Well, wolves are coming here."

It all had an air of finality about it. Says Bangs, "The big thing with the wolf process is that wolves are coming on their own. Unless we prevent them from getting back, they'll get there. It may take them fifty or sixty years."

But they're coming.

. . .

It is September in Yellowstone, and most of the out-of-state visitors have gone. There is a dusting of snow on the peaks, and a cold wind sighing down from them. Clouds have rolled in from the west. In the depth of night, coyotes break the silence with low howls, followed by three short yips. As if in answer to them, the sky flashes with lightning and a crushing blast of thunder. There is a long silence, perhaps of hours. And again they howl. But this time, the noise seems to announce the arrival of elk in camp. Cows are browsing down the hill from my tent. A bull follows them out of the spruce and aspen, bugling its high-pitched challenge to other bulls. The sound seems to echo off the mountain walls. The bull snorts spurts of steam into the cold morning air, and walks menacingly, neck low, head tilted back, nose straight in front of its eyes, boasting, spoiling to chase the cows into a tightly engineered and submissive bunch. For the elk, too, the issue is control.

The sun comes up on the Lamar Valley, a broad grassland edging the meandering Lamar River. Specimen Ridge rolls up a couple of thousand feet, its higher reaches forested with spruce and fir, which thin out into grassy hilltops. The valley might be ten miles wide here. And those little black dots on the valley floor are bison the size of trucks. There were once wolves here, following the bison and elk. There may be wolves again.

12

A SKIRMISH IN ALASKA

There are six to seven thousand wolves in Alaska. They are not listed as endangered or threatened, as they are in the lower forty-eight states, and Alaska Department of Fish and Game officials believe they are increasing and expanding their range. They have traditionally been exploited by trappers and hunters, who in the 1980s were trapping and shooting about a thousand a year for their skins—which are used as rugs or as trim on parkas—or purely for excitement. Until 1989, there was no limit to the number of wolves a hunter could take, and some hunters killed twenty or more a year. The department has encouraged the take, on the general principle that wolves compete with humans for game.

The idea that wolves must be removed to make way for human hunters has long been gospel in Alaska. To some degree, it reflects the experience of hunters and trappers. In the early twentieth century, canine diseases such as rabies and distemper, probably brought to Alaska by the first dog teams, apparently decimated the wolf packs. In the absence of wolves, moose moved into areas of western Alaska and the Brooks Range where native languages had no words for moose. Just as the wolves were recovering, federal officials tried to help Alaskan hunters by liberally poisoning and shooting predators.

When federal predator-control people moved into an area, wolves quickly vanished. Dean Wilson, a fur trapper and buyer from Copper Center, recalls that, in the 1940s, "wolves were quite common. The prey was uncommon at that time. In the 1950s, Frank Glaser, an old U.S. Fish and Wildlife Service predator-control guy, came in, and I guess they put poisons out and snared up wolf kills. They worked there for a couple of winters, and they took down the wolf population substantially. By the late 1950s, we were seeing real substantial increases in moose and caribou."

As a result, in the 1950s and 1960s, much of Alaska had unnaturally low wolf and bear populations, and high populations of moose and caribou. These years were an era of lush hunting for those who looked upon Alaska as a last frontier, a land in which a man or a woman could live apart in nature, dependent on his or her own resources.

The era of predator control continued until 1959, when Alaska became a state and suspended the poisoning. By 1965, a step ahead of prevailing views in the rest of the nation, Alaska even eliminated the bounty that had been paid on wolves.

With an end to the use of poisons, however, wolf and bear populations began to recover. And because they had high numbers of moose and caribou to feast upon, wolves increased rapidly, perhaps even grew to higher densities than had existed before controls. The predator populations probably combined with hunting, cold winters, deep snows, and dry summers to reduce caribou and moose populations. No biologist has precisely apportioned the blame, but most will say hard winters and overhunting by humans were the most important causes of the declines. The Fortymile caribou herd, east of Fairbanks, was estimated to number more than 500,000 animals in the 1920s; by 1975, it had dropped to 6,000. The Western Arctic herd had dropped from several hundred thousand to 75,000 by 1977. Moose declined in the Tanana Flats, near Fairbanks, from 12,000 in the 1960s to 2,800 in 1975. In the 1970s, Alaskan hunters and trappers saw their golden age ending, and they clamored for a resumption of wolf control.

At the same time, biologists in the Department of Fish and Game were claiming to see evidence that the accepted orthodoxies about predator control were wrong. Dick Bishop came to Alaska in the early 1960s and went to work as a biologist for the Alaska Department of Fish and Game. "I subscribed to the predation gospel of the sixties," that predators took only those animals which were bound to die anyway. "I was wrong," he says. "When I went to work on some of these big-game projects, I began to realize there were some inconsistencies in the data. Unless you took into consideration predation, the data just didn't make sense. I was talking to my superior in the department one day. He was really trying to get rid of the bounty on wolves. I said, 'It's a really interesting coincidence that in some of these areas that have no calf survival there are high wolf populations,' and he said, 'My god! Don't tell the public!'"

Alaskan biologists began to feel that the mechanisms of predation described by Errington and other biologists in the lower forty-eight might not apply to wolves in the far north. More and more, it appeared to Bishop that predators were determining the number of prey. When the Fortymile caribou herd failed to rebound, says Bishop, "there was simply wolves. There had to be an element of predation affecting that population. There was just nothing else going on."

Alaskans sought to put their view to the test. In 1976, the Alaska Department of Fish and Game began shooting wolves from airplanes and helicopters in an effort to increase moose and caribou in the Tanana Flats area. Similar wolf controls were launched in other parts of the state as well. Conservation organizations in the lower forty-eight objected to the hunts as a return to the old days of wholesale predator extermination. They viewed the control program as defiance, both of science and of an increasingly held view that wolves were special animals that should not be shot. There were lawsuits and threats of tourist boycotts. Similar hunts began in British Columbia, the Yukon, and Northwest Territories. When television news showed the hunters shooting the wolves, the opposition stiffened. By 1983, then Governor Steve Cowper refused to fund wolf-control measures, and the board stopped authorizing them. For seven years,

Alaskans argued over wolf controls, but the state was unable to move toward consensus.

To build a case that wolf controls were justified, Alaska Department of Fish and Game biologists sought scientific evidence that controls increased moose and caribou numbers. In 1983, department biologists published a review of the wolf-reduction campaign in the Fortymile area, along the Yukon border. Their study concluded that prey populations had been pushed by hunters and bad winters to low densities, but that the wolves and bears that preyed on them didn't begin to decline until several years later, and in their relatively higher densities predators took an even higher proportion of the remaining prey. The study concluded that wolves kept the populations down and declared that, in such cases, "a manager has two choices. These are either wait for a natural recovery of prey while reducing or eliminating harvest, or reduce the number of wolves."

Waiting for nature to recharge the system was not in Alaska's temperament. Hunting was, in the 1960s and 1970s, one of the reasons migrants moved to Alaska. And out-of-state hunters coming for trophy sheep, moose, and caribou brought in money that guides and lodges depended upon. The 1983 study concluded, "Periodic artificial removal of wolves is the most practical option."

But not all—perhaps not even most—of the wolf-control efforts worked. In the Nelchina Basin, Warren Ballard found that grizzly bears accounted for 91 percent of predation on moose calves, while wolves took as little as 4 percent. "Reduction in wolf numbers," Ballard concluded, "failed to greatly improve calf survival." Victor Van Ballenberghe, who studied wolves and prey in Minnesota and in Alaska, and who served on the Alaska Board of Game for three years, says control efforts in the McGrath area similarly showed that removal of wolves did little to increase moose populations. "The only area where they saw a really dramatic response to wolf control," says Van Ballenberghe, "was Game Management Unit 20A, the Tanana Flats area. And all the ingredients were there: the moose population was low, the bear population was low, there wasn't a hunting problem because the hunting season had been closed, and winter weather following the wolf control was mild."

Albert Manville, senior staff wildlife biologist for Defenders of Wildlife, believes even the wolf-control effort in the Tanana Flats area did not help caribou to increase. The caribou population in the area rose from 2,000 in 1976 to 10,700 in 1989. But Manville points out that, except for the winter of 1984–85, all those winters were mild, five of them the five mildest on record. In the next three years, the Delta caribou herd declined to approximately 4,000, and Manville points out that two of those winters were among the most severe on record and that the summers were extremely dry.

Opponents of the wolf hunts argue that, almost everywhere they have been instituted, predator controls have become long-term and intensive, because human hunters want to keep their newly acquired share of the harvest, and because the predators adjust their birth rates to such human assaults by breeding at younger ages and having larger litters. A study of eighty-nine female wolves shot in various parts of Alaska from 1959 to 1966 found that all but ten of them had been pregnant. The shooting and poisoning may have so upset pack and territorial structures that the breeding rate was abnormally high. Douglas Pimlott found that when wolves were aggressively hunted in Algonquin Provincial Park nearly 60 percent of the females gave birth. Pimlott found that only 15 to 30 percent of the individuals in unhunted wolf populations were pups, whereas more than half the individuals in hunted populations were pups. Mech found in Minnesota that, when wolves reached higher density, 66 percent of the pups were males, but in two areas where wolves were hunted, only 27 percent were males. Clearly, wolf control increases the birth rate in surviving wolves. And when control stops, wolf populations may increase with surprising speed. In 1958, when predator control stopped, there were an estimated 120 wolves in Alaska's Nelchina Basin. By 1965, despite illegal airborne hunting of those wolves, Department of Fish and Game biologist Robert Rausch estimated there were between 350 and 400.

For ten years, Alaskans debated wolf controls. Ed Bangs, who was working in Alaska in the 1980s, recalls how emotional it was. When a small plane crashed in the bush, injuring the Department of Fish and Game personnel aboard, he heard people say, "Good, they deserved it for gunning down wolves." The issue was so polarized that in 1986, when Cathie Harms, a young biologist in the Department of Fish and Game, proposed to assemble a public advisory committee to write recommendations to deal with this impasse, a nervous superior hushed her, looked apprehensively over his shoulder, and told her not to use the word "wolf" around the office.

By 1989, a combination of hard winters, dry summers, human hunting pressure, and predators had brought many of Alaska's moose and caribou populations to levels that game officials found alarmingly low. Declared department biologist Robert Boertje, "Moose will occur in low densities in Alaska unless wolf or bear populations are manipulated." The department felt something had to be done. The Board of Game appointed a twelve-member panel of Alaskan citizens to look into the wolf issue and come up with a consensus view of the Alaskan wolves. The panel was chosen to represent the various interest groups—from trappers to birdwatchers—that might have a strong opinion about wolves.

In 1991, the team came up with a report declaring that, as long as this conflict raged, people of extreme views would rush to each new administration to get their interests represented, and there would be no active effort to conserve wolves. It declared, "Wolves have intrinsic value and provide multiple values to society," that "Wolf populations can sustain harvest, but sustainable harvest levels vary," and that Alaska has "a special responsibility to ensure that wolves and their habitat are conserved." It found that wolves can "in some situations keep prey populations at low levels," and that "human intervention can speed recovery of the prey population in some cases." It recommended that human harvesting of a declining moose or caribou population should stop before wolf-control programs were put into play, and that management plans be drafted, with population goals for wolves, bears, caribou, and moose, for each of the twenty-six game-management units in Alaska. By the fall of 1992, the department had prepared specific area management plans for the two game management units close to Anchorage and Fairbanks, and it was ready to recommend them to the Board of Game.

Said David Kelleyhouse, director of the Division of Wildlife Conservation in the Department of Fish and Game, "We have bent over backwards and spent a ton of money incorporating all the views on this. Expectations are high. If this process doesn't work, there's no process that will."

. . .

On a snowy night in November 1992, David Kelleyhouse sits at a table in the bar at Fairbanks' Westmark Hotel, talking with friends about wolf control. Earlier that day, the Alaska Board of Game had begun its deliberations over the first two game-management-unit plans and the proposals contained in them to initiate the first government-operated wolf-removal program in almost ten years. Kelleyhouse is an advocate of predator control, and on the following morning, he will present the proposal to the board. He is feeling expansive. It is clear that he has the votes to win approval.

Kelleyhouse has a jokester's grin, a dimpled chin, and a drooping mustache that somewhat masks the smile and makes him look older and more field-worn. He came to Alaska in 1976, having completed a master's thesis on black-bear ecology at California's Humboldt State University, and went to work for the Fish and Wildlife Service, eliminating introduced fur-farm foxes from Aleutian islands to protect endangered Aleutian Canada geese. This project convinced him that by removing predators, managers could increase prey populations. He says today, "I went back after we had removed those foxes and, man, it was like the Garden of Eden as far as birds. There were petrels and puffins and murrelets and gulls, and they were ground-nesting. Flowers were growing up in the fox trails." He then took a job with the Alaska Department of Fish and Game. "When I started at Alaska Fish and Game," he says, "my very, very first assignment was to go deal with a pack of wolves that had followed the wife of my superior and her dogs." He went on to participate in the studies that concluded wolf controls increased moose and caribou populations. Today, he believes wolf controls are urgently needed. Says Kelleyhouse, "We've got a real emergency situation south of Fairbanks. The Delta caribou herd has collapsed 50 percent in the last three years." The Fortymile caribou herd numbered fewer than twenty-two thousand.

Fish and Game officials look upon themselves as managers. Though they increasingly understand the need to manage whole ecosystems, most of the effort in game divisions goes to hunted species, because hunters still pay most of the bills. The oil boom of the 1970s brought more hunters to Alaska, but in 1980 the federal Alaska Lands Act set aside one hundred million acres of land as national parks, refuges, and preserves, and gave native Americans priority rights to hunt there. Native corporations also closed their lands to nonnative hunters. Sport and part-time subsistence hunters were squeezed into smaller and smaller areas. The Nelchina Basin—by virtue of its road access and proximity to Anchorage, where 40 percent of Alaska's population resides—is heavily hunted. In 1991, there were fourteen thousand applications for hunts in the Nelchina Basin, but the department issued only sixty-five hundred permits. Says Oliver Burris, a former department official and now a spokesman for the Alaska Outdoors Council, which favors wolf controls, "In 1991, the total statewide moose harvest was seven thousand. It would not take much effort at all to produce a harvest of six thousand moose out of the Nelchina Basin alone." Alaskan game officials feel they must do something to increase moose and caribou populations to meet the demand for hunting. And wolf control, they feel, is something they can do.

To control the wolves, department officials would shoot them from airplanes and helicopters. Though the department has encouraged ground-hunting and trapping, such efforts have never taken more than 15 percent of a wolf population; for control to have an impact, there must be a take of at least 40 percent. Some people argue that it would be more effective and more humane to send department technicians out to remove wolf pups from dens and euthanize them with lethal injections, but the public would not stand for it. Some people suggest a return to poisons. Says Kelleyhouse, "It's nonselective. I'm not willing to use toxicants." He would love to be able to use birth control to lower the birth rate of wolves, but there are as yet only unsuccessful efforts to develop a chemical sterilant for dogs; such birth-control technologies for wildlife remain years—perhaps decades—away.

If the department does nothing, and waits for nature to rebuild the herds, Kelleyhouse is convinced the hunters will act on their own. He points to British Columbia, where public outcry also shut down aerial wolf-control efforts. In the absence of government efforts to remove wolves, says Kelleyhouse, "the local people took it into their hands and started lacing the countryside with poisons. It's out of control. I had the exact same kind of threats coming out of Alaska. Once the people decide to break the law, you've really lost it. You have no management of predators."

Kelleyhouse will propose three wolf-control efforts—one in the Nelchina Basin, one in the area of the Fortymile caribou herd, and one in the area of the Delta caribou herd, south of Fairbanks. The plans aim to increase the Delta caribou herd from its current 5,750 animals to between 7,500 and 8,000; increase the Fortymile caribou herd from 20,000 to 60,000 and the Fortymile moose population from 4,000–4,500 to 8,000–10,000; and increase moose in the Nelchina Basin from 22,000 to about 25,000 and to stabilize caribou in the basin at about 40,000 animals, down from its current 45,000. To do this, the department calculates it will have to remove 70 percent of the wolves in each of these areas for three to five years—longer if necessary.

"The reason you can do wolf control in this state," Kelleyhouse says, "is, we have an extensive wolf population and it's like draining water from a well: more will flow in." Once the moose and caribou populations reach target levels, he says, the controls will stop. Kelleyhouse is confident that wolf populations would recover in a year or two. He adds that hunts are being considered only in these three areas, and "maybe we're thinking about a small section in one other area. We don't want to get into a bunch of wolf control."

Kelleyhouse believes that opponents of the plan simply don't understand either the plan or wolf biology, and says that biologists who have worked sufficiently with wolves will agree with him. "Biologists working in the field tend to agree very closely about how wolves work," he says. "Dave Mech would be the first one to tell you what he learned in the sixties is still valid—when prey is abundant and aren't hard-pressed, they aren't regulated by predators. But he'd be the first one to tell you, when prey isn't abundant, predators regulate them.

"There are a few biologists who have not worked with the wolf with hands on who have some opinions, some novel hypotheses, like the multi-equilibrium model of Dr. Haber. Gordon Haber flies over and he looks at these packs, but there's no study design, there's no publication. People like that have very strong opinions about wolves and they've read a lot of literature about wolves, but they haven't actually worked with wolves."

As he says this, Gordon Haber, who has been sitting on the other side of the barroom, comes over to the table and sits down. Haber is dour and intense, an odd combination of observant and brash. He is not someone who laughs easily.

Haber was drawn to wolves as a young man by reading Lois Crisler's *Arctic Wild,* then grew more interested in them when, between academic years at Northern Michigan University, he worked summers as a fire lookout on Isle Royale. In 1966, he went to Alaska, in part to meet Adolph Murie, and he stayed to study the same wolf packs Murie had studied twenty-five years before. Haber settled into a graduate program at the University of British Columbia and made the Mount McKinley packs the subject of his Ph.D. thesis. He has been following the wolves of Denali National Park ever since.

Though he hasn't heard what Kelleyhouse just said, Haber is testy about such criticisms of his work. His research has consisted of observation, rather than manipulation—hence Kelleyhouse's charges that it has little study design and that he hasn't "actually worked with wolves." He hasn't extensively argued his views in the mainline scientific journals. A good deal of his work has gone into critiques of wolf-control proposals in Alaska and British Columbia. By temperament and his own preference, he is a contrary. "I don't work for anybody," he says. Indeed, he funds his own research in Denali National Park. "I'm independent. A lot of what I'm saying is very contrary to the established dogma, but I'm getting all my stuff directly from observation."

He says that the combination of Murie's work and his own constitutes the longest continuous study of a single blood line of wolves, longer even than the study on Isle Royale. "I've seen far more of wolves than any other living person," he says. "That's a fact. I don't say that to brag. You look at my Ph.D. thesis and it has 8,279 hours of observation. Since then, I've added another 2,000-plus. What I'm talking about is not just what I think happens or what I've heard trappers say, but what I've seen. In one winter, I covered over 2,665 miles of tracking the two packs. I know not only the kills they made; I know every single encounter they had with live moose and sheep; and I know their behavior. There's nobody in the business, now or in the past, that has a sample that comes close to that."

He seems to enjoy his reputation for being thorny. Of the Park Service he says, "I've been their primary critic for twenty-seven years. In 1986, they decided, 'We'll get our own wolf-research project so we can get Haber out." That year, the Park Service invited Mech to begin his Alaska study, and Haber was not asked to collaborate, even though they were studying the same packs. Mech's study is now winding down, "but," says Haber, "I'm still there. The Park Service has discovered they're dealing with a mean son of a bitch. But they haven't been able to run me out of there."

Haber believes most wolf and caribou biologists oversimplify the systems they study. He wants managers to see that nature is far more complex than they think. He believes, for example, that caribou herds expand and contract on cycles that take sixty to ninety years to complete. Herds briefly reach unimaginable numbers, collapse as weather changes and food supplies diminish, and then go through long periods of low density. Haber points out that the Fortymile caribou declined in the late 1800s. "Indians were starving to death there in the late 1800s. No amount of hunting could explain that decline." And then the population came back to perhaps more than five hundred thousand in the 1920s. The herd doesn't have a sustainable size, he says: it fluctuates wildly, with or without predation.

Haber has been hired by Wolf Haven to critique the Alaska Department of Fish and Game's plan for wolf controls. His review of the plan led him to conclude, "The biology doesn't make any sense, period." And the essential misunderstanding, he says, is simple: "They feel they can replace the effect of natural predation with human harvest. We can increase the hunting and not have to do any wolf control. None."

He believes that the Delta and Fortymile herds are about to stage their own comebacks. The Western Arctic herd now numbers five hundred thousand. As the larger herds reach peak populations, they will connect with smaller herds through migration and range expansion, and infuse new life into them. All Alaska needs to do, he feels, is wait for this natural increase. And moose could be increased by harvesting older moose and encouraging a younger population that would have higher pregnancy and twinning rates and less vulnerability to predators.

As he sits down across the table from Kelleyhouse, Haber starts right in objecting to the department's wolf-control proposal, saying that the department's population objectives for ungulates in the Nelchina Basin are too high. "Game Management 101," he admonishes Kelleyhouse. It is a shorthand reference to the dome-shaped curve that is taught in wildlife-biology courses and that represents the growth and decline of moose or caribou populations. The highest reproductive rate for ungulates does not come at the top of the dome: reproduction is highest 20 to 40 percent *below* that peak, because at that level there are more individuals of breeding age. He tells Kelleyhouse, "You're not going to get those increases at those larger populations; the highest production is farther back on the curve."

"You're right," says Kelleyhouse. "You want to keep them right up at the top of the curve, or back a bit. I've got a $100,000 or $150,000 budget to determine where they are on that curve. What did Tom Bergerud say on that? We're better off with a herd of five thousand that is producing than with a herd of fifty thousand that's not."

Haber says caribou numbers are already high in the Nelchina Basin. Why kill wolves if the object is to reduce caribou?

Kelleyhouse responds as though Haber has questioned his professional competence: highlighting his sixteen years of experience with wolves, moose, and caribou, he tells Haber that he urged restoration of natural fires to improve moose habitat long before it became accepted practice in the department, and that he directed wolf-control efforts and participated in the research afterward that showed increasing ungulate numbers. "I've spent my whole career working on this," he says.

Haber reiterates that Alaska Department of Fish and Game can't expect to have high production at the large populations the plan calls for. He says Alaska can't aspire to be like Scandinavia, where there are high moose densities in the absence of wolves.

"You're not telling me anything I don't know," says Kelleyhouse, his conciliatory tone starting to thin. "You can't have a Scandinavian situation here, because you're not raising the kind of conifers that the moose like."

Haber continues to needle Kelleyhouse. The Fortymile moose numbered ten to twelve thousand in 1965, but dropped to below twenty-five hundred by 1980 due to severe winters and high grizzly-bear and wolf predation. To expect that region to carry ten thousand moose again is, Haber feels, to ignore weather and other complexities.

By now, Kelleyhouse has tired of Haber's assault. "You know,

Gordon," he says, "we had our best talk at Edmonton, telling stories about times in the woods. When we start talking biology, we'll never get anywhere. It's like arguing religion or politics." He says he'd like to relax now, to have a drink with his friends. He does not wish to continue this debate.

But Haber, deaf to the entreaty, presses his case. "I think the decisions have to be based on some pretty tricky biology. How in the hell can you expect the board to understand it?"

"My biologist is sitting there with the data when it comes to the cut," says Kelleyhouse. With this he draws the conversation to a close, refusing to reply to any more of Haber's arguments.

There is a moment of uncomfortable silence, and across it stretches a gap that divides Alaska. It is not just wolves that inhabit this silence, but the very identities of Alaskans. Some see Alaska as a last frontier, a place where young Americans can go to find lofty independence in a life framed by adventure, risk, and limitless space. Since the oil boom of the 1970s, however, there is also a more settled Alaska, a place less given to risk and more to order, security, and cautious planning, a place in which wilderness is less a challenge than a reassurance. Haber, for all his hours following wolves, lives in Anchorage and is comfortable in the urban setting. He is apt to see rural hunters as people with little respect for biological complexity and the healing intelligence of nature. Kelleyhouse has for years lived in the small town of Tok, many of whose residents hunt and trap for food and winter cash. He is convinced the newly arrived city dwellers have lost their daily association with wildlife and their understanding of nature. Says Kelleyhouse, "I would say the vast majority of people testifying against this are recent arrivals [to Alaska]. They are urban residents. They don't understand what life is like for rural residents."

Frontier Alaska still lives. Greg Neubauer, a placer miner who grew up in the Alaskan bush, sits in his living room in Fairbanks and tells stories of his youth in a mining camp. His father drove a Caterpillar bulldozer from the railhead near Denali one hundred miles across roadless country to Iditarod one winter, with temperatures sixty below zero, in order to make the machine available for him to dig river gravels in the summer. Another year, he caravanned three Cats three hundred miles cross-country into the Brooks Range, a three-month trek in the dark of winter. The paths he cut are both established trails today. Neubauer's mother shot twenty-seven bears in camp, bears that simply came too close and thereby seemed to pose a risk to her children. When Neubauer was a child, a neighbor was killed by a black bear. The neighbor had heard something outside his cabin and opened the door, and there the bear was. When they found the neighbor, his chest was eaten away. Neubauer recalls wolves breaking into the camp cookhouse one day when his father left the door open; they destroyed everything and urinated on what they didn't eat. On the wall behind Neubauer is the coal-black skin of a bear he shot after it damaged his airplane and broke into a case of oil cans, biting into every one and spilling their contents onto the ground.

Such stories advertise Alaska as a place where one may confront a vast and elemental nature. The promise of adventure draws young men and women from the tame and possessive cities of the lower forty-eight to the wilds of Alaska. There is a powerful desire to lead the life of the wilderness family; the young settlers talk about it, dream about it. A few of them even try it.

Not many stick with it. Living off the land in Alaska is not easy: game doesn't wait around for hunters, winters are hard, and subsistence life-styles are lonely. Most of the new arrivals usually end up living in the cities, where the jobs are.

But there is still a powerful desire to lead this life. Though Dick Bishop lives in Fairbanks, he still gets 90 percent of his family's protein from the woods. "There are a lot of people that have a foot in each camp," he says. "They have jobs and live in cities, but they hunt for food, they pick berries and fish." Each fall, they try to shoot a moose to feed them over the winter. Says Bishop, "They feel self-reliant. The part of their lives they appreciate the most is a sample of that life, but they can't devote themselves to it."

Those who hunt say it is harder and harder to stock their winter larder, and many say wolves are to blame for the hard times. Bill Waugaman, a hunter from Fairbanks, complained that in 1992 he tramped an area he had hunted for years and saw thirty-eight sheep and only one bull moose. "I used to see this much game out of the window of my cabin," he says. "We have an emergency there." Says Archie Miller, who has trapped for twenty-eight years on the Upper Yukon Flats: "The wolf population is extremely out of hand. The result is, we've got hardly any caribou, hardly any moose, and hardly any sheep. Take wolves down a peg or two to the point that we can have game for a generation or two. Do it until it gets back the way it was in the fifties." Jim Roland of Fairbanks says, "The predators are taking 97 percent of the game, and the hunters are getting 3 percent. The land will hold ten times what's out there. We just need a new philosophy in Fish and Game." Bill Hager of Fairbanks urges, "Bring back equal allocation: one-third to the bears, one-third to the wolves, one-third to the humans!"

But there is a competing view. Only forty thousand of Alaska's six hundred thousand residents live in rural areas. The vast urban majority consists largely of recent migrants from other states, where, as urban boundaries expand, fewer and fewer people grow up hunting. Today, fewer than 20 percent of Alaskans hunt. The urban majority sees nature less as a pantry than as a spiritual and recreational resource. Says Nicole Whittington-Evans of the Alaska Wildlife Alliance, "Wolves are a part of a natural system. They in fact allow a system to be a healthy system. One of the reasons people move to Anchorage is to be near a system with large predators." In a poll taken by the Alaska Wildlife Alliance and Wolf Haven in 1992, only 8 percent of the responders supported an increase in the number of wolves being killed. Declared an Alaskan woman, "Nature does things really well. We don't have to do better. Living within the limits set by nature is the lesson we have to learn for the future. Wolf control should be a method of last resort."

To settle conflicts over the uses of wildlife, most states established governor-appointed fish-and-game commissions, like Alaska's Board of Game. Commissions are supposed to take the politics out of wildlife decisions, eliminate the influence of market hunters and commercial fishermen, and free the legislature and the governor from having to wade through tedious biological reports in order to make decisions. They are supposed to put wildlife questions into the hands of citizens.

Alaska went other states one better by establishing local advisory committees to report to the Board of Game on local conditions and interests. These committees give a strong voice to the older rural citizens, seldom admitting newcomers or people likely to introduce a contesting view; women are almost completely unrepresented on them. Urban residents often find these committees exasperating. Says former Board of Game member Joel Bennett, "That whole advisory system is poisonous—it's impossible to work with, it's stacked. There's no way they will allow environmentalists to be represented. They're elected by existing members, and whoever shows up is who votes. They view themselves as experts. And, to some extent, the biologists and the board are stopped."

For example, when moose populations crashed in the 1970s, it was probably due in general to overhunting and cold winters. But in the Nelchina Basin, the local advisory committees were convinced that the crash was due to hunter harvest of cow moose. When the moose population recovered and threatened to outstrip the carrying capacity of the land, biologists in the department urged the Board of Game to initiate a cow-moose season. The local hunters, however, managed to get the legislature to pass a law stipulating that the board could not initiate a cow-moose season without concurrence from the local advisory committees. And they would not concur.

The issue is complicated by the sense many rural Alaskans have that environmentalists in the lower forty-eight and the United States government threaten their way of life. Kelleyhouse says the reason the wolf issue came about in the first place was that in 1980, when the federal government set aside a hundred million acres in parks and refuges, and in the process closed them to hunting, it was a blow to the rural life-style. Closure of all these areas has concentrated the hunters more heavily on the remaining areas. Meanwhile, game populations are declining, as is hunter success. The number of hunting licenses sold in Alaska has also been declining, because buying a license doesn't guarantee a hunter the chance to hunt. After buying a license, one must get a permit for a specific hunt, but subsistence rights guaranteed by federal legislation give native Americans priority on most hunts. Kelleyhouse, recalling the setting aside of these federal lands, says he is "bitter at the way it was done. Suddenly people are being told they can't hunt where they hunted all their lives."

The state's tourist industry is now geared more to wildlife viewing than to big-game hunting, and the conflict between the new tourism and the old subsistence life-style is plain to see. At a 1991 Board of Game meeting, when one commissioner suggested that one of the state's management objectives for wolves ought to be wildlife watching, another replied, "If they want to look at wildlife, they can watch it in my freezer." When Al Manville of Defenders of Wildlife first appeared before the Board of Game to argue against wolf controls, a board member addressed him as "the goddamned conservationist from Washington, D.C."

When Walter Hickel ran for governor in 1990, he reviled the federal government and appealed to rural Alaskan tradition. Hickel won, and his appointments to administrative positions in the Department of Fish and Game reflected rural Alaskan values. When Kelleyhouse was elevated from district biologist in Tok to head of the Division of Wildlife Conservation in Juneau, many saw it as an effort to resurrect wolf control. Concluded David Cline of the National Audubon Society, "They're going back in time. They are trying to manage the state as it was in the 1930s and 1940s." Two years later, five of the seven members of the Board of Game were Hickel appointees. In November 1992, not only was the Department of Fish and Game ready to recommend wolf controls; the Board of Game was ready to approve them.

The day after Kelleyhouse and Haber debated wolf control in the Westmark bar, the board authorized wolf-control programs in the area of the Delta caribou herd, the area of the Fortymile caribou herd, and the Nelchina Basin. Department officials would conduct the wolf hunts from helicopters in the Delta and Fortymile areas, but would issue permits to selected private individuals to do the killing from airplanes in the Nelchina Basin.

From the Department of Fish and Game's point of view, the board's action boiled down to a commitment to intensive management. The department's Robert Stephenson said, "We had to do this. I mean, why are we here? To manage low-density populations forever? We're not just here to sit and monitor. It doesn't make much sense to spend $12 million a year on monitoring."

Said Governor Walter Hickel: "You can't just let nature run wild."

But few people outside the department saw it that way. "This is single-species management at its worst," said Melody Bankers of Wolf Haven. Former board chairman Douglas Pope called the department's decisions "myopic" and ascribed them to "upper-echelon wildlife officials . . . lobbied intensively by interest groups in the hunting community." An Alaskan native activist said, "It's an outrage the way they intend to kill them from airplanes and helicopters." Said Tom Dowling, who delivers the mail on a 240-mile round trip between Tok and Delta Junction, "These people have an agenda that they feel our countryside is supposed to be a meatlocker and provide them with unlimited resources for moose. What they're up to is extinction of wolves, is what they're up to."

Outside Alaska, criticism burgeoned into outrage. "How dare you spend my tax dollars to harvest wolves so that hunters can have more game to kill!" wrote an Arizona woman, who added, "Such practices enforce the image of Alaskans as barbaric, stupid, ignorant, backward people." The National Parks and Conservation Association canceled plans for a 1993 conference in Alaska and urged its members to defer travel there. Defenders of Wildlife asked its members to boycott the state until the wolf issue was resolved, and announced plans to sue Alaska and seek a congressional ban on aerial wolf control. The department and the board received forty thousand letters on the issue, some of them threatening board members with violence. When a study by the Alaska Tourist Council predicted that if the wolf-control program was carried out, it would cost the state as much as $235 million in lost tourist revenues, Governor Hickel suspended all aerial wolf-reduction efforts.

Newspapers accused the state of seeking the extinction of wolves. The charge was ill-considered, for the proposed hunts covered only 3.5 percent of the state of Alaska. But the heart of the controversy was not the possibility of extermination—it was the issue of killing.

Much of the outrage expressed against the Alaskan hunts was directed at the proposal for the Nelchina Basin, where hunts were to be carried out by private individuals, not yet named, selected by the Department of Fish and Game. Department officials argued that the arrangement would save them money, but there were hints that there was more to it. During the board's deliberation, board member Jack Didrickson said the arrangement would provide local people with a fine opportunity to harvest wolves.

Sport hunting of wolves is controversial in Alaska. People disagree as to whether it is proper to take the life of so sentient a creature, and debate whether it is proper to take the lives of animals that aren't eaten.

There is also controversy over the methods used to hunt wolves. The wolf is so wary, so likely to flee at the scent or sound of humans, and so dispersed in the vast Alaskan landscape, that it is a rare hunter who can find and approach one on foot. The only way to locate and approach wolves consistently is to use an airplane. But it is a violation both of state law and of the federal Airborne Hunting Act of 1971 to shoot at animals from an airplane or to use aircraft to herd, drive, or chase animals. For years, holding that it is consistent with the Airborne Hunting Act, Alaska has permitted "land-and-shoot" hunting, in which the hunter uses the airplane to find the wolves but must land nearby, get out of the airplane, and approach the wolves on the ground.

In fact, land-and-shoot hunting has often been abused. Hunters shoot wolves from the air, or taxi after them on the ground until they are exhausted—both violations of the Airborne Hunting Act. Sean McGuire, a Fairbanks resident, has witnessed aerial wolf hunts. In one, a pilot tried to drive a wolf out of forests on a hillside into the open valley below, where he might shoot at it. "The wolf wanted to go up, and each time the plane would try to move it down, it would circle back. The plane finally gave up." On another occasion, he heard a four-hour battle on the other side of a mountain ridge—the whining and circling of the plane and gunshots over the sound of the engine. But proof of such hunts was hard to come by. Alaska is a land of open spaces. In winter there are no backpackers or fishermen to question what an airborne hunter is doing in the back country. Says Bennett, "It's a place where you can do things that you wouldn't be allowed to do in other places, or do things without much chance of being caught."

Bennett believes the Department of Fish and Game ignored, and even condoned, such violations, because it viewed the sport hunt as a back-door wolf-control program. He recalls one board member saying, "Hey, if you eliminated land-and-shoot, what are we going to do for wolf control?" Bennett believes, "It was incredibly effective in some areas. In Nelchina, they could send one family of hunters out and be fairly sure of getting one hundred wolves a year." But those who did the hunting publicly denied any wrongdoing.

And then, in March 1989, two national-park rangers overheard a radio conversation between aerial hunters over Kanuti National Wildlife Refuge. They took notes, and produced a transcript of the conversation:

> "Shot at wolves twice . . . let him go for now."
>
> "We had five on the run, shot two. . . ."
>
> "Jimmie got one."
>
> "He wasn't completely dead. . . . We'll go back later. The damn thing jumped up and bit my wing. . . ."
>
> "It's always fun!"
>
> "Jimmie stuck three arrows in him."
>
> "The wolf was still blinking his eyes at us, so I didn't want to take a chance of getting bit, so we'll go back a little bit later."
>
> "He had an arrow up his a—— and he didn't like that one bit."

Investigator Alan Crane later found five skinned wolf carcasses near Old Dummy Lake, inside Kanuti National Wildlife Refuge. One of them contained an arrow shaft and three arrow wounds. In the same area, he found wolf tracks with abrupt changes in directions, including sharp reversals in course, periodically intersected by aircraft ski tracks in a snaking pattern, atypical of a normal landing, taxiing, or takeoff pattern, and with occasionally greater depth of one ski track on the outside of turns, indicating high-speed taxiing. The wolf tracks showed a wolf falling down occasionally, and between leaps were spots of urine, indicating that the wolf was stressed and exhausted. Crane found aircraft and wolf tracks of this nature at five different sites, and all indicated someone had been chasing wolves in a taxiing airplane.

An Anchorage orthopedic surgeon, Jack Frost, and a licensed guide, Charles Wirschem, were charged with violation of the Airborne Hunting Act by using a plane to herd the wolves, with using their airplanes' radios to hunt wolves, and with violating the federal Lacey Act by transporting a wolf that had been illegally killed. The government seized a photograph of Frost and Wirschem on the lake in question in front of their planes, with Wirschem shooting at a wolf. Frost was vilified in the press, which printed transcripts of the radio transmissions. He signed a plea bargain in which he admitted using his aircraft to "disturb" a wolf. He was fined $10,000, his airplane was confiscated, and he lost the right to hunt wolves for two years.

Originally, Jack Frost had settled in Alaska for the hunting. As a boy in State College, Pennsylvania, he tramped the hills outside town, hunting rabbits. The first week of deer season, he remembers, school closed so that fathers could take their sons out into the woods. In winter, when a fresh snow fell, he'd go out and track rabbits and foxes. He came to Alaska in the service, and he stayed.

At forty-seven, he has blond hair framing his bearded face in tightly curled ringlets. He speaks with cautious deliberation. When I met him at a San Francisco hotel, his reserve struck me as a form of apprehension. He seemed to be trying to make sense of a shadow that had fallen over his view of life.

He has hunted moose and caribou all over Alaska, has taken the grand slam—all four kinds of North American wild sheep—and, above all, has hunted wolves. On winter days, he would range in his airplane far from Anchorage. When he found tracks in the snow, he would follow them until he came upon a pack of wolves.

Wolf hunting is not a pursuit for novices. "It's not an easy thing to do to go out and find a wolf in an airplane," says Frost. "It's a skill that takes a long time to learn, and some people never can learn it." Reading tracks from the air while traveling at seventy miles per hour and watching for mountains and treetops is demanding. When snow is soft and deep, the wolves walk in old tracks or stay under the trees, where the going is easier. When wind settles and compacts the snow, the wolves travel on lakes and rivers, but the compacted snow may not reveal their tracks.

It takes skill at flying. Says Frost, "People kill themselves hunting wolves. They fly into trees or they turn too fast. There's a lot of skills—paying attention, flying the plane, landing the plane, shutting off the engine, and jumping out. If the wolf is in timber and going to come out on a lake, you have to coordinate how far away you are from him before you're landing. If you get there too much before him, he may turn and go back. There's a lot of flying judgment.

There are maybe a hundred guys in the state of Alaska who can do it, maybe twenty or forty that do it." Frost and the rest of this small cadre of hunters took most of the sport harvest of wolves in Alaska.

He believes that land-and-shoot hunting is fair chase, because he believes the wolves always have a chance to escape. "Once I spot a wolf that I want, I'm guessing, I would get maybe 20 percent of them," he says. "Lots of times you see wolves where there is no place to land. Lots of times you land and they duck into the bushes and you just don't get them. The first wolf I ever landed on in an airplane was in deep snow, and I jumped out and I shot all the bullets in my rifle. There was a guy in the back seat and he handed me his gun and I shot all the bullets in his rifle. The wolf was 150 yards away and I just missed. It went into the woods."

Frost insists it is not necessary to herd or chase them. He says wolves don't always run away. "In landing on wolves, I have had every reaction from running, to stopping and gawking at the plane, to sitting and looking at it, to trotting over to see what was going on. You can't predict how any wolf is going to react. I have landed thirty yards away on a lake and had them just stand up and look at me." Once he landed below four wolves he had found sleeping on a hillside. Three of them scattered into the trees. "The biggest wolf trotted down the hill to within twenty yards of me. And I missed him with my bow and arrow."

Often enough, he didn't miss. A legally killed wolf must be reported to the department, and the records indicate that Frost killed as many as twenty wolves a year. He says, "I was successful at wolf hunting because I worked hard at it. There have been winters I've flown in excess of 250 hours." Some years, he would take a whole month off from work to hunt wolves.

Frost insists that department officials assured him that land-and-shoot hunting did not run afoul of state or federal law, that landing near a wolf, although it might disturb the animal, would not be regarded as a violation of laws against harassment. "We were told, 'Go ahead and land, and shoot as they are going away.' State fish-and-game enforcement officers said that over and over. They were saying, 'We need to harvest some of these wolves. I mean we *want* you to harvest them.'"

Today, Frost declares he did nothing wrong. He insists, "I've never fired out of an airplane at a wolf," and denies he used the airplane to herd the animals: "They basically charged that we harassed them, driving them to exhaustion, and we didn't do that, either. We went out hunting wolves, we found wolf tracks, we followed the tracks. When we found the wolves, we followed the wolves. We couldn't land where we first saw them. When the wolves came out on a lake where we could land, we landed the airplane as close as we possibly could and jumped out of the airplane and shot the wolves. It was our understanding that that was okay. It was what we were told we could do.

"We thought, if we were not shooting the animals from the air and we were not driving the animals from the ground, the state of Alaska allowed us to take ten wolves, and the definition of take says hunt, kill, pursue, and disturb. If we're allowed to hunt an animal, we ought to be able to pursue him. We ought to be able to disturb him.

"There had never been a court case on this thing. We were the first people ever tried for landing near a quarry. I would have fought it in court if they had simply charged me with the misdemeanor violation of the federal Airborne Hunting Act." But when they charged him with a felony violation of the Lacey Act, for transporting an illegally taken animal, his combativeness faded. Conviction on the Airborne Hunting Act charge would have made conviction on the Lacey Act violation more or less automatic. "Then all of a sudden I'm looking at hundreds of thousands of dollars in attorneys' fees, possibly going to jail for a long time, possibly losing my right to vote and to own arms, and possibly even losing my medical license."

So he pled guilty to the one violation of the Airborne Hunting Act. "The only thing I agreed to in the plea-bargain agreement that I signed was that I used my airplane to disturb a wolf. And I have to say that they probably were disturbed. My plane was there. I didn't intentionally disturb them."

"He is playing with words," says Stephen Cooper, the assistant U.S. attorney who prosecuted Frost. The indictment to which Frost pled guilty charges that Frost "did unlawfully and knowingly engage in using an aircraft for the purpose of harassing and killing a wolf." Cooper says, "He was questioned at great length in the court" as to what the language in the plea-bargain agreement meant. "It was like pulling teeth to get him to say he used the airplane to run the wolves in the direction he wanted so that he could kill them. Eventually the judge got him to answer yes to having used the airplane to herd the wolves to a place where he could shoot. He was driving them by means of an airplane."

Says Frost, "I've always felt I was a law-abiding person. I pay attention to the laws. I try to understand them. All they had to do anywhere along the line is to say, 'It's illegal,' and I wouldn't have done it. I felt I was aware of the law, aware of the issue, and I was doing it within the realm of what was legal."

Cooper concedes, "There's a lot of gray area in that law. If you try to land next to a wolf, are you disturbing the wolf or not? That's very difficult to define." And he expects that, as long as Alaska regards land-and-shoot hunting as a way to control wolves, "all that's going to do is push people close to the line for potential violations of the Airborne Hunting Act. If they try to control wolf populations by trying to get people to do it with land-and-shoot, they're practically asking people to walk on the wrong side of the line."

Partly in response to the outcry over the Kanuti Refuge hunt, the Board of Game stopped authorizing land-and-shoot hunting. Bennett thinks the resurrection of wolf control in 1992 came about in part because the board believed it would never be able to revive land-and-shoot hunting. A poll taken in 1992 showed that 80 percent of Alaskans opposed aerial hunting of wolves. Says Bennett, "Probably the writing was on the wall that land-and-shoot would fall afoul of the Airborne Hunting Act." Without land-and-shoot hunting, the board may have felt that it had lost some of its ability to manage wolves.

For most observers outside the courtroom, the issue was not what was legal, but what was ethical. I suspect that the look of reserve on Frost's face is an expression of his feeling that he is caught in the middle of someone else's debate. "The two most prevalent views about wolves," he says, "are either wolves are a true symbol of the wilderness or a menace to game populations that need to be kept at low levels. I feel the wolf's neither a villain nor a god. It's just another wildlife species. It's a renewable resource that can be reliably harvested. The vast majority of people have one emotion or another. That doesn't leave much room for someone in the middle."

But the debate is far more encompassing than Frost has seen. For most Americans, it seems wrong to shoot a wolf for the sake of the excitement. The debate is not just about wolves; it is about people, and about the nature of killing.

Since the middle of the twentieth century, we have been absorbed with the issue of killing. In the atomic bomb, humankind grasped the power to destroy all life on earth. After 1950, crime and violence grew in our cities. By 1960, we wondered whether we were helpless to prevent war and crime. We asked whether we were a killer species. In the 1960s and 1970s, animal behavioralists and psychologists sought to resolve this question in a wide-ranging debate about aggression, or behavior aimed at hurting another physically or emotionally. The debate was inaugurated by the German animal behavior expert Konrad Lorenz in his book *On Aggression*. Lorenz held that aggression was innate and unavoidable. He was to some degree reiterating the ideas of Sigmund Freud, who wrote, "Men are not gentle friendly creatures wishing for love. . . . A powerful measure of desire for aggression has to be reckoned as part of their instinctual endowment." Like Freud, Lorenz believed that, if aggressive drives were repressed, they would accumulate until they exploded into violence. Lorenz believed that "primitive" people exercised their aggressive drives by making war on neighbors, which allowed them to be loving and gentle and exhibit little aggression toward their friends and families. But in the modern community, he believed, there was no longer a legitimate outlet for aggressive behavior. He wrote, "Present day civilized man suffers from insufficient discharges of his aggressive drive. . . . We are all psychopaths for each of us suffers from the necessity of self-imposed control for the good of the community."

The other side of the debate held that aggression is not innate, that, however genetically disposed we might be to aggressive acts, it takes environmental conditions to elicit them. Critics pointed out that Lorenz based his views on the relatively inflexible behavior of fishes and birds, whereas higher mammals, such as wolves and humans, are not so rigid. They also pointed out that Lorenz's accounts of warlike primitive societies were faulty, and that many human societies experience very little violence.

In the 1970s, an increasing body of research showed clearly that aggression is learned behavior evoked particularly by pain or frustration, and that poor and dispossessed people were more likely to display it. Experimenters showed that both children and adults who see aggressive behavior respond by becoming more aggressive themselves, and that American spectators are more aggressive after watching sporting events.

Bits and pieces of these academic debates trickle into popular understanding. Frost, for example, says, "I think hunting is a part of our nature. And I think, if a man doesn't hunt, those same drives and urges will come out in other areas, and the people that have those drives and urges may become inhumane to other people." Animal-rights advocates who burn down research laboratories, he says, may be simply exercising their hunting instincts. "They want to get something. They want to get that big company." Nonhunters, on the other hand, are apt to side with those who believe aggression is learned, and to say humankind would be better off if it put hunting behind it. That hunters can't feel the suffering of their victims, or that they enjoy killing, disturbs such people. They believe that, whether aggression is innate or not, we are obliged to suppress the urge to kill, because one act of aggression begets others. In the Kanuti Refuge wolf hunt, they see the kind of hateful violence they want to expunge from the human character.

The academic debate about the nature of aggression doesn't resolve the issue for most people. We tend to work out our views from day to day in popular cultural expressions such as movies or song, and in these the nature of killing is still a battleground. Much of our popular culture is redirected or vicarious aggression, invoking the views of Freud and Lorenz by providing harmless substitutes for war as a way of discharging accumulated aggressive impulses. Our spectator sports are violent. Video games encourage our children to tear the head off or rip the heart out of an opponent. Popular music is littered with images of assault. Our children see, on the average, more than one thousand murders a year on television. The purveyors of these entertainments often argue that they are helping displace aggression that would otherwise overwhelm our society, but critics say they are really making us more aggressive.

We seem to be deeply conflicted over this issue—set off not against one another but against ourselves. We rail against abortion yet clamor for the death penalty, preach the Golden Rule yet purchase hand guns.

Frost is caught in this debate, but the wider ethical issue frustrates him. He talks little about the animal and much about the people he sees when he looks through the lens of the wolf. "People have grown up learning about wildlife from Walt Disney movies and the propaganda that is put out by animal-rights groups," he says. "The environmentalists just don't want to allow anybody to kill a wolf." In his opinion, "The general population everywhere is more urban, more likely to be anti-hunting, and think their meat should come off the shelves and not be shot, watched die and butchered in the field, and brought home." He suggests that they are as deeply implicated in death as he is, and that what sets him apart from nonhunters is his own understanding of the mystery of death.

But Frost can't, in the end, pinpoint why killing is essential to him. He says he has tried photography, but it just doesn't satisfy him. "When you come down to why you have to kill," he says, "that's the part I can't come up with, except to say I think it's inbred, it's a genetic part of us."

"Though not everybody feels it," he says, it is a natural thing to do. "You can take a house cat and rear it through ten generations, feeding it dried cat food, and in the eleventh generation you put it down and it sees a canary in a cage and its tail will start twitching. I think that hunting is as deeply inbred in the human species as it is in animals. Many people try to deny that."

At this point, he begins to wander around the edges of the question of why he kills—speaking of the importance of the right to bear arms, and cautioning that he doesn't feel paranoid, that what he is talking about here is not a conviction but a vague speculation. He can't really imagine locking himself in his basement with his guns to defend himself against a siege of criminals or government goons. But, he says, "I don't think it's totally inconceivable in times to come. If the government knows there are lots of arms in private ownership, they could never start loading people in boxcars. That's not science fiction. That happened fifty years ago in Germany. It's happening today," in places like Bosnia and Iraq. "Who's to say that isn't going to happen?"

Even though I cannot be certain, I suspect that what Frost is talking about is not wolves but people. It occurs to me that he's talking in heavily veiled ways about the propriety of killing humans—in self-defense, in war, in the ugliest of eventualities.

The history of the world since 1945 has been shaped a great deal by remorse at the blood our own species has spilled. It has been a search for the sources of our own humanity, and for the sources of goodness and nobility. The great debates of our time have been about how to excise whatever makes us cruel and enlarge what makes us noble. Increasingly critical of our own nature, we would like to perfect what is uneven in it.

We congratulate ourselves on our abhorrence of murder, but perhaps we do so in proportion to our own desire to commit it. We are uncomfortable with the plethora and strangeness of our own species. There is evidence that, the more numerous we grow, the more prone we are to violence. We are divided more and more into separate communities of rich and poor, black and white, orthodox and reform, rural and urban. Like the wolf, we have different ethical standards for strangers. We may only be really moral when we love one another not as a matter of principle but as a matter of mingled breath, shared thoughts, of looking directly into one another's eyes.

I suspect that, when we argue about wolves, we are arguing about love and hate, peace and war, killing and kindness. We are arguing about our own hearts and souls.

In the outcry that followed the Board of Game's authorization of wolf control, Alaska made a half-hearted attempt to show that wolf control was needed. The department invited journalists and environmental leaders to attend a "Wolf Summit" in Fairbanks, at which wolf experts, state officials, and environmental leaders would all discuss the issue. But the wolf-control proponents felt that no one outside Alaska would listen to them anyway. Said Kelleyhouse, "There's no way to satisfy the concerns of animal-rights people." The department did not even provide news media with maps showing the proposed wolf-control zones. In Kelleyhouse's view, "There's a simple fact that the news is only going to carry what it wants to carry. Alaskans have no way to get out what they want to get out to 270 million people. I think it comes down to a state's-rights issue, and I think we have to debate this issue well internally and reach some sort of a decision and proceed."

The Board of Game agreed, and plunged ahead with its plan. In June 1993, the board reimposed the wolf-control effort in the Tanana Flats area. The wolves would be taken only by department personnel, and they would be baited with carcasses and taken with snares, rather than shot from airplanes. But, as a further gesture of defiance, the board voted to end its two-year ban on aerial sport hunting of wolves.

Robert Stephenson tried to be philosophical. "This isn't a big conservation problem. It's not a biological catastrophe. As things rank around the world, this is not a core issue that is going to break the world. I hate to see us blow our emotional budget on it."

But it may be that when we talk about wolves we simply are unable to see what we are really arguing about. We seem to have freighted the wolf with too much meaning for mere discussion to encompass. Said Anne Ruggles, who had been a member of the public advisory team on wolves and a member of the Board of Game, "None of this is biological. We're beyond the biology. We're in the sociological now."

At the Wolf Summit, a man drove up in his truck to the front of the indoor hockey rink in Fairbanks where journalists and hunters, trappers and biologists had gathered to discuss the issue of wolf control. He took out of the back of his truck a freshly killed wolf and carried it into the building, and dumped it on the floor. Blood dripped from the carcass. The bearer of the dead wolf seemed angry. But as state troopers led him away, no one could say exactly what the gesture meant.

13

WOLF-DOGS

Earl Hurst is a forty-eight-year-old bear hunter who lives in Grants Pass, Oregon. A logging-truck driver, he has worked in the woods all his life. In the fall, he puts two or three of his Walker hounds on a two-foot-high box on top of his Toyota pickup, chains them so they won't fall off, and drives down skid roads in the woods. When the dogs catch the scent of a bear, they "open up," or bay. He stops and looses them, and they chase the bear until it goes up a tree. Usually, he then pulls out a video camera to take his trophy in the form of pictures, after which he leashes his dogs, returns to the truck, and lets the badly shaken bear go. "I don't get behind that program of shooting everything you tree," he says. "I just like to hear the dogs."

Early one September morning, a few weeks into the 1991 bear season, Hurst was out with the dogs driving a new logging road thirty miles northwest of Grants Pass, looking for black bears. The road was covered with dust, and he drove slowly, reading track, through the fir and manzanita. He saw what appeared to be the prints of large dogs. Though he had been down that road several times, he had not seen the tracks of these animals before. Guessing that someone had lost his hunting dogs, Hurst thought, "I'll drive down and pick them up, and return them to the owner."

"I rounded this corner," he says, "and there were two of them"—large gray animals which he thought were dogs. He stopped and got out, intending to coax the animals into the back of his pickup.

But as he stepped into the road, a third animal came out of the woods and walked around the back of the truck. By then one of the three hounds on the box on top of the truck was baying. And then the hounds all went quiet and cowered.

The third animal, the one behind the truck, "was bristling and showing his teeth. And its eyes were staring and yellow. I've been around animals all my life, and I can tell an aggressive type of dog almost instantly. I have seen wild dogs that people let roam free. Wild dogs take down deer. And after they kill a deer, they're not very aggressive. These were not like that at all.

"They looked like the pictures of wolves I've seen, but I'd never seen any being that aggressive. They didn't growl or make any noise at all. Their lips were turned up, showing their eye teeth, and the hair on the backs of their heads was standing up. They were really stiff-legged. I remember the eyes. They were driving me insane. It was like they were staring right through me. The first thing that entered my mind was, these were wolves and they were going to attack me and my dogs. They were just flat going to eat me up."

Fearing for his life, Hurst grabbed his gun from the cab of the truck, turned, and shot the one nearest him. It fell at the edge of the road, and the other two animals fled instantly into the woods. He looked at the dead animal. Larger than a coyote, with long legs and a big muzzle, it looked all over to Hurst like a wolf. He got back into his truck and drove to Grants Pass to summon warden Jack Baker, who drove back to the scene with him.

Baker had heard reports of wolves in the area. Three years before, there had been repeated claims of wolf sightings on the Rogue River National Forest, near Crater Lake National Park. People said they had sighted a group of as many as six animals, and that they were not afraid of people and seemed to have been hand-reared. Oregon wildlife officials considered trying to live-capture one to X-ray its skull and send the film to a taxonomist. But when they consulted the U.S. Fish and Wildlife Service, they were told that, if they accidentally killed one and it turned out to be a wolf, that would be a violation of the Endangered Species Act, and they'd be liable for a $10,000 fine and five years in jail. No one cared to take that risk.

While there have been a dozen reports of howls, or of wolves dashing across roads in the Rogue River National Forest, Greg Clevinger, resource staff officer for the Rogue River National Forest in Medford, says he is not getting reports of sightings from the people he would expect to see wolves. "We ask government trappers around here, 'You ever see any wolf tracks or hear wolves?' They say no, they haven't. I've talked to Bob Naney, on the wolf-recovery team in Washington. His feeling is, if you've got wolves, you know you've got them—you get howling, you get tracks." The evidence has not been sufficient to declare these animals wolves or dogs.

Still, officials had been trying to ascertain whether they indeed had wolves. The national forest does not have funding to hire its own crews to go out looking for wolves; it must rely on volunteers. A wolf brigade trained in California by Paul Joslin of Wolf Haven went out to look for wolf tracks and listen for howls, but they heard no howls and—in a winter of almost no snow—turned up no tracks.

When Baker examined the dead animal, he told Hurst, "It sure looks like a wolf. You want me to write you a ticket now, or wait until we find out if it's a hybrid?" Baker collected the carcass and sent it to the forensics lab in Ashland, which forwarded it to the Smithsonian Institution.

While they awaited the results of testing at the Smithsonian, there were additional encounters. Hurst believes the two animals that fled from him appeared a month later on the Middle Fork of Cow Creek, where they walked into someone's campsite and one was shot. The incident was not reported to authorities, because no woodsman wants to risk a fine and imprisonment should someone else decide that what he shot was an endangered species. Says Hurst, "I guess the big male is still running loose. Everybody that hunts that countryside is keeping an eye out, because they don't want to get their dogs chewed up."

Measurements of the skull of the animal Hurst shot led a taxonomist at the Smithsonian to conclude that it was a wolf-dog hybrid. And that is where the question of what's in the woods grows ticklish. According to Clevinger, there are almost a thousand people raising wolves or wolf hybrids in Oregon. Signs posted on trees advertise wolf-hybrid pups for sale. People sell them from the backs of pickup trucks by the road. Hurst says, "From talking to other people in the area, I do know there is some people at Sunny Valley that were raising wolves, and something came up that they were aggressive, and they got rid of them. Since then, I've seen the ads in the paper, usually about once a week, either wanting to buy wolf-dogs or wanting to sell them."

Are humans reseeding habitat once occupied by wolves with hybrid wolf-dogs? If so, says Hurst, "From what I've seen, I don't want anything to do with them."

Wolf hybrids pose considerable problems. There are, according to Randall Lockwood, vice-president of Field Services for the Humane Society of the United States, between one and three hundred thousand wolf hybrids in the United States. And the number is growing. Wolves are still taken from the wild in Canada and Alaska, where they are not endangered, and bred to dogs. There are perhaps hundreds of wolves born in zoos but sold to private owners because the zoos can't afford to keep them, and many of these are also bred to dogs. The growing hybrid population complicates the issue of wolf conservation and clouds our perception of wolves.

People have been breeding dogs to wolves for as long as there have been dogs. Dogs are themselves domesticated wolves, whittled off that wild stock in bits and pieces at various times over the last twenty thousand years. Early speculations suggested that humans took wolves in to help them with hunting. But, says Lockwood, who studied wolves at the St. Louis Zoo and in Alaska, "The idea of humans and wolves running alongside each other, catching prey together, really doesn't fit well with wolf biology. The way wolves hunt is often to cover miles and miles of terrain in a short time. No human group could keep up." Lockwood believes, rather, that wolves were adopted because human and wolf societies were so interchangeable that wolves made good company, and because wolves, with their acute senses and aggressive defense of the pack, were excellent sentinels and warned of the approach of human invaders.

Wolf domestication has gone on more or less continuously. Native Americans repeatedly bred their dogs to wolves. The planter William Byrd observed in 1728 that the Indians of Virginia took wolf pups from dens and kept them as dogs. Botanist Peter Kalm observed in 1750 that Indians in Pennsylvania domesticated wolves. In 1801, fur trader Alexander Henry reported that in North Dakota his dogs went out and ran with wolves for weeks at a time. Wolves would come into their camp when their female dogs went into heat. "Some of my men have amused themselves," he wrote, "by watching their motions in the act of copulating, rushing upon them with an ax or club, when the dog, apprehending no danger, would remain quiet, and the wolf, unable to run off, could be dispatched." In the 1870s, Joseph Grinnell reported that almost all the dogs among the Assiniboins, Crows, and Gros Ventres "appeared to have more or less wolf blood in their veins." Similar cross-breeding must have been going on in Europe and Asia for thousands of years.

The breeders bred for a variety of reasons, and their choices shaped vast differences among dogs. One people might breed their dogs to be shepherds—to be attentive, obedient, and swift. Another might breed watchdogs—large and aggressive creatures that would bark at or attack strangers. Some people bred dogs as toys, others to carry baggage, others to be long-legged hunters. Terriers are aggressive and persistent because they needed such qualities to pursue badgers and foxes into holes. Until the last two hundred years, humans lived in isolated populations, where their dog companions were rigidly selected for local conditions. A handful of mutations added qualities that differentiate dogs from wolves: the upcurved tails, the softer coat, the dark-colored eyes, the wider bodies, the tendency for hind legs to swing alongside rather than directly behind forelegs. But most of the genetic resources that shape the wide range of dog characteristics are present in the wolf. They are simply teased out of the genetic strands and emphasized by generation of artificial selection.

This same molding of wolf genes goes on today, but now dogs are bred less and less for practical ends and more and more to reflect human vanities. It is not at all clear why people breed wolf hybrids, but it probably has much to do with the symbolism of wolves. Wolves are hard to keep, however, and today special permits from the states and the U.S. Fish and Wildlife Service are required. Breeders conceive that they can keep the form of the wolf but trim the lupine behavior and the legal objections by mixing dog genes into the animals. There is no recognized standard for wolf hybrids, as there is for chihuahuas or golden retrievers, and there is no practical purpose, such as shepherding or hunting, that a standard might be shaped around. Lockwood has observed that hybrids in the United States are typically malamute-wolf crosses. In Europe, they are typically collie-wolf or standard poodle-wolf. In the Netherlands, there is a recognized breed called the Saarloos wolf-dog, the result of crossing wolves with German shepherds between the 1920s and 1960s. Movie animal trainers commonly breed Great Pyrenees or husky dogs to the wolves because they feel that is the only way to get an animal that looks like a wolf but performs as reliably as a dog.

Appearances count with wolf hybrids. Says Lockwood, "Wolf-dog hybrids look a lot more like what people think wolves look like than wolves do. Wolves tend to look scrawny. If you see them in the summertime or before they are three years old, they tend to look like skinny, bony, lanky dogs. Most wolves weigh seventy to a hundred pounds. I've seen hybrids in excess of 150 pounds."

If you ask owners what is special about wolf hybrids, they'll tell you different things. Some say they're guard dogs, others say they're extravagantly loyal; some think they have an animal capable of extrasensory perception; some are under the illusion that they're helping to save the genes of the wolf, which they believe is going extinct in the wild. The most common reason has to do with the wolf's sociability, its intense loyalty, its concentrated focus on the members of the pack. Hybrid owners want to train that loyalty on themselves. The wolfish nature of this sociability suggests that the loyalty is natural, not engineered, a gift from nature. And the liveliness of wolf perception suggests acquaintance with the deeper mysteries of life, an acquaintance the owners would like to share.

Lockwood says, "A lot of people mistakenly feel they're being put in touch with nature, closer to the wolf's spirit. They see themselves as part of the environmental movement by owning this little piece of wild nature. For a lot of people it is a religion."

In fact, hybrids don't work out to be good guard dogs. Wolves may be good sentinels, but they are not likely to come to an owner's defense. Twice in one week while Bruce Weide was exercising his captive wolf Koani on a leash in southern California, the wolf was attacked by pit bulls. In both attacks, Weide had to fight the dogs off while the wolf cowered. Weide believed the wolf looked to humans for deliverance, because, in order to keep the wolf, the humans had maintained the roles of dominants. Says Ed Andrews, who has kept captive wolves and wolf hybrids in Washington, "You can't make a guard dog out of a wolf. One of my wolves protected me from other wolves. I would have been killed one day had not my wolf, Cripple-foot, held the other wolves off. But had a man attacked me, he wouldn't have known what to do." Hybrids are no more reliable. The more wolflike a hybrid is, the greater the likelihood that it will either defer to its owner or be so aggressive that it must be kept in a cage.

And though hybrids are capable of intense sociability, the ones that turn into rewarding pets generally do so because their owners are willing to commit a great deal of time and patience to them. Wolves that have been reared in a wolf pack do not take readily to the company of humans. Those who have raised wolves to accept them as companions have taken them from a litter at a very young age and spent long hours with them every day. In effect, such owners have had to work within the wolf's capacity for society, and shape their own lives to the limits of the wolf, one of which is a limited ability to extend itself to strangers. According to Bobbie Holaday, who keeps hybrids in Phoenix, Arizona, "People who take in hybrids are taking up a lifetime commitment. You can't just feed them once a day and leave them alone. When I leave them in the camper alone, they can get kind of destructive. They're a wonderful animal as a companion, but they just don't sit down and play dead."

Many cannot be housebroken. They defecate and urinate not to empty their digestive tracts, but to scent-mark territory, perhaps even to express ambition. They are apt to follow an owner around the yard, replacing the very deposits the owner is shoveling up. If a strange dog has been allowed in the house, they are apt to scent-mark where it has been.

And they are mischievous. Lockwood once sat on a panel of veterinarians who, trying to characterize hybrids, came up with the words "sneaky" and "weird." Lockwood attributes those qualities largely to the fact that wolves are social animals. When a hybrid is left alone all day by a working owner, it gets bored and lonely. It becomes destructive: it chews up furniture to get at the twang of a spring inside, or rips out trees, or digs under fences.

But the appeal of the wild is stronger than the awareness of its drawbacks. The number of wolf hybrids is believed to be increasing. Wolf-hybrid puppies are regularly advertised for sale in the want ads of large city newspapers, and some breeders are making money off the desire for the animals. Says Lockwood, "We are encountering an increasing number of wolf-hybrid puppy mills. In Arkansas, one advertised an April tax special: buy three, get one free. Give them your MasterCard number over the telephone and they'll put a four-week-old puppy in a box and mail it to you. People breed hybrids and sell them for $500 or $600." Much fuss is made about the percentage of wolf genes an animal is reputed to carry: there are 50-percent hybrids and 75-percent hybrids. Lockwood says, "Some come with certificates that say things like 'This animal is 98 percent wolf.'" Some dealers at least ask prospective buyers whether they have an enclosure for the animal, and whether they have the time to spend with it. But many breeders ask no questions.

And they ought to. For it is an animal beset with problems.

Ron Maga is a thirty-four-year-old fireman in Quartz Hill, California. His wife, Jennifer, is a massage therapist. They own two wolf hybrids, which are licensed as such with the county. Says Maga, "They are brother and sister, Peso and Kenai. The father was 100 percent timber wolf, and the mother malamute. A friend brought them from Alaska. He couldn't keep the male because he had two male huskies that would occasionally get in fights and tear each other up, and they were doing this to Peso, too—the dogs wanted to kill the hybrid." So the friend gave Maga the male but kept the female, hoping to breed her to one of his huskies. That was four years ago. About a year later, he also gave Maga Kenai.

"They are trainable," says Maga. "They're very smart. They just look at you like you're some kind of knucklehead if you throw a ball.

We baby them—we bring 'em in at night, and they're house-trained. They're very good.

"But they are not dogs. They are definitely not dogs. They are very family-oriented. They're sensitive. They're emotional. They're almost psychic—they'll pick up on how I feel when I walk in the house. If I'm upset about something, they'll back away. I have a friend who's loud and aggressive and boisterous. They really don't like him. They prefer people who are calm. Whenever he comes over, before he even knocks on the door, the dogs will react. If they're near the door, they'll back off. I'll know it's him before I hear him come up the walk."

They have, says Maga, a special awareness. "I believe our mental status creates a certain frequency that we'll emit; I just think they pick up on that. I think they're very psychic-oriented. They're almost, like, spiritual." Because of that, he says, "I do not want to pen them up. I do not want to chain them up. So they have the run of the backyard. I have a five-foot cinder-block wall all around the yard. I had to put up two and a half feet more of corral fence on top of that, just to keep them in.

"I've been told by the guy I got the dogs from, the worst thing to do is have kids with wolves, because you don't know when they may turn on them. I would bet my dogs would never do that. Their temperament is fantastic. We've had kids in the backyard pulling and scratching and jumping on them, and they don't bother the kids at all. They lick kids to death. They're very, very gentle. One day, a strange dog cornered them both in the backyard and they rolled over and submitted." But since hybrids *have* attacked children, Maga is never perfectly sure that his hybrids' temperament won't change. "I try to feed them dry food," he says. "I've heard, if you start feeding them raw meat, they get more aggressive."

The trouble comes when the Magas go to work: she works full-time, and he is sometimes gone for three or four days straight. "If we're gone," Maga explains, "they dig and tear up things, I think out of spite. There's something about us not being home for a lengthy period of time that aggravates them. They start pushing our buttons. They'll chew sprinkler valves, tiki lights, valve covers. I'll look out the window of the back door in the morning and see they've pulled up a tree. They'll look at me and, if I'm angry, as soon as I think,

'God!,' they'll hide under the deck. I have to admit, they're smarter than I am."

And they get out. "Any type of hybrid wolf is a runner. They love to get out and run. When Kenai was in heat, she would jump the fence. She got out one evening and came back about five in the morning, and I found a rabbit in my front yard. I thought she had killed it and brought it back as a gift for me because she knows she's not supposed to get out."

They had the female spayed, and that seemed to calm her down a little. But if he does not exercise great care, says Maga, "when the sun goes down, they're gone." One day, "my wife put Peso in the garage. He tried chewing through the door and got nails stuck in his mouth." The construction worker who gave Maga the hybrids had another wolf-husky cross which, says Maga, "accidentally got locked in the bathroom of the house they were working on. It chewed through drywall, two-by-fours, lath, and plaster, and got out and ran home that night. It came home all bloody."

Maga is more worried about what people may do to the animals than what they may do to themselves. One night, at dusk, his neighbor stopped a sheriff's deputy who was standing in a nearby field with his gun drawn on the escaped hybrids. The officer thought they were coyotes and was about to shoot them.

"I went through a period when I would come home and say, 'That's it! I gotta get rid of 'em!' I would put an ad in the paper for them, and I would get fifty calls, and the homes were all small. Even though I wanted to get rid of them, I just didn't have the heart to give them to someone else who didn't have a home for them."

The Magas found a home for the male with a woman who had dogs that she worked in films and who planned to use him in television commercials. But Peso got sick, and the woman tried to force the Magas to pay the veterinarian bills. When she found he didn't train as easily as her dogs, she neutered him. That didn't change things. Within a month, she was ready to take him to the pound, so the Magas took Peso back.

"We still have them, and we'll probably keep them," says Maga. "I've never had a dog love me like those wolves have. They're very loyal, but they're a challenge to keep. They need a lot of attention.

"I feel responsible. It's not their fault that the things they do aggravate me—it's just their nature. It has got to be frustrating for them, because somewhere in their genes is a yearning to get out and hunt and run and do the things a wolf does, and I've suppressed that, keeping them in the environment they're in."

Says Maga, "They're not dogs. There's a space between a wolf and a dog, and that's these animals."

As a teenager, Terry Jenkins, now curator at the Folsom City Zoo in California, decided she wanted a wild pet. She bought a wolf from the Folsom Zoo and got a permit to keep it. Like others who acquire exotic pets, Jenkins knew little about it. She found it fascinating, but headstrong and hard to keep. When she looked at other captive wolves, she concluded, "Most of them were scared of everyone and were in little cages in a backyard, and it really seemed to be a tragedy." She thought she might solve the problem by breeding her wolf to a dog, "to produce an animal that looked like a wolf, so it would satisfy that urge to have an exotic pet, but would have the personality and temperament of a dog, so that it would make a good pet." The cross-breeding, she also hoped, would save wolves from having to live in captivity. "It took me a number of years to realize that the wolf hybrid wasn't an answer to that."

Some of the hybrids she produced were intensely loyal and expressive, but others were not good companions. Jenkins had one animal that was very affectionate and appeared to love babies and women, but he tried to kill her. One day she tried to establish herself as his superior and failed: the hybrid lunged at her and bit her repeatedly in the chest, going for her throat. "I shoved my hands down his throat and let him chew on those, and backed out the gate," she says. She never trusted the animal again. "I was quite certain that if I had fallen he would have killed me. He didn't really react like a wolf. A wolf will go through a whole range of body postures and silent expressions before it attacks. Even if they bite you, they'll still back off and give you a chance." At that point, she didn't know what to do with the hybrid. "I called the original owner and she said, 'Shoot him.' " Rather than do that, Jenkins sent the hybrid to a friend in Iowa. But there the animal became extremely aggressive toward women and children. "It was an intense, scary, and obviously very dangerous situation," says Jenkins. When the hybrid got into dominance fights with another male, the lady in Iowa euthanized him.

One of Jenkins' hybrids was a lap wolf: "She loved to be in your lap being scratched." Jenkins would take her into school classes, and she would go around a circle of seated children, licking every face. As she grew older, however, she became shy around strange adults. One day, Jenkins took her out on a leash to show to some people who had come to her house to see the wolf-dog. One couple had an infant, which they placed on the ground. "She had always been gentle with babies. She kissed the baby, but kept looking up at the parents. Then she very carefully reached over the baby's shoulder with her jaws and tried to pick the baby up and move him. The parents moved quickly to save the baby, and the hybrid jumped back and dropped the baby, who was knocked over and started crying. I never let her around a baby or even a little child again. By the time I had children of my own, she was very aggressive to children."

Jenkins' caution was well advised. In 1984, a woman visiting her mother in Reno put her own baby on the floor and left the room. The grandmother's hybrid came in, picked up the unattended baby by the head, and killed it instantly. A wolf hybrid took off a child's arm in New Jersey, and another did the same thing in Montana. In 1988, the Panhandle Animal Welfare Society in Walton, Florida, advertised for adoption a five-year-old neutered male wolf-husky that had been through several owners, and which the shelter believed was such a gentle animal they made it Pet of the Week. A family took him home, but within an hour he jumped a four-foot fence. A neighbor woman found him and put him in her yard. Since he still had the shelter identification tag on, she went inside to call the shelter. While she was on the phone, she heard the animal attack her four-year-old son. She got the hybrid off her child, but he was critically injured. The emergency team had to fight the wolf-dog off with a flashlight while they tried to save the boy. The boy died. In 1993, a child was killed by a hybrid in Vermont.

There are fifty-four million dogs and from one to three million reported dog bites a year in the United States, according to statistics kept by the Humane Society of the United States. There are an average of twenty fatalities inflicted by dogs a year. Between 1986 and 1992, eight wolf-hybrid attacks took human lives.

The hybrid attacks, says Lockwood, "are not slavering, savage attacks. These are not for the most part vicious animals. These were not animals that said, 'Mmmm, I'm going to eat him.' These were animals that were curious, that were inquisitive, that were defensive, or that regarded these children as an interesting inanimate object or as prey. We know a lot of the malamutes are cat killers. In 1991, we had six or seven malamutes and huskies kill children under circumstances that were essentially inquisitive predatory attacks. They regarded a child as they would a rabbit.

"There is this mythology, particularly among the owners of the dogs who bit somebody, that the victim did something to provoke it. The vast majority of dog bites involve people who are not doing something inappropriate at the time. A child puts his arm through a fence: that is stupid, but it's not provocation. Most bites are owned animals injuring the kid next door or a family member."

Lockwood thinks part of the problem is that we have been breeding dogs for aggressive traits for centuries. He shows a picture of a tile from a floor at Pompeii, dated A.D. 79, which says *"cave canem,"* meaning "beware of the dog": "Two thousand years ago, people were taking mean dogs and chaining them to their front doors. This has been going on for generations and generations. To me, this has been one of the major differences between wolf and dog. We have selected our dogs to be far more aggressive, and far less in control of their aggression, than their wild counterparts." Lockwood has watched wolves encounter bears in Alaska and retreat from the danger. By contrast, he says, humans have trained dogs to be so excessively aggressive that they will fight to the death. He has investigated pit-bull attacks and dogfight promoters. "When dogfighters tell me these dogs love to fight or they're just doing what God intended, that's nonsense." He recalls that English breeders enclosed rat terriers with hordes of rats, and a good dog would kill a hundred of them in less than five minutes, and that was the animal they would breed. "A wild canid normally won't kill more than he can eat, but we can breed animals to keep going and going and going. We have bred out the off-button that controlled aggression."

When Lois Crisler left Alaska, she took wolves with her. Living in

Colorado, she sought to make her wolves less indifferent to her by breeding them to dogs, in the belief that the resulting animal would be "courageous, untamable, serious, yet gay," and have "a genius for loving." But in her act of engineering, she merely took the wildness out of the wolf. The hybrids were more aggressive than the remaining wolves. She felt the hybrids contained an inbred schizophrenia, the wildness of wolves and the tameness of dogs, and each pulled at the soul of the animal until it went crazy. Increasingly, she saw in them "a queer, uneasy equilibrium." They were, she feared, filled with anxiety and rage.

The messages Jenkins got from her hybrids, however, were conflicting. The hybrids seemed immensely perceptive and intelligent. Many were loving and joyously shared their feelings—one of them chipped Jenkins' teeth and blackened her eyes just in greeting her effusively. But caring for them was demanding. "Most wolf hybrids have this intense purpose in life, and it doesn't always coincide with yours," she says. She tells of a friend's hybrid that jumped through plate-glass windows to get out. When that animal got older, he also jumped through plate-glass windows to get in. Jenkins had to give her hybrids hours of attention each day, and had to take extra precautions so that they didn't get out or weren't made more aggressive or neurotic by the acts of visitors. The tension between affection and concern was draining. And yet it was addictive. Over a period of twenty-two years, Jenkins cared for 130 different wolf hybrids. "It was my whole life," she says. "That's how involved I was in it." She could not bring herself to euthanize her own animals. And she continued to take other people's hybrids in when they could no longer cope with them, feeling she had the skill and understanding to save them.

Now thirty-eight and a mother, Jenkins has a different view of keeping wild animals. She still has two hybrids, one of which will not allow anyone but Jenkins near her and must be kept in a cage. "She's not a dog," says Jenkins. "She's not a good companion animal." Jenkins acquired the second hybrid to keep the first one company. She can't let them be wolves, but she can't trust them to be dogs. She feels obliged to keep them alive. "I feel a great pain and frustration from this problem," she says. "I hurt daily from it."

• • •

There is a place for dogs, and there is a place for wolves. But whether there's a place for wolf-dog hybrids is an unsettled question. Wolf hybrids are neither house pets nor wild animals. They place different demands on people from those placed by either wolves or dogs.

Where people adjust to those demands, the animals fill their lives. Speaking to a California seminar on wolf hybrids sponsored by the Humane Society of the United States, Lockwood says: "Owning hybrids becomes almost an addiction. I've frequently heard people say, 'Once you've owned a wolf or once you've had a hybrid, there's no going back.' It becomes a kind of codependency—they've altered their life-style to fit the wolf or hybrid. They have a community of like-minded people. They have their own circle of friends, and often their other friends get pushed aside, and it becomes almost a substitute family for them." But when a hybrid tears up their yards, rends their society, divides their families, and weighs them down with guilt and obligation, they want to get rid of it. They look first for wildlife shelters and animal-rescue centers.

The Wildlife Way Station is up a side canyon off Little Tujunga Canyon, in the San Gabriel Mountains, which form the northern rim of the Los Angeles Basin. A narrow road winds up the canyon, through dry hills covered with chamise, scrub oak, manzanita, yucca, and prickly-pear cactus. There are stables everywhere, their corrals framed by groves of eucalyptus, dull gray-green in the intense southern-California sun. The Way Station's side canyon is deep and shaded with sycamores and oaks. Under a live oak, two very wolflike forms stir. They have the piercing yellow eyes, broad foreheads, big jaws, and oversized feet of wolves. They seem oddly colored; they have a reddish tint to their coats, which seems to bleach out when they step into the bright sunshine. Perhaps it is more diagnostic that they are slow-moving, not pacing around their pen, and that they are not shy of visitors but come confidently up to the fence to sniff at them. It would be difficult to say, if you saw them melting into the forest, whether they were wolves or hybrids.

There are eight hundred animals in the facility—lions, tigers, jaguars, macaques—most of them animals people acquired as pets and then tired of. Martine Colette built this park as a refuge for unwanted exotic animals and supported it with private donations. It is part of the Neverland quality of southern California that people take in lions and tigers and bears, but, though wildness is what makes such animals attractive, wildness doesn't thrive in a southern-California yard. If you muster the time and concentration to keep such an animal, you take the wildness out of it. If not, the romance fades, and there is only a nagging responsibility, and you want to get rid of the animal. If ever a refuge for domesticated but unwanted lions and tigers and wolves might take root, it would happen here in southern California, where nature is celebrated and denied with equal fervor.

Wolf hybrids tend to appear in places where people both celebrate and bemoan the loss of nature: there are relatively few in Alaska and Canada, many in California. Colette explains that the Wildlife Way Station takes in wolf hybrids, but does not adopt them out again. "We do send the animals out now and then to a facility. If the hybrids pass for wolves"—that is, if people can feel they are looking at something wild when they view the animal—"they are sometimes taken by zoos or nature-education centers." Otherwise, they are here for life.

The Wildlife Way Station gets dozens of requests each year to take in wolf hybrids. It has at least thirty already on the premises, but it can take only so many of these rejected pets. The refuge performs a kind of triage, accepting only the most wolflike of the hybrids. "People sometimes call and say they've got a hybrid," says compound foreman Audrey Wineland, "but it turns out to be just a dog. We have them send us a picture and give information on age and so on. When we look at the picture, if it looks like a wolf, we'll take it. If it looks like a dog, we can't take it."

There are only a handful of refuges for unwanted hybrids. Nature centers here and there may take in a few. Wolf Haven in Tenino, Washington, has taken in one or two, and the Charles Avery Park in Minneapolis has some. But there aren't enough places for all those now seeking refuge. Carlyn Edison, who rescues hybrids in Austin, Texas, gets three thousand calls a year, but can place only about twenty animals. Bill Chamberlain, of the U.S. Wolf Refuge and Adoption Center in Apache Junction, Arizona, says he has a similar number of inquiries. He takes in hybrids, then tries to socialize them and place them in homes with new owners who have the patience, time, and dedication to care for them. Says Chamberlain, "They're not going to go away. What we need to do is match that population to the number of homes that are able to deal effectively and humanely with them."

The Folsom City Zoo serves as a refuge for unwanted pets and orphaned and injured animals. It has four wolves and two wolf hybrids on exhibit, and a mountain lion that was raised in an apartment with a wolf and a chimpanzee. The zoo tends to respond to fashions in such pets. "When I first came there, in 1982," says Jenkins, "the animal that was most frequently offered to the zoo was raccoons. We literally got hundreds of calls on raccoons. Then we got several calls a week on ferrets. For about the last five years, it's been wolf hybrids, and it's increasing. I'm in the position of having to turn down countless desperately in-need animals. There are always more animals out there than anyone has room for. I really hope they get them euthanized."

Some hybrid owners don't even try the shelters. Instead, they drop the animals off in the woods, expecting that the wolf genes will orchestrate some kind of Jack London story of survival in the wild. But the odds against their survival are enormous: wolves must be taught to hunt and learn to use the landscape. There is no record of domestic hybrids released into the wild successfully occupying the niche of wild wolves. Nevertheless, people continue to put them out in the woods.

The hybrid Earl Hurst shot in Oregon was probably just such an animal. Near Helena, Montana, several released wolf hybrids have been shot by ranchers after killing sheep or cows. Two different "wolf packs" in Idaho hung around campsites and ate food out of bowls before they were recognized as hybrids. In Glacier National Park in 1992, two Tennessee residents were caught by park officials trying to release their hybrids into the wild. Because the released animals didn't know how to hunt, they hung around the campgrounds.

Terry Jenkins regularly hears of hybrids turned loose in the mountains of California. There were tales of wolves in the Warner Mountains after a private wolf-hybrid refuge near Susanville shut down. A man crossing the Sierra Nevada with his dog one winter told Jenkins that a wolf walked into his camp to pay them a visit. Repeated reports of sightings and howlings have come from the vicinity of Ice House, in El Dorado County.

Once a hybrid is turned loose in a setting that might support dispersing wolves, it becomes difficult to say what kind of animal is there. In 1992, an animal thought to be a wolf was hanging around the town of Glacier, Washington. It walked into an empty swimming pool that John Almack of the Washington Department of Game had baited with fish, and Almack went into the pool and darted it. He was reluctant to declare it a hybrid, because wolves were believed to be dispersing into the area, and rarity gave value to even a faint hope. So, after X-raying and measuring it, he put a radio collar on it, drove it into the back country, and released it. Immediately, it homed back to the housing area where it had been caught.

Carter Niemeyer, who traps wolves known to have killed livestock as an Animal Damage Control trapper in Washington, Idaho, Montana, Wyoming, and North Dakota, heard about Almack's wolf. Sent out to look at the animal, he found it so tame that he could take a twenty-minute video of it. "That was about as wolf as a poodle was," says Niemeyer. "The animal was walking down the street. I was just astonished that anyone would collar an animal like that." The Fish and Wildlife Service authorized recapture of the hybrid. The service's agent Jeff Haas persuaded some people to walk their dog into a tennis court, and the hybrid followed it in. Haas simply picked up the animal, put it in his truck, and drove away. Niemeyer traced it back to a woman who claimed she had sold it to another person who let it get away. It now resides in Wolf Haven.

The behavior of the modern hybrids poses questions about what kinds of animals have been responsible for attacks on humans in the past. For example, the most celebrated human-killing wolves of Europe were a pair of wolves collectively called "The Beast of Gévaudan," which appeared in France in 1764 and began killing women and children. Though King Louis XV put out a reward of 6,000 livres for whoever killed the beast, the animals eluded hunters until 1767. In that time, they were said to have taken the lives of more than sixty women and children. When hunters finally killed them, they proved to be enormous animals. One weighed 130 pounds, nearly twice the weight of a typical French wolf. The colors of their coats—one had a reddish tinge, the other a white patch on its throat—were characteristic of dogs, rather than wolves. And the beasts attacked people by biting at their faces, a behavior that suggested either rabid wolves or domestic dogs, though none of the victims contracted rabies. The evidence strongly suggests that the beasts were hybrids.

Hybrids might also account for many of the attacks on livestock that have been blamed on wolves. In the American West, ranch dogs run free, and there are many accounts of ranch dogs and wolves mating. The resulting offspring might, like modern hybrids, lack the caution of wolves. The reports of federal wolf trappers early in this century indicate that they frequently caught cattle-killing hybrids on New Mexican ranches. J. Stokley Ligon's 1924 report on the Predatory Animal and Rodent Control program in New Mexico declared that "in more than 75 per cent of the cases investigation disclosed the fact that dogs or coyotes and in some cases hybrids, wolf dogs, were the offenders." There were several well-documented cases of hybrids attacking livestock in Montana in the 1980s. And today, Ralph Opp, of the Oregon Department of Fish and Wildlife, reports that in a recent six-month period there were six different packs of wolf hybrids attacking livestock in Klamath County.

The possibility that there are hybrids running loose in the West bedevils plans to reintroduce wolves into places from which they have been eradicated. Reintroduction hinges on the premise that there are no wolves remaining in the wild. If people begin seeing hybrids in the woods, reintroduction may be stalled until it is clear what those animals are. If the animals seen are truly wolves, they enjoy the protections of the Endangered Species Act. But if they turn out to be hybrids and they mate with wild wolves, the resulting offspring are not protected, and managers will be unable to tell which wolves are pure and which are not. In southeastern Australia, all the wild dingos appear to be dog-dingo hybrids. Something similar could happen to wolves in North America.

The presence of hybrids in these areas hurts efforts to persuade ranchers and hunters that they have little to fear from wolf-reintroduction efforts. Sheep and cattle killings or attacks on children by hybrids will be blamed on wolves when, in fact, the wolves in the neighborhood may be innocent. Says Lockwood of people who keep or release hybrids, "These people think they somehow are undoing the Little Red Riding Hood syndrome through the keeping of hybrids, but ultimately it backfires."

Because there are so many problems with wolf hybrids, there has been a movement to regulate their breeding and keeping. Nine states forbid the ownership of wolves or first-generation wolf hybrids, and a dozen more require permits or licensing for wolf hybrids or animals that might be taken for wolves. On top of that, an increasing number of cities and counties are outlawing or restricting ownership of wolf hybrids.

In California, ownership of wolves or first-generation crosses requires permits from the U.S. Fish and Wildlife Service and from the California Department of Fish and Game; hybrids other than first-generation crosses with wolves are considered dogs. The law requires every dog owner to vaccinate for rabies and license the animal before it is four months old. However, though the U.S. Department of Agriculture has approved specific vaccines for use with dogs, cats, horses, and ferrets, it has no approved vaccine for wolves: the USDA regards dogs and wolves as different creatures and will not assume that a vaccine effective in dogs will also be effective in wolves. Before approving rabies vaccine for an animal, the USDA requires tests in which at least thirty subject animals are injected with live rabies virus, tested for at least a three-year period, then challenged with the disease for another three-year period. No such testing has been completed on wolves—a test was undertaken at Auburn University but was discontinued when funding ran out. Wolf hybrids fall through the cracks in the law: hybrids must, as dogs, be licensed, and rabies vaccination is part of the licensing process; but, technically, they cannot be licensed, because they cannot be vaccinated for rabies.

Many are vaccinated and licensed anyway. But if a wolf hybrid bites a human, the bite may not be treated as a dog bite. If a cat or dog bites a human, California law requires the animal to undergo a ten-day quarantine, during which a rabies-infected dog or cat will usually shed rabies virus in its saliva or otherwise show symptoms of infection. But no one knows whether a wolf will demonstrate infection in the same short time; bats and skunks, for instance, have been known to take more than ten days to shed the virus. So wild animals that bite humans are generally euthanized and their brains dissected and tested for rabies. Exceptions are made for exhibited and rare wild animals, which may undergo a thirty-day quarantine, but in such cases the human victim will have to undergo treatment for rabies. Unvaccinated dogs or cats exposed to rabid wild animals undergo six-month quarantines at the owner's expense. If a wolf hybrid bites a human, whether or not it has been vaccinated, the animal-control authorities can require either a six-month quarantine at the owner's expense or sacrifice of the animal for testing. In most other states, if a hybrid bites somebody, the animal must be sacrificed and the head presented to the state diagnostic lab for rabies testing.

The American Veterinary Medical Association does not recommend that wolf hybrids be vaccinated, fearing that vaccination will give owners a false sense of security. Many veterinarians won't immunize a hybrid without having the owner sign a release stating he won't hold the vet responsible should the vaccine not work. Companies manufacturing the vaccine say they have no liability if a veterinarian uses the vaccine in ways other than described in their product literature, and the literature does not recommend vaccination of wolves. The AVMA warns that a veterinarian who immunizes a wolf hybrid may have no insurance coverage in the event that the animal contracts rabies.

California requires the owner of a first-generation hybrid to have a permit and to build adequate fencing to keep the animal confined. Increasingly, however, counties and municipalities, seeing that the state has been unable to check whether permit holders have the required fencing, are moving to restrict or outlaw the ownership of hybrids.

Enforcing all these laws poses difficult problems. Once a hybrid owner finds that his pet may be killed or subjected to a six-month quarantine, he may suddenly decide to identify the animal as a dog. A Stanislaus County animal-control officer says that, twice when he has picked up animals identified as hybrids by owners, he told the owners he was going to quarantine their pets for fourteen days, and suddenly "the animals were no longer wolf hybrids." The animal-control officer fears being sued by an owner if he sacrifices an animal that has been represented as a hybrid and the owner changes his story—or by a bite victim if the owner registers a pet as a malamute and after the bite says he told the registering officer it was a wolf-malamute hybrid.

Few health-department officials or animal-control officers can distinguish wolf hybrids from dogs. It takes a trained taxonomist to judge from skull measurements, and a first-generation hybrid's skull may fit in either category. Eye color is no indication, since wolves, malamutes, red Siberian huskies, and weimeraners all have yellow eyes. There is as yet no test of DNA to distinguish definitively between wolves and dogs. When there was great concern over pit-bull attacks, a private company tried to develop a genetic test with which to identify pit bulls and failed.

Dr. Robert Wayne is trying to develop such a test for wolves, but it is a tricky business. Since dogs were bred out of wolf blood lines less than twenty thousand years ago, the genes of dogs and wolves have not diverged much. Dogs were domesticated several times in history, and each domestication presents its own lineage, so a marker present in one dog may not be present in all dogs. Wayne will have to find the DNA sequences that are shared by dogs but not by wolves. Geneticist John Paul Scott believes all the differences we see between dogs and wolves could be accounted for by about twelve mutations. Finding pieces of those twelve fragments of DNA in the immense genome of the dog may be like finding a bottle drifting on the Pacific Ocean. At best, Wayne hopes to come up with combinations that give statistical likelihoods. He will look at the frequency of various gene sequences. "If this one is in 95 percent of dogs and 5 percent of wolves, it gives us a good indication. In the end, what we want to make is a probabilistic statement. We can say an individual, if it has this distribution, is a wolf, say, all but one in a million chances."

Without such a test, identifying hybrids is more an art than a science. One of the leading artists is Monty Sloan, a wolf-behavior specialist at Wolf Park in Indiana. Young, slightly built, and bearded, Sloan started working with wolves at the San Francisco Zoo in 1984, and has worked at Wolf Park since 1988. He does not look anyone straight in the eye. "I think it's something I derived from working with wolves," he explains. "If the wolf doesn't know you, eye contact will result in aggression or fear. I get much more novel responses from dogs and hybrids because I don't look at them until they come over. With wolves and wolf hybrids, you often have to kneel down to get them to come over to you."

Sloan has made a study of differences between wolves and hybrids and has often been called in to judge whether an animal is a wolf or a dog. He says the judgment is seldom simple. "I can't say this animal has wolf in it or this animal doesn't have wolf in it. Anything less than 25 percent wolf will not look wolflike. Very high wolf-content hybrids will tend to be so wolflike in characteristics that you can't tell any dog content. I look at general body build, behavioral characteristics, how they act to a novel person, how they carry themselves. Wolves are very narrow-chested. Hybrids have shorter legs. The back won't be straight across." Microscopic examination usually reveals four bands of color on an individual wolf hair, but only three on a dog's. A hybrid will have pointed ears, whereas a wolf's are more rounded, and the hybrid's ears will be thinner: when the light is behind them, a little pink shows through. Sloan looks also at the animal's behavior. For example, an adult wolf, he says, will roll over in playful submission in front of a pup, but a dog will not necessarily do that.

He seldom finds one distinctive quality. "It's a blend—you can't point to any single characteristic. A lot of the things I see are subtleties. I can't even say what I'm seeing."

Other students of the art look for different qualities. Terry Jenkins sees fur inside the ears, cheek tufts, or dense underfur on the legs in winter as wolf traits. The claws of wolves and wolf hybrids are bigger in diameter than dog claws; Jenkins looks especially at the way an animal moves. "Wolves have fluid movements. Often I've been struck, when seeing a high-percentage wolf walking on a leash, that it was a big cat. The movement is classically smooth." But none of these traits, she concedes, is definitive.

While we argue over the nature of individual animals, the hybrid population continues to grow. Hybrids pose enormous challenges, but few of us have the resources to cope with them. When we release them into the wild, we confound the nature of wolves and challenge our own understanding of nature. We threaten the integrity of ecosystems. We even confound the nature of the dog.

Says Randall Lockwood: "I think we have spent fifteen to twenty thousand years transforming the wolf, through the process of domestication, into an animal that for the most part can live safely, happily, and humanely in human homes. In producing and proliferating wolf hybrids, we take a big step backwards. We are undoing what we have worked twenty thousand years to do."

14

LOOKING FOR SPIRIT

Our interest in wolves expresses the hunger of our imaginations. For many, science is too narrow a view, and wolves are as much spiritual as biological. They say that, to understand wolves, we must go beyond what we can see into realms of spirit. Much of the literature of wolves urges us to hark back to what native Americans said about the animals. A number of people suggest that, if we could but regard the wolf as native Americans did, we would take it to our hearts, see it clearly, and recognize higher powers that stitch us to the cosmos.

Ethnographic studies suggest that native North Americans held wolves in high regard. Plains Indians, particularly, acknowledged a high degree of similarity between humans and wolves, and they celebrated the likenesses in wolf-clan totems, wolf-warrior societies, and hunting techniques that consciously imitated wolf behavior. Modern Eskimos express deep respect for the wolves they see. Might we find new spiritual realms by talking with other cultures about how they live with wolves?

Fort Chipewyan is an Indian village notched into the spruce-forested southwest shore of Lake Athabaska, in northern Alberta. To the west of town is the Peace-Athabaska Delta, a broad, marshy, river-braided plain that stretches for hundreds of miles. The delta forms where the Peace and Athabaska rivers back up against the granitic mass of the Canadian shield. When spring snowmelt comes to the Rocky Mountains, the rivers flood over this plain, producing a watery environment that is home to beaver, muskrat, millions of ducks and geese, the northernmost subspecies of bison, and the largest documented packs of gray wolves in North America.

Today, Fort Chipewyan is a miscellany of low trailers and aluminum-sided houses with steeply pitched roofs. On a knoll at the east end of town is the new Fort Chipewyan Lodge, a prominent, two-story cedar-sided structure overlooking the lake. Those who can afford it, and many who can't, drive up there on weekends to drink in its bar. On the west end of town, also on a knoll, is the 150-year-old Catholic mission, a rambling, red-roofed, wood-frame structure with gables and steeples. The main street is paved from the mission to the lodge, and midway between them a Canadian flag flutters in the stiff breeze out in front of the offices of Wood Buffalo National Park. Unnamed dirt streets run off the paved road, but there is hardly anyone on the streets. Boats and snowmobiles sit glumly in the weeds. Ravens croak from perches in spruce and birch trees. Dogs bark and howl, sometimes in packs, picking up a low moan and passing it along from one end of the silent town to the other.

This town is sixty miles away from the next pavement, unconnected by road to anything, except when winter freezes the Slave River and one can drive on the ice to Fort Smith. The population of nine hundred is almost all native: Crees, Chipewyans, and Métis Indians. It is a mix, not just of native cultures, but of old and new. Traditional Indian music chants from a speaker inside the old folks' home, and at a window an old man stands listening pensively to it. Native trappers sell their furs at the Northern Store, a low, yellow, windowless building on the main street. They are apt to walk out of the store with Green Giant frozen peas, Eggo frozen waffles, Skippy peanut butter, Aunt Jemima syrup, El Molino tortilla chips, Mamma Mia frozen pizza, Shake 'n Bake, and Tang.

In front of the store is a blue wooden bench, coated with dust from the clouds raised by passing trucks and vans, that is the heart of the town. It is here that the old men gather to tell stories, to fish for stories, to wait for stories. Stories thread their lives, tie them to the earth and its creatures and to other human beings. Stories carry the wisdom of elders and give a sense of one's place in the life of the people. I have come here to hear stories about wolves, and to see how wolves live in the spirit of these people.

On the bench, two old men are talking about the unusual August weather. The temperature last night went down to minus two, and tonight it will drop again. For the last three weeks, the clouds have been not the clouds of summer but the clouds of autumn—higher, more anvil-shaped, with rain in them. There is an undercurrent that this is just another one of the changes.

From 1788, when Roderick MacKenzie and the first European fur traders arrived here, the Peace-Athabaska Delta has supported fur trappers, who took the skins of mink, lynx, beaver, otter, and wolf. The delta once yielded two to three hundred thousand muskrat pelts a year. But twenty years ago, when British Columbia raised Bennett Dam on the Peace River, the dam cut off the spring floods that used to spread out over the delta, and much of the delta dried out. Said a Métis Indian trapper, "I see a lake where it used to be good for muskrats. Now it is all straight bush. Everything growed grass and willow. Where we used to fish in three or four feet of water, there is only four inches of water. Now nothing but mud, no fishing grounds there any more. Mink also disappeared, after the muskrat disappeared. Even the frogs left the country."

As the delta dried out, so did traditional culture. The young people no longer go out with the old people to hunt and fish, so they no longer learn the things that their ancestors had to know to stay alive. Many of them will go off to the cities to the south to find employment. Others will stay in Fort Chipewyan, living on government handouts and the scorn of the outside society, buying their food at the Northern Store, and drinking to fill the emptiness of life. Domestic violence runs high; a Mountie says that half the complaints he responds to are of men being beaten by their wives, and that alcohol is the root of just about all the complaints. A new set of forces shapes their culture.

The men on the bench aren't finding stories; they share only gossip—so-and-so went out on his boat to Fort MacMurray for supplies and hasn't come back, so-and-so owes me money. The old stories are gone, but new stories aren't yet born. The wind off the delta is not bringing words to mend the world.

For days, I hovered around the blue bench and eavesdropped. Whenever I approached as if to sit down and enter the stream of conversation, however, the men would suddenly disperse, like birds flushed from a field. One morning, I sidled to within a yard of the bench and took two men by surprise. One of them rose suddenly and scuttled down the street. The other turned and, as he caught sight of me, looked suddenly stricken. He got to his feet, tapped his watchless wrist, mumbled something that might have been "Oh, look how late it is," and scurried down the street.

When I talked to people in the town about wolves, most of them had little to say to me. The kindest of them pled ignorance and referred me to the trappers who still go out in winter to harvest furs. So I looked up as many of them as I could find in town. I discovered that the wolf is not admired here.

Archie Simpson is a seventy-six-year-old Cree who lives in a small house in Fort Chipewyan. But for his graying hair, he hardly shows his age. He is a compactly built man, with searching eyes and small straight teeth. He grew up in the Peace-Athabaska Delta before much of it was set aside as Wood Buffalo National Park, and has hunted and trapped over more of it than any living man. His government-permitted trapping area is in the park, and he still goes out for weeks at a time in the middle of winter. He is regarded as the dean of trappers and wise in the ways of wolves.

"Wolf have an easy life," he says. "They just running around for the exercise. That's what they are, I think." They follow the herds of caribou or bison. He says he has seen as many as sixty wolves together following the caribou, has watched them hunt and kill bison and moose, and has even driven wolves off a moose so that he could take some of the meat for his dogs.

Simpson has chased wolves on his snowmobile and shot them for the bounty, which was paid until 1954, and the fur, which he sold to commercial dealers. "I see forty in one bunch. Fresh tracks, I follow.

At night, after the snow's over, they're following buffalo. I seen sixteen buffalo got killed by these wolves. So I chased 'em on a lake, Lake Claire, for the fur now. I got about ten wolves in two days. They were beautiful wolves. Barren-lands wolf."

Sometimes he shoots them simply because they steal his furs. "The wolf here is about thirty in a pack. Sometimes they're alone. When they're alone, that's when you got to watch. They'll go to your trap lines. You see, if they hit your trail, they'll follow it. They're hungry—that's why they'll follow you. As soon as a single wolf hits your trap line, they just follow your trap line, taking the fur out of your traps. One time, I took eleven lynx. The wolf took 'em all. If there were no wolf I'd have all that money."

Sometimes he shoots because he feels threatened. "One time, first time I went in the bush with my dad in the fall, I was pretty young that time. I was going to go to the creek. I hitch up my dogs, and after I hit all my traps, I was riding and going pretty good. It was a prairie before you hit the lake. Lots of buffalo. My dogs stop and look toward the shore. They [normally] don't stop for nothing. I could hear wolves coming to me, across the prairie. I had a good gun and could shoot about eleven or twelve altogether. I tied up my dogs. I counted seven wolves. Just right, I thought. The wolves are just coming. They don't look like they're going to stop. The first one, I shot him. He fall among my dogs. That big black wolf was here to that spruce [thirty yards]. They still keep coming. I shot five. I let two go.

"I didn't know what was gonna happen if I didn't have a gun. That was too damn close. After I shoot 'em, I was shaking. We skin 'em. They're in good shape. You could see not starving. But the fur was nothing on it. No bounty. Nothing." His brother took the skins and hung them up, but forgot about them, and so didn't cure them or ever sell them.

"A few years after, I went again in the fall. I used dogs. I got on Lake Six in the cabin there. I stopped my dogs. The sun was just going down. I took a pail down the hill and I was going to get some ice. I went out on the lake and I was standing there looking around. I look down and I see something sitting on the point. I look and look and look. I started chopping the ice. At the same time, he moved. As soon as I stopped, he just lay down flat. As soon as I start again, he started coming trotting. I stop and he stop. I start and he start. I ran as fast as I could to my dogs. When I got to my dogs, I took my gun. I look around on the lake: no wolf. I look under the tree. There the wolf was sitting. I shot and I killed him. A big wolf, a big gray wolf. I think it really meant it, that wolf. Lucky I got away.

"Once, riding a Skidoo, coming home from Little Buffalo Lake, I see wolf tracks on the lake, but I didn't know he was coming behind me. I stopped. I looked and he was sitting right behind me. I start and he follow. I stop and he stop. I start and he's coming. So I trick him. When he was coming, I took my gun out, just turn my Skidoo around, and I chase him. And I caught him right in a snow drift.

"I never tasted the meat, just threw the carcasses away. Dogs won't eat wolf. They'll eat lynx. Lynx is good meat; I like that meat. Dogs will eat martin. But they won't eat wolf.

"They're pretty smart, the wolf. I can call a moose in; I can't call a wolf in. If you miss once, he never come back. When you catch one, the rest go away. They go someplace else, they don't come back. If they see a person, they don't stay there. They've got to go. Even they hear a Skidoo, they'll go."

Forty other natives have rights to trap in the area, but few of them exercise their rights. Simpson says, "There used to be camps all over the place. But now nothing. They're all here in town.

"That Bennett Dam, what it done, where people kill a thousand rats, it's all brush. My father-in-law cut bush in his trap line eight miles from here. One time I took twenty-one hundred muskrats out of that place. Now there's no water. Anyplace you go, you see dust flying. Any lake you go, you could see young ducks swimming. Now you won't see that." And prices for furs have gone down: he used to get $1,000 for a good lynx, but now he's lucky to get $300.

Simpson sees how it has affected the culture. "I used to go to the old people when I was young. That's the way I learned everything hunting. Today, it's lots different. If they go out on a Skidoo and they go a long way and it broke down, some of 'em, they can't even make a fire and they can't make a bed. Some of 'em don't know how to use the ax. Lots of things finished from the old way. Even the killing of a moose, they throw the hide away. They should learn the old way. They'll be helping themselves."

Simpson thinks about his father: "My father was not a working man, just trapping and hunting and fishing. My dad used to tell me how to hunt the animal: You gotta do this. You gotta do that. You gotta follow the wind. If the moose go away from the wind and you go behind it, you'll never get it. You gotta watch when you're going after moose. You gotta follow the wind. But now nothing. Moose don't follow the wind. They don't move that way. Even the animals are changing. Walking, walking like that"—his hands indicate that the moose is turning without regard to the direction of the wind—"eating, eating, then they lay down. Everything has changed. I don't know why. Everything has changed. The world has changed."

The bison are also changing. "In the springtime, all the young buffalo are born. During the fall, you don't see the young ones—the wolves kill all the young ones. If there are old cows, they won't have no young ones. How in the hell's there going to be the buffalo?"

The buffalo in the park are declining in numbers, and biologists note that they have been infected with brucellosis and tuberculosis. Some wildlife managers urge that some of the disease-free buffalo be brought in and quarantined, and the rest of the buffalo killed. The healthy captives would later be put back into the wild. Simpson believes the buffalo are declining for other reasons. "I've been trapping a long time and I never seen any dead buffalo, dead for nothing. If I see a dead buffalo, it's dead for the wolf. I told them, I don't believe the disease. Their disease is the wolf. If they take away the wolf, the population of buffalo might come up again."

Still, he doesn't want to see wildlife managers shoot buffalo. "I like to protect the buffalo and I like to protect the wolf. I don't like to see killing all the buffalo and starving the wolf. I hate like to poison the wolf. I hate like to shoot 'em clean out. They're feeding themself. They're like some people looking for a job. They're looking for food. If they down all the buffalo, I want 'em to move the wolf where they could make their living.

"Move the wolf. Move them somewhere else where they could live. If they kill all these buffalo and the wolf are living on that, where are they going to go, the wolf? They might go to the town and kill the children. If the wolf get to town, they'll kill anybody."

. . .

Carl Granath's mother was an Indian, his father a Swedish trapper. Carl married a Métis woman in Fort Chipewyan and has been trapping lynx, fox, martin, and otter since 1966. His trapping area is thirty miles from town in rocky country with a lot of lakes and muskegs and forests of pine and spruce. He goes out by boat in October, before it freezes, to clear brush so that he can get a snowmobile over his trap line in winter. Though he returns to town by December, he then goes out again and stays most of the winter on his trap line, coming into town on his Skidoo every few weeks to see his wife and children.

"I like it. I look forward to the winter." He sees wolves out along his trap lines from time to time. "I don't think they are dangerous. Years ago, when I was driving dogs, I would always run into them, especially in winter and after dark. They would howl to let you know they're around. They never bother me.

"In the 1980s, I had one wolf hanging around the cabin. I saw him hundreds of times. He was kind of a silvery color, like a silver fox. I had thrown out scraps to him. I had dogs, and the same wolf would come in and be playing with my dogs—just playing—he never bothered the dogs. I started off one day on the Skidoo to check the lines, and I went a half-mile and remembered something I'd forgotten back at the cabin. I walked back and saw the wolf playing with the dogs. He used to go up and down my lines and sometimes take furs. He might take one or two lynx at a time. He got to be like a pest. It was pretty expensive. I'd get pretty mad at him. I had in my mind, if I seen him out on the ice, I'd shoot him. But I never did.

"I respected him. I never bothered him; I didn't try to shoot him. When I used to come to town after I had no dogs, this wolf would follow me. And he'd stay on the outskirts of town, and when I went back, he'd follow me back. I thought I must be his friend or something. That wolf was around for five years."

But the fifth year the wolf followed Granath to town proved its last. "People complain about wolves getting too close to town," says Granath. "People must have complained, and Fish and Wildlife put out some poison at the town dump. They must have gotten it, because Fish and Wildlife said they got a wolf at the dump, a silver gray wolf. And he never showed up after that. I kind of missed him after all the years he was around.

"I know people today, lots of them are scared of wolves. They won't go anywhere after dark. I don't know what they are scared of out there."

Andrew Campbell is a big, powerfully built man, with a dour expression and a hint of the accent of his father, who came to Canada from Scotland. His mother was a Métis Indian. He is sixty-five years old, born the year the first airplane came into the north country. He lives at the mouth of a small canyon on the west side of town, just below the grounds of the Catholic mission. "I've been here all my life," he says, "been with wolves."

As a boy of eight or nine, he would go out winter trapping with his father, and his father would send him back to town alone with the dog sled to get supplies. "He would send me into town from a cabin thirty miles out. My mom wouldn't let me go back, because she was afraid the wolves would get me."

He has a trap line and a cabin inside the park, seven miles from his home in Fort Chipewyan. "I love it out there. I still trap, but you can't make a living off trapping now—there's not enough animals, and the price of things is too high. Look at the Skidoo I got out here—that's $4,000. The gas is an expense. So you go out on your trap lines, it's going to cost you a hundred dollars." And the Northern Company fur buyers "put your fur way down low and the groceries they put way up high."

Campbell regards wolves as an additional adversity. "I don't like the wolf. A wolf is a bad animal. They'll kill any animal. They take the dogs from right in front of the house. Every year, we lose dogs right here in the yard to wolves. They're pretty quiet when they come here, because there's people all over town. That wolf will know whether you're asleep or you're awake.

"When they kill a dog here, they pick it up and carry it away. The wolf will break an ordinary dog chain—they rip the collar right off. They don't leave any blood. I don't know how. I got my dog out here, and a wolf came and cut her right in half. Half the dog was there, and the wolf took the other half away. And there's no blood. You explain that. They must drain the blood out of 'em. When they cut 'em in half like that, they must suck the blood right out of 'em. There's not time to lick the blood up, because I'm out there.

"One time, forty years ago, fifteen come after me. I didn't know whether they were after me or after my dogs. I was on the dog sled in a little muskeg, tracking caribou. All of a sudden, the dogs give a jump. They almost knocked me off. There were wolves coming down the hill. Even though it was daytime and a clear sky, these wolves were coming this way. The dogs were running. I stopped the dogs and grabbed the rifle. There was a black wolf there and I shot her. I stayed in the bush and made a fire. Those wolves went clear around me. They surrounded me. They started howling. I had this fire going so they wouldn't come any closer. I grabbed my rifle, because they were too close.

"Lots of time, they come awful close. I don't know what they were after, me or the dogs. But I never got hurt."

He does concede, "I don't know anybody to be attacked." But he believes wolves are laying waste to the rest of the wildlife. "There's an awful pile of wolves here. They kill just about every day. Around Claire Lake, you see, them wolves are living right off the buffalo. They just live with them, following them around. They kill anything they can. They get a herd of caribou and they'd run. These wolves"—he holds two fingers of his left hand down—"they lay here in the front. This other wolf"—he holds down the index finger of his right hand—"will grab her and hang on. This wolf, he's got poison in his teeth. I've lived here sixty-five years and I know something. That's my study. The disease from the poison in the teeth gets into the caribou. He gets weak. They say they're killing the sick and the weak. Well, sure. That caribou is sick already because the wolf's bit him.*

"They'll eat them alive, on the run. Sometime you'll see the whole thigh eaten. Their guts has fallen out of their back parts while they're standing. Sometimes the buffalo is lying there two weeks before it dies."

*While this sounds like the folklore of a man who considers wolves supernaturally evil, biologist Tom Bergerud has suggested that in fact prey may die from *pasturela*-bacteria infections caused by wolf bites.

This talk of wolves and loss turns quickly to talk of hardship and treachery. "If you kill the buffalo off, the wolves are going to clean this whole country out," he says. "This country won't be worth anything." Campbell's face reddens and his voice rises as he talks about politicians and environmentalists: "A lot of people like to save the wolf. It's not the people they're worried about, it's just their business. They're not worried about the people in the North. By God, if they were in my shoes all the time and seen the tough times I seen, they'd know what I'm talking about!"

Jerry Bourke is a fifty-year-old Chipewyan Métis. "I've been out in the woods most of my life," he says. "I spent most of my life on the trap line. I didn't go to school much. I went out with my dad. He got sick, and after that I went out with my uncle. My dad or my uncle told me about animals."

He has a trap line a hundred miles north of Fort Chipewyan. "You're right in a corner where there's nothing at all. It's just animals and you. Some years, there's so many wolves." He finds wolf-killed moose and caribou. Though he finds wolf tracks around his camp and believes the wolves are attracted by his dogs, he has never lost a dog to one. "I hear that other trappers lose dogs to wolves. I don't know what they do. I never shoot wolves. I never shoot anything unless I make money on it or I need meat." Even when there was a $40 bounty on wolves, he wouldn't shoot one. "It wasn't worth it. It was too much work to clean a wolf.

"In the olden days, they'd see wolves more. Since they started using Skidoos, they don't see wolves much. The wolves hear you coming a long way away, and they run. Any wild animal hates Skidoos. Only moose don't. A moose will stand three hundred yards away while you drive by. A caribou will run."

In January 1979, Bourke was attacked by a wolf. "It was at night, and it was snowing hard. I had been going all day on the Skidoo, and I had run out of gas on the edge of Collins Lake, near where I had a trap and a snare. I was cold. I decided I'd jump off my Skidoo and run to the trap, and by the time I got back I'd be warm." Although he had two rifles on the Skidoo, he says, he didn't carry either firearm as he ran. The trap was over a low ridge, and as he came over the ridge, he saw wolves. They were eating two lynxes he had caught, one in the trap, one in the snare.

"There was twelve of them. One started to bark and come after me. I got out on the ice. I didn't have any gun with me. The wolves were coming after me. I broke a stick off a tree.

"The wolf grabbed me just above the knee. I was wearing a Ski-doo suit and jeans and underwear. The wolf tore that suit and the jeans just like a piece of cotton. I hit him over the head with the stick and that drove him away, and I run as fast as I can. I was scared. I ran to my Skidoo.

"I gassed up as fast as I could. What wolves hate is a Skidoo. Any wild animal hates Skidoos. I gassed up the Skidoo and started it, and when I started the Skidoo they all run off."

Bourke could see later, "I had kind of cornered and surprised them, coming down over that rise.

"It didn't bother me, because I had been attacked by a bear before. I had been bitten by a dog before. I have spent all my life in the bush; I'm not afraid of animals." The next day, when he and his brother came upon a pack of twenty-two wolves, he didn't even think about the attack.

These are not healing stories. They express no kinship between wolf and human, but tell at best of an understanding tolerance in the human, at worst of a frailty that armors itself in hatred. Where, I wonder, are the healing stories?

Clearly, they exist somewhere in the native world today. Robert Stephenson, who studied wolves on Alaska's North Slope, found the local natives had great respect for the wolves. The Koyukon of Alaska's Yukon Basin put a piece of fat into the mouth of a wolf that had been killed and spoke to its spirit. The Dogribs, who lived between Great Slave and Great Bear lakes, would not kill wolves. Nor would the Hans, who lived in the Yukon. Eskimos generally respect wolves. The Nunamiut Eskimos, says Stephenson, believe an animal has a powerful spirit. "You act with respect for it because you want to please its spirit so you can have more to eat. They'll tell you their father told them never brag, 'I'm going to go over there and shoot a wolf.' It will bring you bad luck. In Anaktuvuk, after they skin an animal, they cut through the vertebrae behind the spine, because it will let the spirit go. You'll see little acts and rituals. A lot of 'em probably do it and don't know why."

But when Stephenson compared the views of Eskimos, who live on the open tundra, where they see wolves, to the views of Indians who live in forested areas like Fort Chipewyan, where they do not, he found different attitudes. "Indians have less use for wolves than Eskimos," says Stephenson. "Indians don't see wolves very often. Mostly they see kills. When you live in the interior, you find these kills—a calf killed by a bear, or a moose killed by wolves in winter. If you don't see the animal, and you just see what it does, that doesn't give you a very favorable impression. They see wolves as a competitor and don't have a lot of use for them. They complain about them a whole lot. Old-timers tell you that's the only bad animal in Alaska—you better shoot them when you run into them."

Stephenson's generalization seems a fair one. Like the Eskimos, the Indians of the plains saw wolves coming and going, and they, too, spoke admiringly of them, told stories of wolves and humans sharing a common language, dressed as wolves when they went into battle, dreamed of wolves when they wanted vision. Indians of the forested east were less likely to celebrate a sense of kinship with the wolf. Fort Chipewyan is at the northern edge of the forest, and most of its inhabitants take a dim view of wolves.

Still, there are people in Fort Chipewyan who look for a sense of kinship with the wolf. One of them is Lloyd Antoine, a forty-seven-year-old Cree Indian. Round faced, his mouth set on the edge of a smile, he has the calm affability of a man who sees the good in other people. Though he grew up in Fort Chipewyan, he went away to college. He was out of the village for ten years, going to school and working in Edmonton, so he knows something of the outside world. After ten years there, however, he decided the city was no place for him. "You're totally alone, and there are thousands of strangers around you," he says. "You've got to be rude. You got to look the other way a lot of times. In the city, people have an inability to feel." He came back to Fort Chipewyan hoping to raise his children in the closeness and community of native culture.

As a boy, Antoine had worked alongside his father and six brothers on a trap line in Wood Buffalo National Park. He points out a sixty-square-mile area on a map at the offices of the Cree Indian Band, where he works. Where blue shapes indicate lakes, he says, as a result of Bennett Dam's going in "there are no lakes now. My brothers and I used to trap right here, about 1972 to 1978," he says, pointing at one of the lakes. "I took out twelve hundred muskrats and my brother took out two thousand in one three-week period. All of this now, there's no water, there's just grass. All of this is pretty much gone as far as muskrat. Beaver and otter's gone, because there's no water. The fox are pretty well gone, because they live on muskrat, too. Buffalo used to come through our trap lines. We used to see them by the thousand. They'd trample the muskrat lodges, and the wolves would follow them." Now the buffalo are disappearing, and there aren't many wolves in his area.

Antoine still has rights to work the trap line in the park, but he observes that fewer and fewer people go out hunting or trapping. "It's pretty grim," says Antoine. "It costs you more money to get out there than you can get out of it. People just can't make a go of it any more. When you run a trap line here, it's because you want to be out there. You want to have a sense of being a person."

Antoine goes out, not to make a living, but to try to preserve vestiges of the old ways: he is trying to get back to something he believes his family once had, and the wolf is part of it. "The wolf was my dad's spiritual brother. It kind of passed down from my grandfather. This was an animal that takes care of you. If you hear a wolf, you stop and listen. It's trying to tell you something: it's time to settle down or to go for a hunt; it could be telling you that a big wind is coming."

In December 1965, Antoine and his brothers were out trapping with their father. "We were running the north end of our trap line. All eight of us were in there." They were working individually on different parts of the trap line for part of the day, then coming together for lunch. "I went out alone. Right in the middle of a lake, there was an island, and there was a wolf in there, and all day long this one wolf kept howling. I felt uneasy. My brothers came where we would congregate for tea about noon before we would go out again. One of my brothers said, 'Did you hear that wolf?' I said, 'Yes.' He said, 'That's kind of a bad omen. It's not too often during the day you hear a wolf.' Maybe we surrounded him and he couldn't get out—they are really private animals." They concluded that there was more to it than that. "We said there's something else you can't understand about it." A few days later, his father died. "A long time after, my brother told me the brother wolf was calling my father that day and telling him it was time to go. My father knew from that trip he didn't have a long time to live."

Antoine didn't hear the meaning himself. "I've been exposed to too many worlds," he says. "I've kind of lost it. I did not have that kind of exposure" to the old ways. He only began to think about what his father had tried to teach him after he returned from living in the city. One day, six years after he came back, "It hit me that this is the way things are—that the animals do talk to people, that it's not just some kind of a storybook."

Other families may have special relationships with other animals; in his family, it's the wolf. "It's something very private," he says. "We don't tell." Maintaining the relationship doesn't require ceremonies or taboos, and wolves may still be shot or trapped. Says Antoine, "If a wolf is there, you catch him. It's a means of making a living." In the past, they ate the meat and they used the sinews for sewing and the teeth for ornament. More recently, they sold the fur to buyers in Edmonton or Winnipeg. "When I caught a nice wolf, I got $275 for it," he says. "I put food on the table. I was happy."

That you could catch such a wolf owed to the wolf's willingness, says Antoine, "to lay down its life so you could live." An animal who comes into your gunsights does so as a gift. "You have this reverence. It means a lot to hear the wolf—it goes beyond having food on the table. You're thankful. You're going to be taken care of. It's a way of life. It's a whole circle."

So, when he's out on the delta, he listens for wolves. "If you hear the wolf, you feel good. When we are out there hunting and hear the wolf, I tell my children, 'Stop and listen. That's your grandfather talking.'"

However, they seldom hear it. He feels that his own two oldest sons, who work in Fort MacMurray, have lost touch with the land. "They're ignorant of a lot of things they would have known if the trapping were still alive. They're making fantastic money, but they're losing a lot not being out there—they lose a sense of serenity, a sense of being part of nature, being part of the land. We still have that knowledge, but it's getting away from us. You go to a bank and get some money and go to a grocery store to get some food. That's not a way of life. And yet we practice it."

Says Antoine, "Unless we smarten up and go back to pursue some of the things we are very fast losing, we will not be what they call in our language 'the people.' We'll be lost. We'll be walking around like zombies."

They have lost more than the spring floods and the muskrats: they have lost the stories. Their fathers and mothers learned about spiritual relationships when the old people told them stories as children. The stories told, for example of Wisacisa, a person who lived with the animals and could speak their language. They made the listener feel a kinship with the wolf, the goose, the moose. "It was a relaxing kind of coexistence," says Antoine. "In Cree it's got a whole meaning—it helps you understand some of the meaning of life." The stories were shared only when listener and teller were likely to be attentive to each other. "The only time I was told fables in my family was when everything had been quieted down and we had been fed."

But in the new culture of the town, there is no time for stories. "We just don't have the opportunity, because of the way we are. When it came time to share our culture, you detached yourself from thinking about going to work. If you were able to go out to our old traditional summer camping grounds with one of the old people, you might hear the old stories, but not here in town. This has got nothing to do with our old way of life when you come here. This is a place you can watch TV."

Back at the lodge that night, I turned on the television and watched rock singer Michael Jackson, in black leather and silver studs, surrounded by a circus of alienated beings, singing "I'm Bad" in the subways of New York City. Someone moonwalked on roller skates. Dancers glared like caged animals. It was a lightless, treeless, waterless, futuristic world, about as far away as you can get from Fort Chipewyan. Satellite dishes were bringing this signal into homes all over town.

We need stories that tell us how we are tied to the wider world. It is as much a need as food and air and water. In the citified world, however, we throw up walls around us to keep from bumping into one another. We are starved for those stories, impoverished of the thread of connection. That is one reason we want to put wolves back into the wild. If we again see wolves in the wild, we may share stories about them. And that, we hope, may restore us.

For now, few North Americans or Europeans see wolves in the wild, and our stories about them are personal and idiosyncratic and far removed from the real animal that would, in Eskimo or Plains Indian cultures, seem to validate the telling. We profess to love wolves or to hate them, but we don't see the same animal, or share the same meaning. When we speak of wolves, it is not entirely clear what we love or hate.

In 1994, the movie *Wolf,* written by novelist Jim Harrison and directed by Mike Nichols, was released. Harrison, who spurns the city because he believes civilization stifles the human spirit, recalls that one night in his backwoods Michigan cabin, the approach of car headlights set him into a rage, and he tore a door off his cabin and ran out into the night. "There was something inside me," recalled Harrison. He had used the image of the wolf entering his spirit before in his writings, and he saw in that image a wildness that could revivify the human soul. He eventually wrote a screenplay about the idea, in which a middle-aged man is bitten by a wolf and something of the wolf enters the man to heal his spirit. He becomes supernaturally strong and predatory. Director Mike Nichols, however, saw the story differently, and thought the wolf-bitten hero lost his humanity. For two and a half years, writer and director quarreled over whether the wildness kindled by the wolf bite was healing or dissolute. After writing five drafts of the script, Harrison withdrew from the project, and Nichols' old performing partner, Elaine May, finished the script.

That argument is exactly the kind of thing that is happening in Fort Chipewyan. Each person you talk to sees a different wolf. One wolf strikes fear into the heart, another is companionable and playful, another rapacious and another knowing and sympathetic. In Fort Chipewyan, as in the rest of the world, there are many different levels of culture: three different Indian groups, the ritual and discipline of the Catholic church, the soulless technocracy of industrialism, the unearthly fantasies of rock video, and a smattering of ecological science. There is a constant shuffling and streaking of messages. Everybody draws on the inventions of other cultures. We are no longer a single people, united by our stories. Each of us invents our own wolf.

If a spiritual view of wolves survives on the Peace-Athabaska Delta, no one but Antoine shared it with me. I am inclined to doubt that it was simply because I was an outsider. Spiritual life comes from what one does day by day; it is not something one turns on and off, like television. If wolves once had spiritual meaning, it was because those who felt it lived with wolves day in and day out. The people of Fort Chipewyan no longer live with wolves. And trying to acquire traditional views of a creature we no longer live with, as if we were acquiring an old car or an antique table, will not work.

On my last day in Fort Chipewyan, I walked past an empty schoolyard. The empty playground swings swayed in the breeze. A taffy-colored mongrel pointed its nose to the gray sky, squinted its eyes, and howled, low and soft, musically. A quiet pleasure rippled down the sides of its body. Its tail wagged slowly back and forth, a metronome set on adagio. It seemed to be replying to some message drifting on the cold air from the delta. But what? The distant howl of an ancestry close at hand? Stories on the wind?

15

BRIDGING THE GAP

If we are ever to deal adequately with wolves, we will have to overcome our long history of estrangement. North American culture, like the European cultures that gave rise to it, is a forest culture. Our innate sense of good land embraces deep shade and groves of trees. Like that of the people of Fort Chipewyan, our outlook has grown in the presence of wolves but with wolves rarely in view.

We are also a pluralistic culture, with diverse and sometimes discordant views of things. We look for consensus, and science is increasingly the framework by which we seek agreement. Only in the last fifty years have we had the technology and the inclination to study wolves as they really live in the wild, and science hasn't yet managed to summarize their complexity in a way that reflects our own. Despite an impressive amount of study of wolves, a great deal more is needed—as is more effort to communicate the uncertainties of science to the general public.

Science is only one way of seeing things, and many of us will look at wolves in other ways. If we have no wolves in view, we shall go on inventing them and seeing them as shadows of ourselves. Today, neither our science nor our myth yet gives us an adequate basis for dealing with wolves. The wolf is more complex and varied than the biologist yet sees, and we humans are too varied and disparate to be served by a single mythology. Can we ever develop a common view of a wild creature that exacts a cost but fails to repay us with frequent encounters?

Pat Tucker hopes we can. Every day, she clips a long leash on Koani and takes her out for a walk in the hills above Missoula, Montana. Koani is a black wolf. Her legs are long, her head is broad, her posture tense and alert, her yellow-eyed gaze penetrating. At eleven months, she weighs eighty-five pounds, and to walk her is an athletic event. Tucker runs to keep up with Koani until the wolf stops suddenly to look long and hard at children playing down the hill, or cars passing on the freeway, or a beetle crawling up a grass stem. Tucker stands by patiently. Tucker is tall, green-eyed, self-effacing, and so good-humored that laughter sneaks out of her every few moments. She is without a predatory thought. Her attention wanders before the wolf's, in part because she has seen all this before, in part because it is human to let one's gaze move on rapidly. When Tucker's attention passes to something else, Koani lunges off and Tucker is caught unawares, jerked like a windblown leaf, stumbling, free arm flailing, trying to dig in her heels, laughing at the silliness of it all.

Koani gets at least three hours of walking a day—one of many accommodations Tucker must make to the wolf. Koani's captivity is a project of converging purposes. A film-production company wished to make a television movie about wolves and needed someone to rear a tame wolf so that they could film her growth. Tucker, an environmental educator working for the National Wildlife Federation, and long interested in wolf conservation, figured she could use the wolf in her environmental-education programs. She and her filmmaker husband, Bruce Weide, now take Koani to schools around Montana. They show slides and bits of Weide's videos about wolves, and they talk about wolf behavior and ecology. Says Tucker, "We go in, have them do a howl, and get them excited about wolves. These are the people who are going to be inheriting the ranches." Changing people's attitudes when they are young is a critical part of restoring wolves to Montana.

A live wolf is a potent teacher. "It's amazing to watch the impact the real animal has on people," says Tucker. "You can show them pictures and pictures, but it doesn't really hit them until they see the real animal, and they see her in the same room, and it's not fangs dripping with blood or chewing off my arm."

While visiting relatives in California, Tucker and Weide take Koani to nature centers and outdoor outfitters to talk about wolf reintroduction in Yellowstone and to raise money for the Wolf Education Center, an Idaho project that sponsors Koani's public appearances. On a January night in 1993, they arrive to give a talk about wolves at the Adventure 16 store in Solana Beach.

Adventure 16 sells tents and boots, freeze-dried food, walking shorts, and Gore-Tex raingear. It is a long, narrow store with backpacks and sleeping bags hanging on the walls. By the time Tucker and Weide arrive, there are two hundred people, mostly men and women in their twenties and thirties, waiting. Folding chairs have been set up, and a number of small children sit on the floor in the aisle.

The presentation starts without the wolf. Tucker tells the audience, "I want the people who see her to know something about what she really ought to be doing, instead of running around on a leash." For that reason, she and Weide will talk for a while about wolves before bringing Koani out.

They talk about the wolf as a symbol of wilderness. Says Weide, "The stories that come from myth and folklore are guided by the society you live in. The wolf is a symbol of wildness, and whether that's viewed as positive or negative depends on the society you grow up in." He reads a quote from Luther Standing Bear, an Oglala Sioux: "We do not think of the great open plains . . . as wild. . . . Only to the white man was nature a wilderness." Weide also recounts the story of Little Red Riding Hood and stresses Little Red's mother telling her, "Whatsoever you do, don't talk to strangers." He has a deep, sly voice with sexual overtones for the wolf when it encounters Red Riding Hood: "Hey, little girl, what's your name?" He has an old scratchy voice for the wolf's grandma voice: "The better to see you with." In the end he says, "The reason for that grisly ending wasn't to warn children that wolves were bad. It was to warn them not to talk to strangers." The wolf in the story is simply a masked human, its wildness the predatory nature of the human heart.

They also talk about the harder edges of wolf biology. "There are people who think wolves are nice creatures that eat nothing but sick, weak mice," says Tucker. "Wolves are primarily dependent on large, hoofed animals." And she wants people to understand how tough it is for a wolf to make a living. "Most of us in this audience weigh more than a wolf. Think about putting on one-inch fangs and running up and attacking a moose. This is a dangerous business, going out and hunting."

She explains that when settlers shot out the deer and elk and replaced them with cows and sheep the wolves ate what the ranchers provided. She recounts the history of poisoning, trapping, and cruel killing that followed. "A lot of Montanans feel wolves are going to kill nothing but livestock if they get back there," she says. "It's kind of ironic that the wolf is a symbol of wildness, but to them it's becoming a symbol of control by bureaucrats over their lives." She pleads for understanding of the ranchers' problems: "We should not be trying to tell ranchers that wolves will not kill livestock." She shows a photo of a dead calf, its haunches eaten out by wolves. "Even if you get compensated, there's still a feeling of violation." She compares it to going out to one's car in the morning and finding that someone has smashed a window and torn out the radio. "I think we really need to acknowledge those feelings."

But, she points out, only about one wolf in ninety in Minnesota ends up killing livestock. And Weide observes that no human has been known to have been killed by a healthy wolf in North America.

Tucker is not doctrinaire. "Nobody knows everything about wolves," she says. In fact, "You can find an exception to everything you say about wolves. Wolves are individuals. Packs do different things at different times."

And so the catechism goes, until at last Tucker stands on the table in the front of the room for, she says, "some audience involvement and a commercial break. I'm going to demonstrate a howl. You've gotta get your bellies into it. It's like yoga. People have said I sound like a sick cow, but I've had wolves answer me." She howls a long, alto howl, keening up and then down. Children in the audience giggle nervously. But when she finishes there is a respectful, almost awed silence, and then an earnest applause.

Now all two hundred howl along with her. It is more than entertainment. The audience is giving voice to something they hope lies deep inside them, something unrestrained, something noble, something ancient and wild. These are people who go out into the woods under backpacks and over paddles, who spend much of their lives seeking wildness.

At last, Weide leads Koani into the room on a leash. The wolf moves down the aisle, calm and affable, but aloof. She doesn't rush up to people wagging her tail. She is, if anything, unnerved by all this attention, and lies down on the floor, shrinking and submissive, then rolls onto her side and lets people stroke her. She becomes immovable, and Weide tugs at her leash, dragging her across the floor. Finally, so that they can get the show on the road, he pulls on the leash, and Tucker gets down and pushes the undignified and very unthreatening Koani's flanks, and they wrestle her, like a sack of flour, to the front of the room. Koani is anything but the Big Bad Wolf of European fables and Wyoming sheep ranchers' nightmares. She is so rumpled and domestic that Tucker must remind the audience, "Don't go away thinking a wolf will make a great pet. They don't. We've done a lot of things to alter our lives to keep Koani. A wild animal belongs in the wild. They're a large, wonderful predator, and we really should celebrate that."

This is the wolf—the symbol of wildness. It has destructive urges and a bite we have feared for centuries. It is the visible form of our own darker nature. But here it licks faces and lies passively on the floor, submitting weakly to the wide eyes and tentative touch of children. That's exactly what these people hoped wildness would mean.

If Koani is the wolf we want to see, is she the wolf that we want in the wild? Tucker herself is not sure. "She's an ambassador from wolves to us; she left her culture to come to ours." Koani must be submissive and must forgo a wolfly penchant for travel. The whole aggressive side of wolf nature—competing for status and killing for food—must be stifled in her. Some "ambassador wolves" rebel. They become sour on children and public appearances; they escape and wander into the gunsights of people who fear them. So far in her young life, Koani has never shown an inclination to hurt a child, but on her daily walks, Tucker never lets go of the leash. "I've had a few good falls," she says, "but I don't even want to make that instinctive choice to let go."

The logistics of travel on these educational trips are exhausting. When they travel in the van, Koani rides in a cage, which she dislikes so much that she will not drink inside it. They must stop frequently and let her out to drink, exercise, and explore. It may take an hour or more to coax and wrestle her back in. After appearing in a schoolroom or a lecture hall, they must walk her awhile, then trick and beg her back into the van.

Tucker frets over whether it is the right thing to do to a wolf. "With captive wolves, there's no way you can really give them what they need." Ideally, a captive ought to have other wolves for company. "But if you have other wolves, you need an army to take them out, so that they can smell new scents and see new things. They're just not a great animal to have in captivity. I have mixed emotions. Hopefully, these captive wolves provide some important educational value, but I sure don't think it's a great thing for them."

Biology and the mythology are, at least in the body of Koani, at odds. Tucker has doubts about what kind of understanding the audience is getting. "It is a mixed message, taking an animal like this in on a leash and telling them an animal like this can be dangerous," she says.

"Wilderness without animals is dead," Lois Crisler wrote. "Animals without wilderness are a closed book." Biologically, a lone captive wolf makes no sense. It lives almost wholly in a symbolic world. And Koani is so tame, so shy and yielding, that only a limited range of symbols is being offered. That's good if the goal is to get people past the apprehension they learned in the stories of Little Red Riding Hood and the Big Bad Wolf. It is better than no wolf at all, but it's not the whole wolf.

It is our inability to see the whole wolf that makes the creature so mysterious, so powerful, so controversial. It is the fact that we conceive of nature simultaneously on moral and biological levels that makes conservation—particularly of other species—so difficult a challenge. Bridging the gap between moral and biological views is one of the great challenges to us as a species.

We cannot help viewing wolves in symbolic terms, but symbols change. The wolf was once widely seen as a symbol of the depravity of wildness; it is now to many a symbol of the nobility of nature. Largely by the use of symbols, we nearly eradicated the wolf. Largely by manipulating symbols, we may yet save it. As symbols change, wolves are returning to their former haunts in Montana, North Dakota, Idaho, Washington, Wyoming, Minnesota, Michigan, and Wisconsin. Their return will also benefit other species, for wolves need vast amounts of wild habitat, relatively unimpaired by human uses, and large tracts of wild land will provide sanctuary for hundreds of other species that make up healthy ecosystems.

Many have argued that the reason for protecting wolves is to save ecosystems, but there is little likelihood that a whole ecosystem will collapse without wolves. Where the wolf has been eliminated, bears, mountain lions, coyotes, and other predators assumed some of its biological functions. Says John Theberge, "You can't say that wolves are essential to wilderness ecosystems as far as function. We had far more mammalian species in the Paleocene, and the ecosystems did fine without saber-toothed tigers and the megafauna. Nothing is essential to an ecosystem—the ecosystem will still go on capturing energy. It will adjust. It will be different. But we ought to conserve for the marvelous complexity that evolved."

We ought to save wolves not simply to keep ecosystems marvelous, but to keep the integrity of an unfolding process. Wolves are part of the machinery of evolution, and we should value them highly because they affect the evolution of other creatures. Remove them, and we may alter the direction of evolution.

The greatest harm we do to the world is that we oversimplify it. We look upon evolution as past tense, entertaining the vanity that both nature and humankind are finished works, stone statues, complete and unchangeable. But, of course, neither nature nor humankind is unchanging. Evolution is still unfolding, and since its direction is focused and deflected by small and seemingly insignificant things, we ought not to destroy any of the parts.

This is not just a question of ecological health; it may also affect our mental health. It is part of our design as a species to be mindful of other creatures. Wolves are an essential likeness to ourselves, a mirror in which we can examine ourselves as we can with no other creature. We see in them reflections of our own good or evil, our own selfless love and our own perplexing violence. We see ourselves as we are and as we might be. The loss of a creature we think upon and dream about could well deprive our children of chances to attain depth and range and complexity.

The very act of saving wolves may help to shape our character, and to incorporate limits in our behavior before our own brashness leads the world to tragedy. The shadow of the mushroom-shaped cloud, the hole in the ozone layer, the radioactive fallout from Chernobyl, the accumulation of greenhouse gases in the atmosphere—these are powerful indications that we must accept limits. Saving wolves does not mean giving up all control, but trying to accept the need for self-control. It means giving up the idea that each of us can take from the world whatever we desire.

It is also an effort to see that all of us have the potential for good and evil installed within us, that life turns in some ways on violent acts, and that we need to address the conflicts within our own nature that lead to violence. As long as we recognize as evil only things that exist outside our skins, we shall fail to solve our problems.

Saving wolves is to some extent a gesture, but gestures have enormous meaning, and more power than words alone to change our minds. Returning wolves to the landscape and reestablishing natural ecosystems are gestures that lead us to greater consideration of the natural world and our dependence upon it. The immense variety and individuality of wolves offers us an opportunity to see the true complexity of life. At least this complexity will require us to design responses to wolves that entertain their wide range of character and their striking likenesses to ourselves. Wolves require us to look at the world through science and spirit simultaneously and to integrate thought and feeling. If we can do that, there is hope, not just for wolves, but for humankind.

Appendixes

APPENDIX 1

FOR FURTHER READING

The literature of wolves is surprisingly large. In the course of my research, I found and read more than eighty popular books and hundreds of scientific articles about wolves. What follows is chiefly a summary of the popular books I found most insightful and most thorough.

L. David Mech's *The Wolf: The Ecology and Behavior of an Endangered Species* (Garden City, N.Y.: Natural History Press, 1970) is still the most comprehensive work about the biology of the gray wolf. Since its publication, a great deal more research has been done, and it may no longer be possible to assemble in a single readable text all that is known about wolves. Most of the other original works about wolves have focused upon local studies. Durward L. Allen's *The Wolves of Minong: Their Vital Role in a Wild Community* (Boston: Houghton Mifflin, 1979), recently reprinted by the University of Michigan Press, is a valuable and very readable account of the research conducted on Isle Royale prior to 1976. Adolph Murie's classic study of wolf predation in Alaska, *The Wolves of Mount McKinley*

(Washington, D.C.: U.S. Government Printing Office, 1944), has been reprinted by the University of Washington Press. Its meticulous science and beautiful writing could well serve as a model to aspiring scientists today; a half-century after it was originally published, it is still rewarding reading. Ludwig Carbyn, S. M. Oosenbrug, and D. W. Anions' *Wolves, Bison . . . and the Dynamics Related to the Peace-Athabasca Delta in Canada's Wood Buffalo National Park* (Edmonton: Canadian Circumpolar Institute, University of Alberta, 1993), gives a view of wolves preying on bison in a relatively open country. L. David Mech's *The Arctic Wolf: Living with the Pack* (Stillwater, Minn.: Voyageur Press, 1988), and Jim Brandenburg's *White Wolf: Living with an Arctic Legend* (Minocqua, Wisc.: NorthWord Press, 1988), both provide accounts of wolves on Canada's Ellesmere Island, where wolves prey upon musk oxen and snowshoe hares. Douglas Pimlott's *The Ecology of the Timber Wolf in Algonquin Provincial Park,* co-authored with J. A. Shannen and G. B. Kolenosky, Ontario Fish and Wildlife Research Branch Report 87 (Ontario Ministry of Natural Resources, 1969), gave a detailed view of the kinds of results that came from the first decade of field studies. It has long been out of print. Pimlott also co-authored, with R. J. Rutter, a popular work, *The World of the Wolf* (New York: J. B. Lippincott, 1968), which is still one of the best introductions to wolf biology and conservation, but, alas, is also out of print. It is to be hoped that some enterprising publisher or natural-history association will one day bring them back.

Last, but by no means least, I would add to this list of books about the biology of wolves Barry Lopez's fine *Of Wolves and Men* (New York: Charles Scribner's Sons, 1978). I place it last on this list only because it focuses on much more than wolf biology, and places the science into a wider human perspective. Lopez showed that we cannot see the creature apart from our own imaginations, which caused many scientists to dismiss it as literary and discomforting. Nevertheless, it presents a broad and scientifically accurate discussion of the life and ecology of the wolf.

Several interesting and well-considered works focus on the study of captive wolves. Eric Zimen's *The Wolf: A Species in Danger* (New York: Delacorte Press, 1980) is an extensive account of a German researcher's studies of a captive pack, with discussion of some of the conservation problems in Europe in the 1970s. Michael W. Fox's *The Behavior of Wolves, Dogs and Related Canids* (New York: Harper and Row, 1971) provides an interesting overview of the body language and social interaction of captive wolves.

Three excellent works chronicle the demise and/or return of wolves to specific parts of the United States. Jan DeBlieu's *Meant to Be Wild* (Golden, Col.: Fulcrum Press, 1991) provides a thoughtful and detailed chronicle of the return of red wolves to North Carolina. Rick Bass' very engaging narrative *The Ninemile Wolves* (New York: Random House, 1992) tells the story of the first years of the expansion of wolves south from Glacier National Park into western Montana. David E. Brown's *The Wolf in the Southwest* (Tucson, Ariz.: University of Arizona Press, 1983) recounts the story of the eradication of the Mexican wolf in the Southwest, and presents a detailed picture of the work of the Predatory Animal and Rodent Control service.

No library collection devoted to wolves would be complete without the works of Lois Crisler. Her *Arctic Wild* (New York: Harper and Row, 1958) is a trove of insights into the mind and character of the wolf; though it is no longer much cited by wolf scientists in their works, earlier researchers confessed a debt to it. I find it difficult to resist the temptation to quote long, eloquent sections of Crisler's deeply felt and acutely perceptive prose. Much less often cited in the wolf literature—because it focuses on the wolves Crisler took from Alaska and bred to dogs in Colorado—is *Captive Wild* (New York: Harper and Row, 1968). A gloomy—but no less eloquent—cautionary tale about what we do when we try to take the wildness out of life, it is a work anyone thinking of owning a wolf or a wolf hybrid ought to read before bringing home a puppy.

Several excellent books discuss competing theories of predation. Paul Errington's *Of Predation and Life* (Ames, Iowa: University of Iowa Press, 1967) summarizes his studies of minks and muskrats on Iowa marshes, and develops the general theory of predation that underlies most of today's biological studies. Chapters in Durward Allen's *Our Wildlife Legacy* (New York: Funk and Wagnalls, 1954) provide a spirited and well-reasoned critique of the persecution of predators that was still going on when it was published. There is not yet a popular account of the revisionist view that predators indeed control prey populations; those who wish to find that contending view will have to look in the scientific literature. A good place to start might be William C. Gasaway, Rodney D. Boertje, Daniel V. Grangaard, David G. Kelleyhouse, Robert O. Stephenson, and Douglas G. Larsen, "The Role of Predation in Limiting Moose at Low Densities in Alaska and Yukon and Implications for Conservation," *Wildlife Monograph,* vol. 120 (1992), or William C. Gasaway, Robert O. Stephenson, James L. Davis, Peter E. K. Shepherd, and Oliver Burris, "Interrelationships of Wolves, Prey and Man in Interior Alaska," *Wildlife Monograph,* vol. 84 (July 1983), or Robert Hayes, "An Experimental Design to Test Wolf Regulation of Ungulates in the Aishihik Area, Southwest Yukon," a report prepared by the Yukon Fish and Wildlife Branch in December 1992.

A few excellent works focus on the ways wolves are interpreted in myth and legend. Far above the rest is Lopez's *Of Wolves and Men,* which looks in especially fine detail at the ways Plains Indians saw themselves in wolf behavior and describes how humans imitated and celebrated wolves in ritual and daily life. A second work—available only in French but well worth the effort—is Daniel Bernard's *L'Homme et le Loup* (Paris: Berger-Levrault, 1981), which describes in rich detail European traditions of wolf legend, wolf lore, and wolf hunting.

Stanley Young and Edward Goldman's *The Wolves of North America* (Washington, D.C.: The American Wildlife Institute, 1944) deserves mention here. As a collection of trappers' lore and explorers' tales it is still rich, colorful, and enjoyable reading. But it views the wolf chiefly as an economic factor in the lives of stock ranchers and deer hunters, and its anecdotal approach is far from reliable. If one reads Stanley Young's *Last of the Loners* (New York: Macmillan, 1970), a collection of highly lurid anthropomorphic accounts of legendary stock-killing wolves of the West, one is less likely to trust Young at all.

APPENDIX 2

PLACES TO SEE WOLVES

As of 1994, the International Species Information System, a listing of the animals in member zoological collections, listed about three dozen zoos holding North American species of gray wolf, and a number of others holding red wolves. Dozens of additional zoos and nature centers not participating in this breeding registry also keep wolves. Wolves are therefore relatively common in North American zoos, but they are rarely displayed in such a way as to give a fair idea of the nature of the animal. A wolf's most insistent natural quality is its motion, and the long-distance travel of a wild wolf is prohibited by the moats and cyclone fences of exhibits. Also, wolves tend to sleep most of the day in a zoo enclosure, and that makes them far less interesting to visitors.

The large metropolitan zoos have increasingly concentrated upon the breeding of endangered species as their chief mission, and in consequence they have been phasing out their timber-wolf exhibits. Explains Doris Applebaum, curator of mammals at the Detroit Zoo,

"They're not that endangered. And if you want to breed timber wolves, it is difficult to find a place to send the offspring." Zoos have, where possible, substituted red wolves or Mexican wolves for forest and tundra subspecies of gray wolf. The San Diego Zoo no longer displays gray wolves, nor does the San Francisco Zoo, the Philadelphia Zoo, or the Detroit Zoo. Where wolves are still displayed, they tend to be put into large "wolf-woods" exhibits, where there is much space and lots of cover. Such exhibits tend to hide the animals from the visitors, and perhaps make the wolves less popular zoo attractions.

There are a handful of centers devoted exclusively to the study or educational display of captive wolves which welcome visitors and provide interpretive programs.

The International Wolf Center in Ely, Minnesota, has live wolves on display during summer months, and it is the permanent year-round home to the "Wolves and Humans" exhibit that helped interest the American public in returning wolves to Yellowstone. A membership organization, the center serves as an educational resource aimed at the conservation and understanding of wolves. It publishes a quarterly magazine which focuses on wolf research, conservation, and education.

Working with the International Wolf Center, Vermilion Community College, in Ely, Minnesota, conducts wolf-study weekends in winter months and week-long wolf-research expeditions in summer. On the winter weekends, participants go out at night to try to get wolves to reply to their howls, and ski or snowshoe to remote lakes to inspect fresh wolf kills. On some weekends, participants may take brief airplane flights to view previously located radio-collared wolves.

Wolf Haven, in Tenino, Washington, started as a home for wolves whose owners could not reorganize their lives to care for the animals, and by 1994 had thirty-nine captive wolves in spacious pens, full of trees and shrubs that give the wolves space and cover to get out of sight if they aren't in the mood to deal with the daily human visitors. There are small white Arctic wolves and large gray-brindled plains wolves. Wolf Haven tries to pair all the wolves in the pens, to give them the close companionship of a mate. Tubal ligations and vasectomies are performed on the wolves to prevent breeding. Says

Paul Joslin, director of research and education, "The intention is not to perpetuate the wolf-puppy mill." About twenty thousand visitors tour Wolf Haven each year.

Wolf Haven has broadened its objectives to assume leadership in several areas of wolf conservation. It commissioned biologist Gordon Haber to critique proposals for wolf-control programs in British Columbia and Alaska, and sent spokespeople to testify at hearings on wolf controls. Joslin has trained howling brigades to go out and look for tracks and listen for howls wherever the status of wolves in the wild is uncertain. In 1992 and 1993, Wolf Haven–trained howling brigades searched for wolves in Washington, Idaho, Montana, Oregon, Arizona, and Mexico. Wolf Haven sponsored an International Wolf Symposium, attended by most of the wolf biologists in the world, in Edmonton, Alberta, in 1992.

Wolf Park, in Battle Ground, Indiana, was started in 1972 as a center for the study of wolf behavior by Purdue University ethologist Erich Klinghammer. Scientists at Wolf Park have developed a wolf ethogram, an outline of the typical gestures and expressions of wolves, which can serve behavioral scientists as a common language for wolf study. Monty Sloan of Wolf Park has become a leading expert in distinguishing wolves from wolf hybrids. Wolf Park publishes *Wolf,* a quarterly newsmagazine about wolf conservation and research issues worldwide. It offers six-day seminars on wolf behavior several times a year. In summer months, Wolf Park admits the public to evening wolf talks at which visitors sit in bleachers in front of a large wolf pen containing a captive pack that is socialized enough to have lost most of the wolf's innate shyness of humans. Visitors will find that here the wolves look as hard at the visitors as the visitors look at the wolves, and there is a level of interplay unusual to the display of the animals.

The Wild Canid Research and Survival Center in Eureka, Missouri, was started in 1971 by the noted zoo director and television star Marlin Perkins. Focusing chiefly upon the breeding of endangered varieties of wolf, such as the Mexican wolf and the red wolf, it provides tours, seminars, and campfire programs by prior reservation on alternate Saturdays from August through January, but is closed to the public during the months when the wolves are breeding and rearing young.

The North American Wolf Foundation, a wolf-education organization in Ipswich, Massachusetts, offers structured presentations to the public on weekend afternoons, and by reservation to groups larger than twenty on weekdays. It draws about twenty thousand visitors a year.

The Wildlife Science Center in Forest Lake, Minnesota, dates back to a private facility established in 1976 for the study of the behavior and psychology of captive wolves. Studies there explored such things as seasonal hormone changes among female wolves, the effectiveness of drugs used to immobilize wolves in the field, and methods of persuading wild wolves brought into captivity to eat prepared foods. Studies of captive wolf packs at the center yielded much of our knowledge of territoriality, dominance hierarchies, and other behavior. Research is still conducted at the facility, but since 1989, as the Wildlife Science Center, about half its effort has gone to education, and fifty to seventy-five thousand visitors come by each year, most of them students from Minneapolis-area schools. It is open to the public by prior arrangement.

The Julian Science Center is a fledgling wolf-research-and-education center in the mountains east of San Diego, California. Founder Paul Kenis hopes to develop a facility at which Mexican wolves and other endangered canids are captive-bred, and to which the public can come for extensive study of wolves. It is currently open to the public only by appointment.

In 1994, Wolf Song of Alaska, a membership wolf-education organization in Anchorage, Alaska, was hoping to begin work on a large facility, perhaps thirty-five miles from Anchorage, at which it would ultimately display as many as sixty wolves of different subspecies from around the world. Wolf Song hoped to stay out of the political controversies over wolf controls in Alaska, and to concentrate on education. It hoped to open a facility by 1996.

There are also many places where one may try to see wolves in the wild. In Canada, there are ranger-led campfire programs or nighttime wolf howls conducted at Jasper National Park, Prince Albert National Park, Riding Mountain National Park, and Algonquin Provincial Park, but the likelihood of seeing wolves in the dense forests of these parks in summer is low. In Ontario's Algonquin Provincial Park, for example, despite a century of wolf-tourist sharing of the environment, wolves haven't lost their shyness. Visitors get only quick and interrupted glimpses. The exceptions are few and far between: In the early 1960s, some men were playing baseball on a diamond scratched out into the old airfield at Mew Lake when a wolf came out of the forest and stole a base. In the early 1970s, a wolf loitered around the campground at Whitefish Lake. In 1987, a wolf began to run up to groups of human wolf-howlers at night. It stood on hind legs with its front paws on car doors and it chewed bumpers; it tugged on a man's collar and pulled a woman's hair. None of this seemed to be aggressive behavior, but one night two boys chased the wolf. As it ran by a fire, it came within arm's reach of a sixteen-year-old girl, who shone a flashlight in its face. The wolf bit her, breaking the skin though not seriously injuring her. Then it let go, scratched at a nearby tent, and picked up a shoe, and shambled off with it. The wolf was destroyed and tested for rabies, but it proved not to be rabid. Those are the only three records in the park of wolves associating with people.

The best odds of actually seeing a wild wolf today are probably to be had by backpacking into the wild country of Denali National Park in Alaska. It is open country, but difficult to travel in. Fogs and rainstorms obscure the view, and the presence of grizzly bears requires additional care and planning on the part of the visitor. But care and effort may reward the patient seeker. Some park rangers there estimate that as many as 15 percent of back-country travelers see wolves. Gordon Haber reports that, as more and more backpackers visit Denali, the wolves are becoming increasingly habituated to humans. "It's an everyday event for wolves to walk up to people in the back country or campgrounds and sit down three feet from them," says Haber. "They pick up a book and walk off with it. They sniff a hand. People will say they looked over their shoulders and there was a wolf sniffing at their heels. It's a touching kind of relationship. The wolves are totally at ease. It's like, 'I see you as a friend.'"

One of the hopes for reintroduction is that we will have more encounters with real wolves. Yellowstone and White Sands alike offer broad grassy valleys and hillsides on which people may some day wait, as Robert Stephenson did with Nunamiut Eskimos, to watch wolves.

APPENDIX 3

SUBSPECIES OF GRAY WOLF

E. R. Hall's *Mammals of North America* (New York: John Wiley and Sons, 1981) recognized twenty-four subspecies of gray wolf (*Canis lupus*) in North America:

Canis lupus alces—a large moose-hunting wolf from the Kenai Peninsula of Alaska

Canis lupus arctos—a white wolf occurring on Arctic islands, including Ellesmere Island

Canis lupus baileyi—the Mexican wolf, the smallest and southernmost wolf of the Sierra Madre of Mexico and adjoining American states

Canis lupus beothucus—Newfoundland wolf, a white wolf now extinct and known only from four skulls and a single skin

Canis lupus bernardi—the Banks Island tundra wolf from Banks and Victoria islands in Canada's Northwest Territories

Canis lupus columbianus—the large wolf of most of British Columbia

Canis lupus crassodon—a medium-size wolf from Vancouver Island

Canis lupus fuscus—of the Cascade Mountains, from British Columbia south to Oregon

Canis lupus griseoalbus—from central Manitoba and Saskatchewan

Canis lupus hudsonicus—a medium-size tundra wolf living mostly on caribou west and north of Hudson Bay

Canis lupus irremotus—the Northern Rocky Mountain wolf

Canis lupus labradorius—a medium-size wolf from Labrador and northern Quebec

Canis lupus ligoni—a smaller wolf from southeastern Alaska

Canis lupus lycaon—the Eastern timber wolf, which once inhabited eastern North America from Hudson Bay to Florida, west to Minnesota and Ontario

Canis lupus mackenzii—the Mackenzie wolf, of the Arctic coast of Northwest Territories

Canis lupus manningi—a small light-colored wolf on Baffin Island and possibly neighboring islands

Canis lupus mogollonensis—a small wolf of northern Arizona and New Mexico, now extinct

Canis lupus monstrabilis—a small species inhabiting West Texas and northeastern Mexico, now extinct

Canis lupus nubilus—the buffalo wolf of the Great Plains that ranged from Manitoba and Saskatchewan to Texas

Canis lupus occidentalis—a very large wolf from northern Alberta and southern Northwest Territories

Canis lupus orion—the Greenland wolf

Canis lupus pambasileus—a large wolf of interior Alaska

Canis lupus tundrarum—the Alaskan tundra wolf, a large wolf of Alaska's Arctic coast

Canis lupus youngi—a medium-size wolf that once inhabited the Southern Rocky Mountains of Utah and Colorado but is now extinct

In 1992, in a paper entitled "Another Look at Wolf Taxonomy" presented at the International Wolf Symposium in Edmonton, Alberta, Ronald Nowak of the U.S. Fish and Wildlife Service suggested combining these twenty-four subspecies into five, as follows:

1. *arctos* would include *bernardi* and *orion*
2. *baileyi*
3. *lycaon*
4. *nubilus* would include *beothucus, crassodon, fuscus, hudsonicus, irremotus, labradorius, ligoni, manningi, mogollonensis, monstrabilis,* and *youngi*
5. *occidentalis* would include *alces, columbianus, griseoalbus, mackenzii, pambasileus,* and *tundrarum*

V. E. Sokolov and O. L. Rossolimo in "Taxonomy and Variability," a paper included in D. I. Bibikov's *The Wolf: History, Systematics, Morphology and Ecology* (Moscow: U.S.S.R. Academy of Science, 1985), recognized nine subspecies of *Canis lupus* in the Old World:

Canis lupus albus—the large wolf of Eurasian tundra from Finland to Kamchatka
Canis lupus campestris—a small wolf of the deserts and steppes of Central Asia
Canis lupus chanco (sometimes referred to by other authorities as *Canis lupus laniger*—the Tibetan wolf)—a medium-size wolf of China, Manchuria, Mongolia, Tibet, and southwestern Russia
Canis lupus cubanensis—in the Caucasus, Turkey, and Iran
Canis lupus desertorum—in Kazakhstan
Canis lupus hattai—an extinct species from the Japanese island of Hokkaido
Canis lupus hodophilax—an extinct species from the other Japanese islands
Canis lupus lupus—the common wolf of Europe and forested Russia
Canis lupus pallipes—a small wolf from India and the Middle East

Other authorities might have added *Canis lupus arabs,* a small, light-colored wolf from southern Arabia, which Sokolov and Rossolimo regarded as a synonym of *pallipes*; *Canis lupus lupaster,* in Egypt, which many authorities regard as a subspecies of *Canis aureus,* the golden jackal; and *Canis lupus communis,* from the Ural Mountains of Russia and Siberia.

Nowak suggested combining *campestris, chanco,* and *desertorum* with *lupus.* His analysis would also recognize *pallipes, cubanensis, albus, communis,* and *hattai.* He did not analyze *arabs, lupaster,* or *hodophilax.*

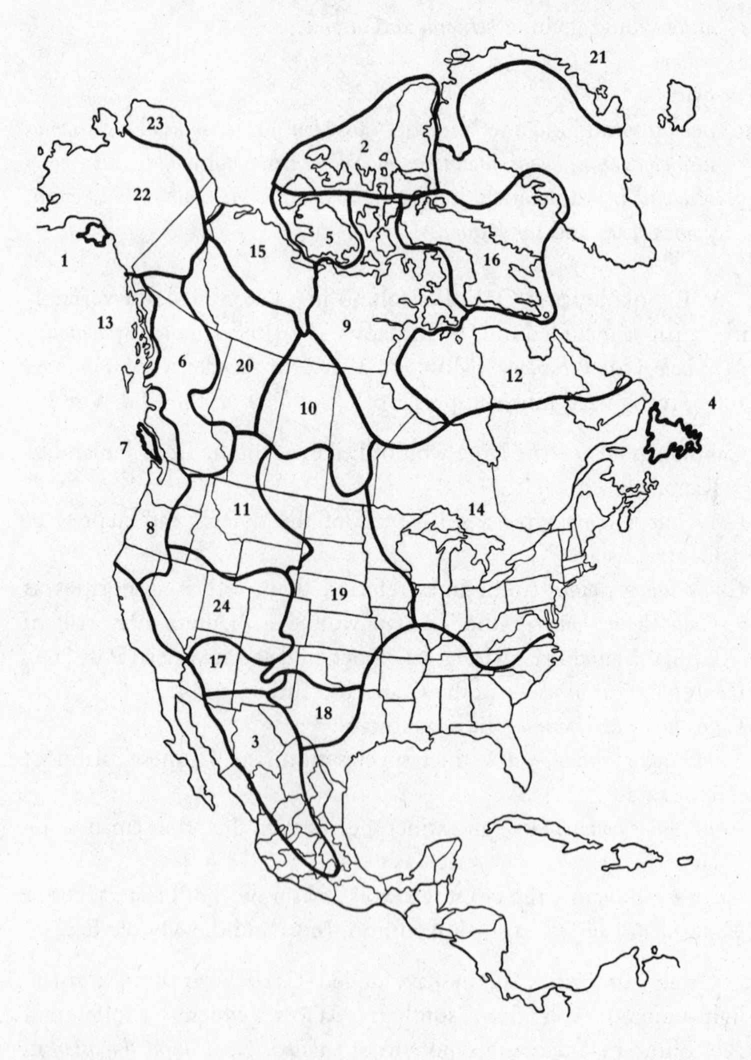

Key to Map 1: New World subspecies of *Canis lupus* recognized by Hall (1981):

1. alces
2. arctos
3. baileyi
4. beothucus
5. bernardi
6. columbianus
7. crassodon
8. fuscus
9. griseoalbus
10. hudsonicus
11. irremotus
12. labradorius
13. ligoni
14. lycaon
15. mackenzii
16. manningi
17. mogollonensis
18. monstrabilis
19. nubilus
20. occidentalis
21. orion
22. pambasileus
23. tundrarum
24. youngi

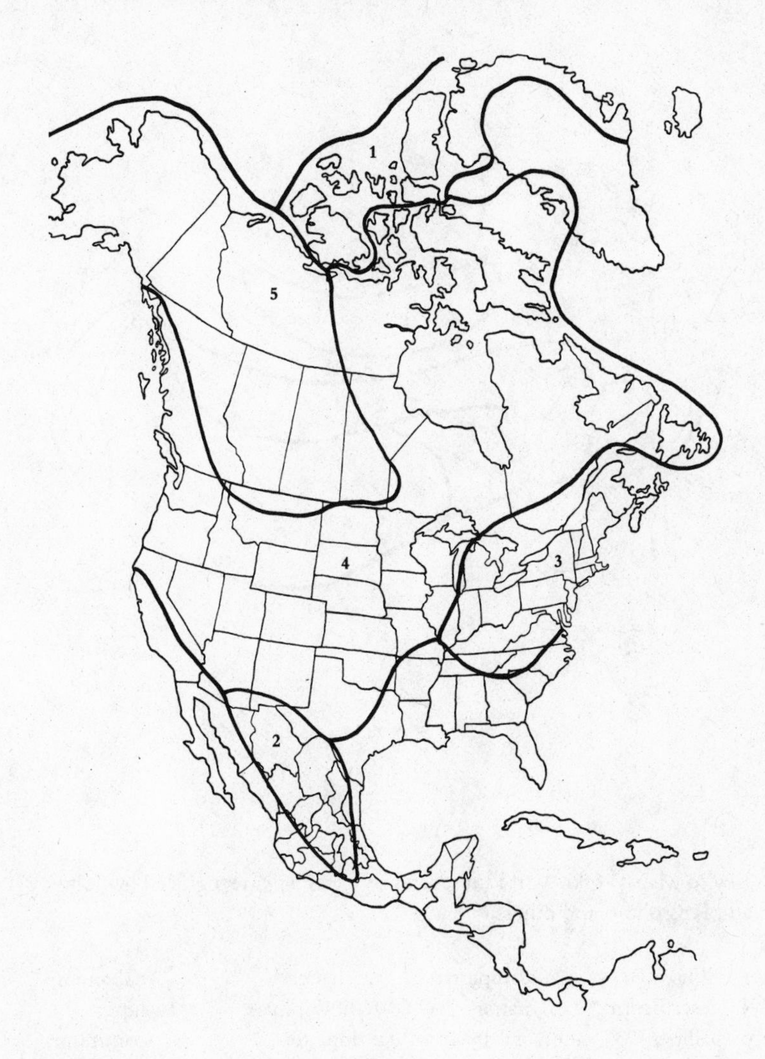

Key to Map 2: New World subspecies of *Canis lupus* as suggested by Ronald Nowak (1992):

1. arctos
2. baileyi
3. lycaon
4. nubilus
5. occidentalis

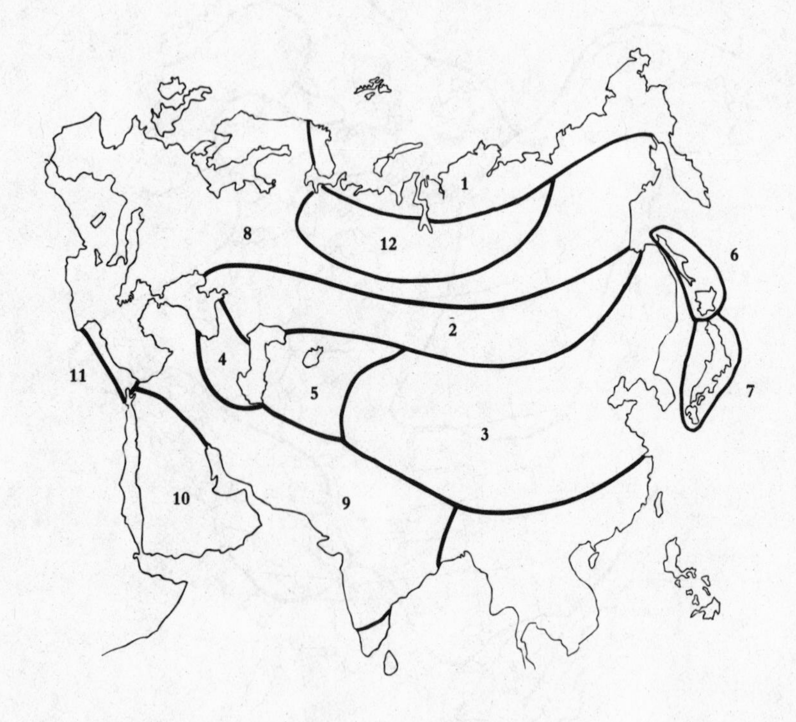

Key to Map 3. Old World subspecies of *Canis lupus* recognized by Sokolov and Rossolimo and others:

1. albus	2. campestris	3. chanco	4. cubanensis
5. desertorum	6. hattai	7. hodophilax	8. lupus
9. pallipes	10. arabs	11. lupaster	12. communis

Maps based on Ronald Nowak (1992), redrawn by Peter Steinhart by permission of Ronald Nowak (1992).

INDEX